CONTENTS

Page No.

EXECUTIVE SUMMARY

EX.1	PURPOSE	EX-1
EX.2	OVERVIEW	EX-1
EX.3	KEY THEMES	EX-2

CHAPTER 1—INTELLIGENCE

1.1	INTRODUCTION	1-1
1.2	SCOPE	1-1
1.3	DEFINITION AND RESPONSIBILITIES OF INTELLIGENCE	1-1
1.3.1	Defining Intelligence	1-1
1.3.2	Responsibilities of Intelligence	1-2
1.4	OBJECTIVES OF INTELLIGENCE	1-3
1.5	LEVELS OF INTELLIGENCE	1-4
1.5.1	Strategic Intelligence	1-4
1.5.2	Operational Intelligence	1-5
1.5.3	Tactical Intelligence	1-5
1.6	CATEGORIES OF INTELLIGENCE PRODUCTS	1-5
1.6.1	Indications and Warning Intelligence	1-5
1.6.2	Basic Intelligence	1-5
1.6.3	Current Intelligence	1-6
1.6.4	Target Intelligence	1-6
1.6.5	Scientific and Technical Intelligence	1-6
1.6.6	Counterintelligence	1-6
1.6.7	Estimative Intelligence	1-6
1.6.8	Identity Intelligence	1-6
1.6.9	Additional Categories of Intelligence Products	1-6
1.7	INTELLIGENCE THEORY AND PRACTICE	1-7
1.7.1	Relationship of Data, Information, and Intelligence	1-7
1.7.2	The Intelligence Process	1-8
1.7.3	Developing Intelligence	1-9
1.7.4	Developing Understanding	1-10
1.7.5	Characteristics of Intelligence Excellence	1-12
1.7.6	Intelligence Practice	1-13
1.8	INTELLIGENCE OPERATIONS	1-14

MAR 2014

		Page No.
1.9	THE COMMANDER'S ROLE IN THE INTELLIGENCE PROCESS	1-15

CHAPTER 2—FUNDAMENTALS OF NAVAL INTELLIGENCE

2.1	INTRODUCTION	2-1
2.2	THE MARITIME CONTEXT	2-1
2.2.1	The Maritime Operational Environment	2-2
2.2.2	U.S. Navy Mission and Naval Force Operational Advantages	2-3
2.2.3	Core Capabilities of the U.S. Navy	2-4
2.3	DEFINING NAVAL INTELLIGENCE	2-7
2.3.1	Naval Intelligence—What it Is	2-7
2.3.2	Naval Intelligence—What it is Not	2-8
2.4	FUNCTIONS OF NAVAL INTELLIGENCE	2-10
2.4.1	Intelligence Preparation of the Operational Environment	2-10
2.4.2	Indications and Warning	2-12
2.4.3	Current Intelligence	2-12
2.4.4	Targeting	2-12
2.4.5	Intelligence Assessment	2-13
2.4.6	Security	2-13
2.4.7	Lead and Manage the Intelligence Process	2-14
2.5	PRINCIPLES OF NAVAL INTELLIGENCE	2-14
2.5.1	Know the Adversary and the Operational Environment	2-14
2.5.2	Commander's Needs are Paramount	2-15
2.5.3	Ensure Unity of Intelligence Effort	2-15
2.5.4	Plan for Combat and Crisis Response	2-15
2.5.5	Use an All-Source Approach	2-16
2.6	CHARACTERISTICS OF NAVAL INTELLIGENCE EXCELLENCE	2-16
2.7	NAVAL INTELLIGENCE PROFESSIONAL CHARACTERISTICS	2-17
2.8	NAVAL INTELLIGENCE AND OPERATIONAL ART	2-19

CHAPTER 3—NAVAL INTELLIGENCE OPERATIONS

3.1	INTRODUCTION	3-1
3.2	THE INTELLIGENCE DISCIPLINES	3-1
3.2.1	Counterintelligence	3-3
3.2.2	Geospatial Intelligence	3-3
3.2.3	Human Intelligence	3-3
3.2.4	Measurement and Signature Intelligence	3-3
3.2.5	Open-Source Intelligence	3-4
3.2.6	Signals Intelligence	3-4
3.2.7	Technical Intelligence	3-4
3.3	THE INTELLIGENCE PROCESS	3-5

Page No.

3.4	TASKING, COLLECTION, PROCESSING, EXPLOITATION, AND DISSEMINATION AND THE INTELLIGENCE PROCESS	3-5
3.5	PLANNING AND DIRECTION	3-6
3.5.1	Intelligence Requirements Overview	3-6
3.5.2	Procedures for Requesting Higher Echelon Intelligence Support	3-11
3.5.3	Augmentation and Federation Requirements	3-11
3.5.4	Direction	3-13
3.5.5	The Importance of Planning and Direction	3-14
3.5.6	Evaluation and Feedback for Planning and Direction	3-14
3.6	COLLECTION	3-15
3.6.1	Collection Management	3-15
3.6.2	Collection Management Responsibilities	3-18
3.6.3	Principles of Collection Management	3-19
3.6.4	Collection Requirements Management	3-22
3.6.5	Collection Operations Management	3-30
3.6.6	Collection Management Best Practices	3-34
3.6.7	Evaluation and Feedback for Collection	3-34
3.7	PROCESSING AND EXPLOITATION	3-35
3.7.1	Processing and Exploitation Considerations	3-35
3.7.2	Evaluation and Feedback for Processing and Exploitation	3-36
3.8	ANALYSIS AND PRODUCTION	3-37
3.8.1	Sources and Methods	3-38
3.8.2	Conversion of Data and Information into Intelligence	3-39
3.8.3	Production	3-45
3.8.4	Analysis and Production Considerations	3-52
3.8.5	Evaluation and Feedback for Analysis and Production	3-53
3.9	DISSEMINATION AND INTEGRATION	3-54
3.9.1	Form	3-55
3.9.2	Dissemination Methods	3-55
3.9.3	Intelligence Integration	3-57
3.9.4	Evaluation and Feedback for Dissemination and Integration	3-58
3.10	VALUATION AND FEEDBACK	3-58

CHAPTER 4—INTELLIGENCE PREPARATION OF THE MARITIME OPERATIONAL ENVIRONMENT

4.1	INTELLIGENCE PREPARATION OF THE MARITIME OPERATIONAL ENVIRONMENT	4-1
4.2	GENERAL INTELLIGENCE PREPARATION OF THE OPERATIONAL ENVIRONMENT CONSIDERATIONS	4-2
4.2.1	Intelligence Preparation of the Operational Environment and the Intelligence Process	4-2
4.2.2	Intelligence Preparation of the Operational Environment Relationship to the Levels of War	4-4

		Page No.
4.2.3	Intelligence Preparation of the Operational Environment Considerations across the Range of Naval Operations	4-5
4.3	DEFINE THE MARITIME OPERATIONAL ENVIRONMENT	4-6
4.3.1	Identify the Naval Force's Area of Operations	4-7
4.3.2	Analyze the Mission and Joint Force Commander's Intent	4-7
4.3.3	Determine the Significant Characteristics of the Maritime Operational Environment	4-7
4.3.4	Establish the Limits of the Naval Force's Area of Interest	4-16
4.3.5	Determine the Level of Detail Required and Feasible within the Time Available	4-18
4.3.6	Determine Intelligence and Information Gaps, Shortfalls, and Priorities	4-19
4.3.7	Collect Material and Submit Requests for Information to Support Further Analysis	4-19
4.4	DESCRIBE THE IMPACT OF THE MARITIME OPERATIONAL ENVIRONMENT	4-20
4.4.1	Develop a Geospatial Perspective of the Maritime Operational Environment	4-20
4.4.2	Leveraging Intelligence Disciplines	4-26
4.4.3	Evaluate the Impact of the Maritime Domain	4-27
4.4.4	Develop a Systems Perspective of the Maritime Operational Environment	4-28
4.4.5	Describe the Impact of the Maritime Operational Environment on Adversary and Friendly Capabilities and Broad Courses of Action	4-37
4.5	EVALUATE THE ADVERSARY	4-39
4.5.1	Update or Create Adversary Models	4-40
4.5.2	Identification of High-Value Targets	4-45
4.5.3	Determine the Current Adversary Situation	4-46
4.5.4	Identify Adversary Capabilities and Vulnerabilities	4-48
4.5.5	Identify Adversary Centers of Gravity and Decisive Points	4-48
4.6	DETERMINE ADVERSARY COURSES OF ACTION	4-54
4.6.1	Identify the Adversary's Likely Objectives and Desired End State	4-54
4.6.2	Identify the Full Set of Adversary Courses of Action	4-55
4.6.3	Evaluate and Prioritize Each Course of Action	4-56
4.6.4	Develop Each Course of Action in the Amount of Detail Time Allows	4-56
4.6.5	Identify Initial Collection Requirements	4-62
4.7	APPLICATION	4-65

CHAPTER 5—NAVAL INTELLIGENCE SUPPORT TO PLANNING, EXECUTING, AND ASSESSING OPERATIONS

5.1	INTRODUCTION	5-1
5.2	JOINT CONTEXT	5-1
5.3	NAVAL INTELLIGENCE SUPPORT TO PLANNING	5-1
5.3.1	The Navy Planning Process Overview	5-1
5.3.2	Mission Analysis	5-2
5.3.3	Course of Action Development	5-5
5.3.4	Course of Action Analysis (Wargaming)	5-7
5.3.5	Course of Action Comparison and Decision	5-10
5.3.6	Plans and Orders Development	5-11
5.3.7	Transition	5-13
5.3.8	The Navy Planning Process in a Time-Constrained Environment	5-14

			Page No.

5.3.9	Naval Intelligence Support to the Navy Planning Process	5-15
5.4	**NAVAL INTELLIGENCE SUPPORT TO EXECUTION**	5-16
5.4.1	Operation Plan Phasing	5-16
5.4.2	Intelligence Support to Operation Plan Phases	5-18
5.4.3	Intelligence Support to Naval Tactical Operations	5-20
5.5	**NAVAL INTELLIGENCE SUPPORT TO ASSESSMENT**	5-22
5.5.1	Intelligence Support to the Assessment Process	5-22
5.5.2	Tactical Level Assessment	5-23

APPENDIX A—NAVAL INTELLIGENCE READING LIST

A.1	U.S. NAVAL INTELLIGENCE HISTORICAL WORKS	A-1
A.1.1	General Histories	A-1
A.1.2	Naval Intelligence Up to World War II	A-1
A.1.3	Cold War	A-2
A.1.4	Post-Cold War	A-2
A.2	NAVAL INTELLIGENCE BIOGRAPHICAL OR AUTOBIOGRAPHICAL WORKS	A-2
A.3	HISTORICAL OR SCHOLARLY WORKS ON THE NAVY AND WARFARE	A-3
A.4	GENERAL INTELLIGENCE WORKS	A-3
A.4.1	Scholarly Works on Intelligence and Intelligence Tradecraft	A-3
A.4.2	Historical Works	A-3

APPENDIX B—NAVAL INTELLIGENCE CASE STUDY

B.1	INTELLIGENCE CASE STUDY—OPERATION ICEBERG (THE INVASION OF OKINAWA)	B-1
B.2	INTRODUCTION	B-2
B.3	INTELLIGENCE DURING THE PLANNING PHASE	B-2
B.4	INTELLIGENCE EN ROUTE TO OKINAWA	B-7
B.5	INTELLIGENCE AT OKINAWA	B-8
B.6	INTELLIGENCE IN SUPPORTING COMMANDS	B-11

APPENDIX C—NATIONAL INTELLIGENCE COMMUNITY

APPENDIX D—INTELLIGENCE DISCIPLINES

D.1	INTRODUCTION	D-1
D.2	COUNTERINTELLIGENCE (CI)	D-1
D.2.1	Definition	D-1
D.2.2	Discussion	D-1
D.2.3	Fleet Counterintelligence Considerations	D-2

		Page No.
D.2.4	Guidance	D-2
D.3	**GEOSPATIAL INTELLIGENCE (GEOINT)**	**D-3**
D.3.1	Definition	D-3
D.3.2	Discussion	D-3
D.3.3	Fleet Geospatial Intelligence Considerations	D-5
D.3.4	Guidance	D-5
D.4	**HUMAN INTELLIGENCE (HUMINT)**	**D-6**
D.4.1	Definition	D-6
D.4.2	Discussion	D-6
D.4.3	Fleet Human Intelligence Considerations	D-7
D.4.4	Guidance	D-7
D.5	**MEASUREMENT AND SIGNATURE INTELLIGENCE (MASINT)**	**D-8**
D.5.1	Definition	D-8
D.5.2	Discussion	D-8
D.5.3	Fleet Measurement and Signature Intelligence Considerations	D-9
D.5.4	Guidance	D-9
D.6	**OPEN-SOURCE INTELLIGENCE (OSINT)**	**D-10**
D.6.1	Definition	D-10
D.6.2	Discussion	D-10
D.6.3	Fleet Open-Source Intelligence Considerations	D-11
D.6.4	Guidance	D-11
D.7	**SIGNALS INTELLIGENCE (SIGINT)**	**D-11**
D.7.1	Definition	D-11
D.7.2	Discussion	D-11
D.7.3	Fleet Signals Intelligence Considerations	D-12
D.7.4	Guidance	D-12
D.8	**TECHNICAL INTELLIGENCE (TECHINT)**	**D-13**
D.8.1	Definition	D-13
D.8.2	Discussion	D-13
D.8.3	Fleet Technical Intelligence Considerations	D-14
D.8.4	Guidance	D-14

APPENDIX E—UNDERSTANDING TASKING, COLLECTION, PROCESSING, EXPLOITATION, AND DISSEMINATION

E.1	OVERVIEW	E-1
E.2	TASKING	E-1
E.3	DISSEMINATION	E-4
E.4	THE COMPOSITE PROCESS	E-4
E.5	TCPED—WHAT IT IS NOT	E-8
E.6	SUMMARIZING TCPED	E-11

APPENDIX F—ANALYTIC TRADECRAFT

		Page No.
F.1	ANALYTIC CONCEPTS	F-1
F.1.1	Analytic Strategies	F-2
F.1.2	Analytic Methodologies	F-4
F.1.3	Analytic Principles	F-4
F.2	THE ANALYTIC PROCESS	F-6
F.2.1	Defining the Analytic Problem	F-6
F.2.2	Generating Hypotheses	F-7
F.2.3	Gather Information and Determine Information Needs	F-8
F.2.4	Evaluate Sources	F-9
F.2.5	Evaluate Hypotheses and Select the Most Probable Hypothesis	F-9
F.2.6	Production and Packaging	F-10
F.2.7	Evaluation and Feedback	F-11
F.2.8	Ongoing Monitoring	F-11
F.3	ANALYTIC TECHNIQUES	F-11
F.3.1	Brainstorming	F-11
F.3.2	Chronologies and Timelines	F-12
F.3.3	Matrices	F-14
F.3.4	Network Analysis	F-16
F.3.5	Flow Charting	F-20
F.3.6	Scenario Analysis	F-21
F.3.7	Indicators	F-21
F.3.8	Analysis of Competing Hypotheses	F-22
F.3.9	Red Teaming and Devil's Advocacy	F-24
F.3.10	Key Assumptions Check	F-24
F.4	ANALYTIC PITFALLS	F-24
F.4.1	Biases of Evaluation of Sources	F-25
F.4.2	Biases in Perception of Cause and Effect	F-26
F.4.3	Biases in Estimating Probabilities	F-27
F.4.4	Other Biases	F-27
F.5	ANALYTIC INTEGRITY STANDARDS	F-28
F.6	INTELLIGENCE ANALYSIS—AN ANALYST'S PERSPECTIVE	F-29

APPENDIX G—NAVY INTELLIGENCE MISSION-ESSENTIAL TASKS

G.1	THE INTELLIGENCE PROCESS AND MISSION-ESSENTIAL TASKS	G-1
G.2	DEFINING TERMS	G-1
G.3	NAVY MISSION-ESSENTIAL TASKS CONCEPT	G-1
G.4	OPERATIONAL LEVEL INTELLIGENCE TASKS AND THE INTELLIGENCE PROCESS	G-4
G.5	TACTICAL LEVEL INTELLIGENCE TASKS AND THE INTELLIGENCE PROCESS	G-5

Page No.

APPENDIX H—INTELLIGENCE PREPARATION OF THE MARITIME OPERATIONAL ENVIRONMENT WORKSHEETS

H.1 DEFINE THE MARITIME OPERATIONAL ENVIRONMENT H-1

H.2 DETERMINE THE ADVERSARY CENTER OF GRAVITY H-1

H.3 DETERMINE THE ADVERSARY CENTER OF GRAVITY HISTORICAL EXAMPLE H-3

H.4 DETERMINE ADVERSARY COURSES OF ACTION H-5

H.5 EVALUATE AND PRIORITIZE EACH ADVERSARY COURSE OF ACTION H-6

H.6 ADVERSARY COURSES OF ACTION WORKSHEET EXAMPLE H-8

APPENDIX I—INTELLIGENCE ESTIMATE

I.1 INTELLIGENCE ESTIMATE TEMPLATE I-1

APPENDIX J—FORMAT FOR ANNEX B: INTELLIGENCE

J.1 ANNEX B TEMPLATE J-1

NWP 2-0

LIST OF ILLUSTRATIONS

Page No.

CHAPTER 1—INTELLIGENCE

Figure 1-1.	Relationship of Data, Information, and Intelligence	1-8
Figure 1-2.	The Intelligence Process	1-9
Figure 1-3.	Key Considerations of the Intelligence Process	1-10
Figure 1-4.	The Information Hierarchy	1-11
Figure 1-5.	Developing Intelligence	1-11
Figure 1-6.	Characteristics of Intelligence Excellence	1-12

CHAPTER 2—FUNDAMENTALS OF NAVAL INTELLIGENCE

Figure 2-1.	Understanding the Maritime Operational Environment	2-3
Figure 2-2.	Intelligence Preparation of the Battlespace and Aspects of the Operational Environment	2-11
Figure 2-3.	Operational Art	2-21

CHAPTER 3—NAVAL INTELLIGENCE OPERATIONS

Figure 3-1.	Intelligence Disciplines, Subcategories, and Sources	3-2
Figure 3-2.	The Intelligence Process	3-5
Figure 3-3.	Intelligence Planning and Direction Activities	3-7
Figure 3-4.	Relationship between Intelligence Requirements and Information Requirements	3-8
Figure 3-5.	Intelligence Request Flow, Noncrisis	3-12
Figure 3-6.	The Collection Management Cycle	3-16
Figure 3-7.	Simplified Collection Requirements Management Flow	3-17
Figure 3-8.	Collection Management Principles	3-19
Figure 3-9.	Collection Management Considerations	3-21
Figure 3-10.	Elements of Collection Requirements Management	3-23
Figure 3-11.	Sample Intelligence Requirements Management Log	3-23
Figure 3-12.	Sample Tailored Intelligence Requirements Log for Carrier Strike Group	3-24
Figure 3-13.	Sample Collection Plan Format	3-25
Figure 3-14.	Asset and/or Resource Availability and Capability Factors	3-26
Figure 3-15.	Collection Timeliness	3-28
Figure 3-16.	Collection Tasking Worksheet	3-29
Figure 3-17.	Guidelines for Requesting National Resource Collection	3-31
Figure 3-18.	Collection Operations Management	3-32
Figure 3-19.	Sample Synchronization Matrix for a 24-Hour Period	3-33
Figure 3-20.	Collection Management Best Practices	3-34
Figure 3-21.	Processing and Exploitation Activities	3-36
Figure 3-22.	Distinction between Sources and Methods	3-39
Figure 3-23.	Notional Intelligence Data Processing Example	3-40
Figure 3-24.	Analysis and Production Activities	3-41
Figure 3-25.	Alphanumeric Confidence Scale	3-43
Figure 3-26.	Low-Moderate-High Confidence Scale	3-43
Figure 3-27.	Types of Intelligence Data	3-44

MAR 2014

Page No.

Figure 3-28. Intelligence Products 3-46
Figure 3-29. General Military Intelligence Elements 3-50
Figure 3-30. Attributes of Good Intelligence 3-54

CHAPTER 4—INTELLIGENCE PREPARATION OF THE MARITIME OPERATIONAL ENVIRONMENT

Figure 4-1. Intelligence Preparation of the Maritime Operational Environment—The Process 4-2
Figure 4-2. Intelligence Preparation of the Operational Environment and the Intelligence Estimate 4-4
Figure 4-3. Intelligence Preparation of the Maritime Operational Environment—Step One 4-6
Figure 4-4. Geospatial Areas and Maritime Domain Factors 4-9
Figure 4-5. The Geospatial Domains in Fleet Intelligence Preparation of the Battlespace 4-10
Figure 4-6. The Political, Military, Economic, Social, Information, and Infrastructure "Systems Perspective" with Interacting Subsystems 4-11
Figure 4-7. Relationship between Geospatial Domains and Political, Military, Economic, Social, Information, and Infrastructure Subsystems 4-11
Figure 4-8. Dimensions of the Information Environment 4-12
Figure 4-9. Activities Across the Information Environment 4-13
Figure 4-10. Intersection of Operational Environment Domains, Systems, and Dimensions 4-15
Figure 4-11. The Domains, Systems, and Dimensions of the Operational Environment 4-16
Figure 4-12. Relationship of Area of Operations, Influence, and Interest 4-17
Figure 4-13. Intelligence Preparation of the Maritime Operational Environment—Step Two 4-21
Figure 4-14. Assessed Ability of Intelligence Disciplines to Collect Against the Geospatial Environment 4-26
Figure 4-15. Assessed Ability of Intelligence Disciplines to Collect Against the Information Environment 4-28
Figure 4-16. Maritime Modified Combined Obstacle Overlay 4-29
Figure 4-17. Systems Perspective of the Operational Environment 4-30
Figure 4-18. Sample Political, Military, Economic, Social, Information, and Infrastructure Nodes 4-32
Figure 4-19. Association Matrix 4-33
Figure 4-20. Example of a Network Analysis Diagram 4-34
Figure 4-21. Measures of Node Centrality 4-35
Figure 4-22. Assessed Ability of Intelligence Disciplines to Collect Against the Political, Military, Economic, Social, Information, and Infrastructure Systems 4-37
Figure 4-23. Summarizing the Key Elements of Operational Factors 4-38
Figure 4-24. Operational Factors—Space, Time, and Force 4-39
Figure 4-25. Example Intelligence Preparation of the Operational Environment Step Two Assessment Product 4-40
Figure 4-26. Intelligence Preparation of the Maritime Operational Environment—Step Three 4-41
Figure 4-27. Adversary Template Components 4-42
Figure 4-28. Naval Adversary Template 4-42
Figure 4-29. Threat Template Depicting a Systems Network Perspective 4-43
Figure 4-30. Time-Event Matrix 4-45
Figure 4-31. Target Value Matrix 4-47
Figure 4-32. Order of Battle Factors 4-47
Figure 4-33. Examples of Adversary Capability Statements 4-49
Figure 4-34. Centers of Gravity Characteristics 4-49
Figure 4-35. Center of Gravity Identification and Analysis Process 4-51
Figure 4-36. Analyzing Critical Factors 4-53
Figure 4-37. Intelligence Preparation of the Maritime Operational Environment—Step Four 4-54
Figure 4-38. Course of Action Evaluation Criteria 4-56
Figure 4-39. Adversary Course of Action Prioritization Process 4-57

NWP 2-0

Page No.

Figure 4-40. Constructing a Situation Template 4-58
Figure 4-41. Geospatial Situation Template 4-59
Figure 4-42. Systems Situation Template 4-60
Figure 4-43. Situation Matrix 4-61
Figure 4-44. Constructing an Event Template 4-63
Figure 4-45. Event Template 4-64
Figure 4-46. Constructing an Event Matrix 4-65
Figure 4-47. Event Matrix 4-66

CHAPTER 5—NAVAL INTELLIGENCE SUPPORT TO PLANNING, EXECUTING, AND ASSESSING OPERATIONS

Figure 5-1. The Navy Planning Process 5-2
Figure 5-2. Example of a Course of Action Sketch and Statement for a Joint Force Maritime Component Commander or Navy Component Commander 5-6
Figure 5-3. Example of a Course of Action Narrative for a Joint Force Maritime Component Commander or Navy Component Commander 5-7
Figure 5-4. Examples of High-Payoff Targets, High-Value Targets, and Time-Sensitive Targets 5-10
Figure 5-5. Naval Intelligence Support to the Navy Planning Process Overview 5-15
Figure 5-6. Operation Plan Phasing Model 5-17
Figure 5-7. Notional Operation Plan Phases versus Level of Military Effort 5-18
Figure 5-8. Naval Mission Areas and Associated Commands 5-21
Figure 5-9. Assessment Levels and Measures 5-23

APPENDIX B—NAVAL INTELLIGENCE CASE STUDY

Figure B-1. Operation ICEBERG—The Invasion of Okinawa B-3

APPENDIX C—NATIONAL INTELLIGENCE COMMUNITY

Figure C-1. The National Intelligence Community C-2
Figure C-2. National Intelligence Leadership Structure C-3

APPENDIX E—UNDERSTANDING TASKING, COLLECTION, PROCESSING, EXPLOITATION, AND DISSEMINATION

Figure E-1. The Responsibilities of Intelligence, the Intelligence Process, and the Relationship of Data, Information, and Intelligence E-2
Figure E-2. Tasking of Intelligence Disciplines E-3
Figure E-3. Tasking, Collection, Processing, Exploitation, and Dissemination Tasking E-3
Figure E-4. Tasking, Collection, Processing, Exploitation, and Dissemination Dissemination E-5
Figure E-5. Tasking, Collection, Processing, Exploitation, and Dissemination Component Characteristics and Definitions E-6
Figure E-6. The Doctrinal Depiction of Tasking, Collection, Processing, Exploitation, and Dissemination E-7
Figure E-7. Tasking, Collection, Processing, Exploitation, and Dissemination Within the Intelligence Process E-8
Figure E-8. The Difference Between Tasking, Collection, Processing, Exploitation, and Dissemination-Based Information and Analysis-Based Intelligence E-10
Figure E-9. A Comparison of Tasking, Collection, Processing, Exploitation, and Dissemination and Intelligence Processes E-10

MAR 2014

Page No.

APPENDIX F—ANALYTIC TRADECRAFT

Figure F-1.	Analytic Principles	F-5
Figure F-2.	The Analytic Process	F-6
Figure F-3.	Issue Redefinition Example	F-7
Figure F-4.	Brainstorming Rules	F-12
Figure F-5.	Chronology Example	F-13
Figure F-6.	Annotated Timeline Example	F-15
Figure F-7.	Matrix Example	F-17
Figure F-8.	Network Analysis Example	F-19
Figure F-9.	Flow Chart Example	F-20
Figure F-10.	Indicators Example	F-23
Figure F-11.	Analysis of Competing Hypotheses	F-23
Figure F-12.	Hierarchy of Intelligence Analysis Types	F-32

APPENDIX G—NAVY INTELLIGENCE MISSION-ESSENTIAL TASKS

Figure G-1.	Definitions of Navy Mission-Essential Tasks Associated Terms	G-2
Figure G-2.	Example of Navy Tactical Tasks with Associated Measures	G-3
Figure G-3.	Universal Joint Task List Operational Level Intelligence Tasks	G-4
Figure G-4.	Navy Tactical Level Intelligence Tasks	G-6

DEPARTMENT OF THE NAVY

NAME OF ACTIVITY
STREET ADDRESS
CITY, STATE XXXXX-XXXX

5219
Code/Serial
Date

FROM: *(Name, Grade or Title, Activity, Location)*
TO: *(Primary Review Authority)*

SUBJECT: ROUTINE CHANGE RECOMMENDATION TO *(Publication Short Title, Revision/Edition, Change Number, Publication Long Title)*

ENCL: *(List Attached Tables, Figures, etc.)*

1. The following changes are recommended for NTTP X-XX, Rev. X, Change X:

 a. CHANGE: (Page 1-1, Paragraph 1.1.1, Line 1)
Replace "...the ~~National Command Authority~~ President and Secretary of Defense establish~~es~~ procedures for the..."
REASON: SECNAVINST ####, dated ####, instructing the term "National Command Authority" be replaced with "President and Secretary of Defense."

 b. ADD: (Page 2-1, Paragraph 2.2, Line 4)
Add sentence at end of paragraph "See Figure 2-1."
REASON: Sentence will refer reader to enclosed illustration.
Add Figure 2-1 (see enclosure) where appropriate.
REASON: Enclosed figure helps clarify text in Paragraph 2.2.

 c. DELETE: (Page 4-2, Paragraph 4.2.2, Line 3)
Remove "Navy Tactical Support Activity."
"...~~Navy Tactical Support Activity, and~~ the Navy Warfare Development Command ~~are~~ is responsible for..."
REASON: Activity has been deactivated.

2. Point of contact for this action is *(name, grade or title, telephone, e-mail address)*.

(SIGNATURE)
NAME

Copy to:
COMUSFLTFORCOM
COMUSPACFLT
COMNAVWARDEVCOM

Routine Change Recommendation Letter Format

NWP 2-0

FOREWORD

Navy Warfare Publication (NWP) 2-0, Naval Intelligence, is the capstone doctrinal publication for the Navy Intelligence Community. This publication will be used as the foundation of the Navy Intelligence doctrine development effort in the creation of a complete series of publications that will provide Naval Intelligence professionals with foundational concepts and practices that characterize effective intelligence support to naval operations.

NWP 2-0 provides a foundation for the conduct of intelligence to commanders and Naval Intelligence professionals with an overview of the unique characteristics of intelligence support to maritime operations at the operational and tactical levels of war.

While the publication is written for the Naval Intelligence professional, it is also intended for use by maritime operators, Joint, Agency, and Coalition partners who would like to obtain a greater appreciation for enduring naval intelligence principles and practices that are critical to fleet operational success. The concepts and standards described in this capstone document are integral to the three pillars of Information Dominance – Assured Command and Control, Battlespace Awareness, and Integrated Fires. These foundational concepts are critical to the conduct of intelligence and the Information Dominance warfighting discipline across the fleet.

E. L. TRAIN
Rear Admiral, U.S. Navy
Commander, Office of Naval Intelligence

MAR 2014

NWP 2-0

EXECUTIVE SUMMARY

EX.1 PURPOSE

Navy Warfare Publication (NWP) 2-0, Naval Intelligence, is the capstone publication in a series of doctrinal publications that address topics associated with intelligence support to naval operations. As such, it provides an overarching framework for those other publications by defining key foundational concepts related to both the theory and practice of intelligence in supporting naval decision makers at the operational and tactical levels of war. The publication is designed to first provide a general introduction to the subject of intelligence and discuss the unique attributes of naval intelligence. These concepts are then applied to a discussion of specific naval intelligence operations and how the process of intelligence preparation of the operational environment is tailored to support operations in the maritime environment. The publication culminates with an overview of how naval intelligence supports the planning, execution, and assessment of operations that will serve as the basis for future doctrinal publications that discuss intelligence support to specific naval operations and missions.

EX.2 OVERVIEW

The following is an overview of the main topics covered in each chapter of NWP 2-0.

Chapter 1—Intelligence. This chapter defines key foundational concepts of the intelligence discipline with an emphasis on concepts that are most applicable to military intelligence theory and practice. It provides the definition, responsibilities, and objectives of intelligence, emphasizing the fact that intelligence is a tool to reduce the commander's uncertainty about the adversary and relevant aspects of the operational environment to enable the planning and conduct of operations. The characteristics of intelligence as they apply to the various levels of war are discussed as are key terms related to the types and categories of intelligence. This chapter also discusses the theory and practice of intelligence and emphasizes the intelligence process as a means of understanding that intelligence consists of a set of highly interrelated operational activities that serve to support the planning, execution, and assessment of operations. The chapter concludes with a discussion of the commander's role in the intelligence process, highlighting the fact that intelligence is an inherent and essential part of command which deserves the same level of attention and focus provided to other operations.

Chapter 2—Fundamentals of Naval Intelligence. This chapter begins with an overview of the maritime environment and a discussion of the particular features of naval operations since understanding the context of the naval service is essential to understanding the culture of naval intelligence. Naval intelligence is then defined and its unique functions, principles, and characteristics of excellence are discussed in detail. The principles of naval intelligence are: to know the adversary and the operational environment, recognition that the commander's needs are paramount, ensuring unity of intelligence effort, planning for combat and crisis response, and use of an all-source approach in addressing intelligence requirements. The chapter concludes with a discussion of operational art and the application of this concept to the synchronization and integration of intelligence operations to meet the needs of naval decision makers. This chapter is supplemented with two appendices. Appendix A provides a reading list for naval intelligence professionals that provides greater depth on the concepts provided in this NWP and appendix B presents a case study on naval intelligence support to Operation ICEBERG, the United States (U.S.) invasion of Okinawa, that highlights many of the themes contained in this publication.

Chapter 3—Naval Intelligence Operations. The focus of this chapter is on applying the intelligence process to meet the needs of commanders, staffs, and supported organizations. It begins by introducing the intelligence disciplines and describing how each one fits within the larger national intelligence community framework. The chapter then discusses the unique attributes of each intelligence operation—planning and direction; collection; processing and exploitation; analysis and production; dissemination and integration—and the interrelationships among these operations. It emphasizes the need for clear and active planning in intelligence operations as this is the key to ensuring synchronized, integrated, efficient, and effective intelligence operations. Given evaluation and feedback is a continuous process that is applicable to all intelligence operations, the unique evaluation and feedback considerations for each component of the intelligence process are addressed. This chapter is supplemented by five appendixes. Appendix C describes the national intelligence community, and appendix D provides an extended discussion on the intelligence disciplines and their potential effectiveness at addressing issues of concern to naval intelligence. Appendix E describes how tasking, collection, processing, exploitation, and dissemination (TCPED) are modified to produce actionable intelligence. Appendix F contains a discussion of analytic tradecraft. Since analysis is considered a key skillset of the naval intelligence professional, this appendix discusses analytic theory, discusses various analytic techniques to address a variety of naval intelligence problems, and provides insights on cognitive biases and other pitfalls that are inimical to sound analytic practice. Appendix G discusses how the intelligence process serves as the basis for naval intelligence mission-essential tasks at the operational and tactical levels.

Chapter 4—Intelligence Preparation of the Maritime Operational Environment. Chapter 4 begins with a discussion of the concepts behind intelligence preparation of the operational environment (IPOE) and the relationship of IPOE to the intelligence process, the levels of war, and the range of naval operations. IPOE is a structured analytic process that occurs in four steps and is focused on determining adversary courses of action (COAs) and understanding relevant aspects of the operational environment that are key to the planning, conduct, and assessment of operations. The chapter then covers the main steps of the IPOE process: define the maritime operational environment, describe the impact of the maritime operational environment, evaluate the adversary, and determine adversary COAs. Each step in the IPOE process is discussed in detail and tailored for considerations specific to the maritime operational environment. Appendix H supplements this chapter with worksheets that naval intelligence professionals can use to present their IPOE analysis to naval planners and operators.

Chapter 5—Naval Intelligence Support to Planning, Executing, and Assessing Operations. This chapter begins with a discussion of the joint context for naval operations and then discusses specific intelligence support considerations for the planning, execution, and assessment of naval operations at the operational and tactical levels of war. The section on planning discusses how intelligence supports the Navy planning process (NPP) from mission analysis through transition in both deliberate and crisis planning environments. The section on support to execution discusses naval intelligence support considerations for the various phases of operations, from shaping to enable civil authority, and introduces concepts associated with intelligence support to naval missions at the tactical level. The chapter concludes with a discussion of naval intelligence support to assessment and includes an overview of the three-phase battle damage assessment (BDA) process. Appendix I, which provides a template for an intelligence estimate, and appendix J, which provides a format for the Annex B (Intelligence) of an operation plan or order, supplement this chapter.

EX.3 KEY THEMES

The key themes highlighted in this publication are as follows:

1. The commander's intelligence needs are paramount.

2. Commanders have definitive responsibilities for intelligence operations under their purview.

3. The concept of operational art is wholly applicable to intelligence, which is a form of operations. Naval intelligence professionals not only need to have mastery over a diverse and broad range of knowledge concerning the adversary, the intelligence process, the national intelligence community, intelligence disciplines, and the operational environment but they must also be experts in the synchronization and integration of complex intelligence operations to meet the needs of their commanders.

4. While intelligence work is an art, there are enduring principles and characteristics of excellence which, if mastered and adhered to, will result in high quality intelligence products and support.

5. Intelligence operations work best when they are centrally planned and directed and executed in a decentralized manner.

6. Collaboration is fundamental to the practice of naval intelligence. Naval intelligence is part of a national intelligence community that has significant collection, processing, exploitation, and analytic resources that can be leveraged to meet the needs of naval commanders. Naval intelligence also makes unique and important contributions to the national intelligence framework.

7. Naval intelligence leaders must foster an open, collaborative environment where ideas from any level and source are respected and evaluated on their merits. This type of environment is needed for innovative analytic thought and a culture of continuous process improvement.

8. Planning and direction are critical to efficient and effective intelligence operations. Given the long lead times associated with some collection, processing, and analytic operations, naval intelligence professionals must become experts at anticipating the needs of their commanders and proactive in submitting requirements.

9. Naval intelligence professional qualities include adaptability; intellectual honesty; teamwork; the ability to communicate clearly, concisely, and accurately; curiosity; and resourcefulness.

10. Clear, concise, and specific priority intelligence requirements are key to successful intelligence operations and should result from frank and consistent dialog among the commander, operations, plans, and intelligence personnel.

11. Intelligence is a scarce resource, so intelligence must be driven by definitive requirements that address the needs of the commander.

12. Intelligence that is anticipatory, objective, complete, accurate, timely, useable, relevant, and available meets the standards for intelligence excellence. While there is tension between some of these characteristics, such as timeliness and completeness, naval intelligence professionals must balance these characteristics to deliver quality intelligence to their commanders.

13. Naval intelligence professionals must be able to articulate the end state—What does the commander require? When is it required? In what format is it required?—and then visualize how to synchronize all intelligence operations to achieve that end state. If the end state if not achievable or will require additional resources or time, the commander must be immediately notified.

14. An all-source (multidiscipline, multimethod) approach is the best approach to addressing intelligence problems.

INTENTIONALLY BLANK

CHAPTER 1

Intelligence

By comparing a variety of information, we are frequently enabled to investigate facts, which were so intricate or hidden, that no single clue could have led to the knowledge of them . . . intelligence becomes interesting which but from its connection and collateral circumstances, would not be important.

General George Washington
Letter to James Lovell
April 1, 1782

By "intelligence" we mean every sort of information about the enemy and his country—the basis, in short, of our own plans and operations.

Clausewitz
On War, 1832

1.1 INTRODUCTION

Intelligence directly supports successful planning and execution of operations in the maritime domain. This publication provides naval intelligence professionals with foundational concepts and practices that characterize effective intelligence operations and support. To that end, the publication begins with a general overview of intelligence that defines key terms and establishes a conceptual framework for the remainder of the publication.

1.2 SCOPE

While intelligence is a broad subject with many facets, the scope of this publication is on addressing the application of operational art to the successful management and conduct of intelligence operations in the maritime domain. To that end, the focus of this publication is on providing commanders and naval intelligence professionals with an overview of the unique characteristics of intelligence support to maritime operations at the operational and tactical levels of war. Although naval intelligence is involved in a wide array of activities that support national and theater level requirements, the scope of this publication is on naval intelligence support to fleet operations, which are inclusive of all naval operating forces. While written primarily for the naval intelligence professional, this publication is also intended for operators and Service, joint, agency, and coalition partners who would like to obtain a greater appreciation for the enduring naval intelligence principles and practices that are critical to fleet operational success.

1.3 DEFINITION AND RESPONSIBILITIES OF INTELLIGENCE

1.3.1 Defining Intelligence

intelligence. The product resulting from the collection, processing, integration, evaluation, analysis, and interpretation of available information concerning foreign nations, hostile or potentially hostile forces or elements, or areas of actual or potential operations. The term is also applied to the activity which results in the product and to the organizations engaged in such activity. (Joint Publication (JP) 1-02. Source: JP 2-0)

The definition of intelligence has evolved over time. The origin of the term comes from the 16th century when the connotation of the term "intelligence" shifted to encompass the gathering of information by secret means, typically by spies who were employed by rulers or various government offices. So, at a fundamental level, intelligence is a term that applies to the discovery of secret information by secret means. While this is a generalization, particularly in an age when information about many topics of national and military interest is readily available in open sources, the focus of intelligence remains penetrating the veil of secrecy that surrounds adversary capabilities, plans, and intentions. Given this understanding, intelligence can be defined as a process whereby "information which has been systematically collected for a definite purpose and from a definite point of view, and which has been subjected to critical analysis . . . and careful evaluation and collation so that it is no longer a mere mass of facts but a coordinated and organized body of pertinent information." This is consistent with the joint definition of intelligence although the joint definition of intelligence uses the term to mean not only the activities and organizations involved in the intelligence effort but also as the product resulting from those organizations' activities.

This full understanding of this joint definition is used for three reasons. First, there are many types of intelligence, such as the intelligence provided to national level decision makers to support the development of policy. Second, this definition provides a more expansive view of intelligence as being concerned with understanding of the operational environment.[1] While there are many elements within the operational environment that are outside the purview of intelligence, there are many factors beyond knowledge of the threat, such as identification of potential landing zones to support foreign humanitarian assistance (FHA) operations, that require the use of intelligence resources and processes. Third, the joint definition encompasses the view that intelligence may refer to the intelligence process, the product of that process, and the organizations involved in producing intelligence.[2]

1.3.2 Responsibilities of Intelligence

Joint doctrine states that the primary role of intelligence is "to provide information and assessments to facilitate accomplishment of the mission" and that its most important role is "to assist commanders and their staffs with analysis of key aspects of the operational environment to assist them in their decision-making process."[3] The development and communication of knowledge is a key characteristic of intelligence that makes it distinct from more general forms of information. Within a military context, these primary functions and roles of intelligence can be broken down into more specific responsibilities:

1. Inform the commander. This includes analysis of the adversary and other relevant aspects of the operational environment and the production of threat assessments on a continuing basis to support the commander in creating and exploiting opportunities to accomplish friendly force objectives.

2. Describe the operational environment. The operational environment is complex and includes such factors as the geopolitical situation; the impact of geographic, oceanographic, and weather conditions on force movement and maneuver; and understanding the relationship between different nodes—societal, political, religious, infrastructure, etc.—within areas of concern. Given the vast number of factors that comprise the operational environment, naval intelligence professionals must work closely with their commanders, staff, and supported organizations to determine the relevant aspects of the operational environment that are the focus of intelligence operations.

3. Identify, define, and nominate objectives. All aspects of military planning are dependent on the determination of clearly defined, achievable, and measurable objectives. To properly fulfill this responsibility, the intelligence team must understand the command's responsibilities and mission; the commander's intent; the full range of means available to collect, exploit, analyze, and disseminate

---

[1] Operational environment is defined as "A composite of the conditions, circumstances, and influences that affect the employment of capabilities and bear on the decisions of the commander." From JP 1-02, Department of Defense Dictionary of Military and Associated Terms.
[2] This view was first put forward by Sherman Kent, a Central Intelligence Agency (CIA) official who was known as "the father of intelligence analysis" and this triune view of the meaning of "intelligence" is recognized throughout government and academia. Lowenthal, M. M., Intelligence: From Secrets to Policy, 2009, p. 8.
[3] JP 2-0, Joint Intelligence.

intelligence; the adversary; the weather; and other relevant characteristics of the operational environment. Intelligence analyzes these factors to develop an understanding of adversary intentions, objectives, and centers of gravity, critical vulnerabilities, and other limitations that may be exploited to defeat them, as well as most likely and most dangerous adversary courses of action (COAs). From this understanding, objectives are recommended to and approved by the commander and then continuously monitored and reviewed to account for changes in the situation or commander's intent.

4. Support the planning and execution of operations. Commanders and staffs at all levels require intelligence to plan, direct, conduct, and assess operations once the objectives, nature, and scope of operations has been defined. Intelligence plays a crucial role in identifying and nominating objectives and targets and determining the means, operations, and tactics required to achieve mission success.

5. Counter adversary deception and surprise. Intelligence does not produce certainty regarding adversary capabilities and intentions. Cognitive biases, active adversary deception efforts, and other factors always create a degree of uncertainty in any intelligence estimate or product. Intelligence professionals must always be sensitive to the possibility they are being deceived and must counter this potentiality through use of multiple collection methods and proven analytical methods such as red teaming, devil's advocates, and alternative hypotheses.

6. Support friendly deception efforts. By attaining understanding of adversary capabilities, mindset, and decision-making processes, intelligence provides critical support to operational personnel engaged in developing and executing friendly force deception efforts. Once deception efforts are employed, intelligence assesses the effectiveness of those efforts by monitoring adversary responses.

7. Assess the effects of operations. Intelligence plays a key role in evaluating the impact of friendly force operations on the adversary and in assessing other aspects of the operational environment relevant to the command and its mission. The goals of assessment are to determine whether friendly operations are producing desired or undesired effects, whether objectives have been achieved, when unanticipated opportunities can be seized, or when modifications to planned operations to respond to adversary actions or changes in the operational environment are required. While the assessment process is the responsibility of the Navy staff operations directorate (N-3) or operations director/department head, intelligence plays a major and critical role in contributing to the assessment process.

These seven responsibilities drive the full range of intelligence activities and products that are the focus and end result of intelligence operations. These responsibilities constitute, in summary, both a shield and a weapon for the commander. When properly employed, intelligence provides a shield that protects against surprise and safeguards plans. It serves as a weapon by providing the commander with the information needed to plan and successfully execute operations and the information needed to contend with and exploit emergent circumstances in the operational environment.

1.4 OBJECTIVES OF INTELLIGENCE

Intelligence has two equally important objectives. First, intelligence attempts to reduce uncertainty—the Clausewitzian "fog and friction" of war—by providing accurate, relevant, and timely knowledge about the adversary and the operational environment (see chapter 4). Second, intelligence assists with the protection of friendly forces through counterintelligence efforts. Counterintelligence (CI), as defined in joint doctrine, refers to "[i]nformation gathered and activities conducted to identify, deceive, exploit, disrupt, or protect against espionage, other intelligence activities, sabotage, or assassinations conducted for or on behalf of foreign powers, organizations or persons or their agents, or international terrorist organizations or activities." Reducing uncertainty consists of four interrelated tasks that are elaborated on in the chapter 4 discussion on maritime intelligence preparation of the operational environment (IPOE). These four tasks include:

1. Identification and analysis of existing conditions and adversary capabilities

2. Estimating possible adversary courses of action (ACOAs) or potential changes to the operational environment based on the preceding analysis

3. Assisting with the determination of friendly vulnerabilities that the adversary could exploit

4. Assisting in the development and evaluation of friendly COAs.

The intelligence process works continuously to meet these objectives by providing a methodology for addressing intelligence shortfalls and focusing on confirming or refuting the fragmentary information that pervades the operational environment.

1.5 LEVELS OF INTELLIGENCE

In general, there are three levels of intelligence: strategic, operational, and tactical. These three levels correspond with the unique intelligence support requirements of the three levels of war, which are defined in joint doctrine as:

1. Strategic level of war—The level of war at which a nation, often as a member of a group of nations, determines national or multinational (alliance or coalition) strategic security objectives and guidance, then develops and uses national resources to achieve those objectives.

2. Operational level of war—The level of war at which campaigns and major operations are planned, conducted, and sustained to achieve strategic objectives within theaters or other operational areas.

3. Tactical level of war—The level of war at which battles and engagements are planned and executed to achieve military objectives assigned to tactical units or task forces.

While the speed and ease of data flow often blurs the distinction between the levels of war, these distinctions are still very significant to the intelligence professional for two primary reasons. First, the level of the decision maker drives unique intelligence requirements which have substantial impacts on such things as the content, format, and timeliness of the intelligence produced. Second, the construct of levels of warfare is an important concept as it aids in the visualization of the vertical and horizontal flow of requirements, tasking, raw intelligence data, and finished intelligence products.

An understanding of how this flow occurs is critical for all intelligence professionals since leveraging collection, processing, exploitation, analysis, and dissemination assets across each level is key to meeting the intelligence requirements of commanders and decision makers at all levels. While it is true that levels of command, size of units, types of equipment, or types of forces or components are not associated with a particular level of intelligence operations, understanding who has operational control over intelligence capabilities; how to communicate requirements; what the capabilities and limitations of these resources are; and how these capabilities can best satisfy requirements are fundamental elements of the operational art associated with synchronizing intelligence operations across national, operational, and tactical echelons to address a commander's unique intelligence needs.

1.5.1 Strategic Intelligence

Note

Within the strategic level of war, intelligence serves two customers: senior civilian and military leaders at the national level, and combatant commanders within theaters of operations.

1.5.1.1 National Strategic Intelligence

Intelligence at this level supports formulating national strategy and policy decisions and includes monitoring of the international situation for signs of imminent or emerging threats, the development of long-term military plans and decisions on force disposition, support for the development of weapons systems and force structure requirements, and support to strategic operations. While there are many time-sensitive strategic-level intelligence concerns, particularly with regard to strategic threats, the strategic level is also the primary consumer of longer-term intelligence analysis that supports policy, planning, weapons development, and acquisition efforts. National strategic intelligence capabilities, such as intelligence and communications satellites, are controlled at this level.

1.5.1.2 Theater-Strategic Intelligence

Intelligence at this level supports the formulation of theater-level strategy and is concerned with joint operations across the range of military operations. It includes determining when, where, and in what strength the adversary will conduct theater-level campaigns and strategic unified operations.

1.5.2 Operational Intelligence

Operational intelligence focuses on theater-level requirements such as supporting the planning and conduct of campaigns; monitoring events (to include man-made or natural disasters); determining the location, capabilities, limitations, and intentions of adversary forces in theater; and identifying adversary centers of gravity.

Note

> When used in this context within this publication, the term "operational intelligence" describes intelligence support to the operational level of war. For naval intelligence professionals, the understanding of the term "operational intelligence," to mean an all-source intelligence analysis process is referred to as operational intelligence (OPINTEL) and is discussed in greater detail in chapter 3.

1.5.3 Tactical Intelligence

Tactical Intelligence supports planning and conducting the full range of missions at the tactical level and focuses on understanding the operational environment; detection and warning of imminent threats; and determining the locations, capabilities, and intent of adversary forces within the area of operations(AO)/area of interest (AOI) of the commander.

1.6 CATEGORIES OF INTELLIGENCE PRODUCTS

While the intelligence produced to meet requirements is unique, all intelligence can be classified as belonging to a type or combination of types based on its purpose and focus. Joint doctrine defines six categories of intelligence products, which are:

1.6.1 Indications and Warning Intelligence

This is extremely time-sensitive intelligence that is designed to provide warning of an imminent threat or establishes a set of monitored indicators which, if observed to occur, provide advanced warning of potential adversary decisions to initiate hostile action.

1.6.2 Basic Intelligence

This typically consists of intelligence that is of an encyclopedic, descriptive, and comparative nature that has been compiled in advance and may be stored in a database. Examples of basic intelligence would include the general military intelligence (GMI) found in the Modernized Integrated Database (MIDB), leadership biographies, and country studies. While this type of intelligence tends to be the most accurate and reliable, it is not time-sensitive and is typically used as a starting point in the development of more relevant and timely products for operational and tactical-level commanders. Joint doctrine also refers to this as GMI, which encompasses a broad range of intelligence topics of military significance. The term basic intelligence is used here as it is more expansive and includes the compilation of intelligence on a wide variety of topics that are not exclusively military in nature.

1.6.3 Current Intelligence

This time-sensitive intelligence describes the current situation such as the conditions in the operational environment, adversary force dispositions, and movements. It is usually more specific than basic intelligence but also harder to obtain and less reliable.

1.6.4 Target Intelligence

An intelligence process focused on understanding adversary capabilities, disposition, personnel, units, facilities, systems, nodes, and links for the purposes of selecting and prioritizing targets to satisfy operational objectives and matching the appropriate response to them based on an understanding of the operational environment, requirements, and friendly capabilities.

1.6.5 Scientific and Technical Intelligence

Scientific and technical intelligence (S&TI) is concerned with understanding foreign developments in pure and applied science that have warfighting implications and also on understanding characteristics, capabilities, limitations, and vulnerabilities of adversary weapon systems.

1.6.6 Counterintelligence

Analyzes the threats posed by foreign intelligence and security services and the intelligence activities of nonstate actors such as organized crime, terrorist groups, and drug traffickers. CI analysis incorporates all-source information and the results of CI investigations and operations support a multidiscipline analysis of the force protection threat.

1.6.7 Estimative Intelligence

This intelligence is focused on potentialities and provides a prediction of future events, conditions, or activities. It requires evaluation of the past (basic intelligence) with an understanding of the unfolding situation (current intelligence) to determine possible futures and winnow those to produce predictive assessments such as a forecast concerning when, where, how, and if an adversary will attack or defend. Estimative intelligence is the most important type of intelligence to the commander, as it drives decisions, but it is also the most demanding to produce and inherently the least reliable type of intelligence.

1.6.8 Identity Intelligence

Identity intelligence results from the fusion of identity attributes (biologic, biographic, behavioral, and reputational information related to individuals) and other information and intelligence associated with those attributes collected across all intelligence disciplines. Identity intelligence utilizes enabling intelligence activities, like biometrics-enabled intelligence, forensics-enabled intelligence, and document and media exploitation, to discover the existence of unknown potential threat actors by connecting individuals to other persons, places, events, or materials, analyzing patterns of life, and characterizing their level of potential threats to U.S. interests.

1.6.9 Additional Categories of Intelligence Products

In addition to the six types of intelligence found in joint doctrine, there are seven additional categories that are typically used to further characterize the different types of intelligence products. While most products incorporate a combination of these focus areas, they provide an additional means of specifying the type of intelligence that is being collected or produced. These seven additional categories are:

1. Political Intelligence: Concerned with understanding political developments within nations or in the relations among nations. The latter meaning is also sometimes referred to as diplomatic intelligence.

2. Geographic Intelligence: Concerned with understanding the unique geography or hydrography of specific locations and interpreting how that geography can influence adversary decisionmaking, force disposition, and movement at all levels of war.

3. Economic Intelligence: Also known as commercial or financial intelligence, it is concerned with understanding the financial activities and strength of nations, corporations, organizations, and individuals and assessing the geopolitical and economic impacts of actual or anticipated activities.

4. Military Intelligence: Concerned with the capabilities, limitations, dispositions, and movements of military forces, to include terrorist and paramilitary organizations.

5. Biographical Intelligence: Also known as personalities intelligence, it is focused on developing biographies of foreign leaders in multiple spheres (i.e., political, military, academic, organizational, commercial) to obtain a better understanding of their motivations, decision-making capabilities and methods.

6. Sociological and Cultural Intelligence: Focused on understanding the sociological and cultural factors that influence societies and organizations and how these factors influence decisionmaking.

7. Medical Intelligence: Foreign medical, bioscientific, and environmental information that is of interest to strategic planning and to military medical planning and operations for the conservation of the fighting strength of friendly forces and the formation of assessments of foreign medical capabilities in both military and civilian sectors. Medical intelligence also encompasses adversarial medical capabilities, foreign healthcare infrastructure for military and civilian populations, and environmental health (such as any form of pollutant, any type of chemical or biological warfare agent, and hazardous plant and animal life).

1.7 INTELLIGENCE THEORY AND PRACTICE

Having defined intelligence and its responsibilities, objectives, levels, types, and categories it is next important to discuss, in a general way, the theory and practice of how intelligence is developed. The discussion below details a general theory about how data is transformed into intelligence, which contributes to the understanding of the operational environment necessary for effective decisionmaking. It then discusses the process for conducting that transformation, the intelligence process, which represents an iterative model for the development of intelligence. How this model is applied in attaining deeper levels of knowledge about the adversary and the operational environment is also discussed. Finally, a brief overview of the complicating factors that often work against the intelligence process is provided.

1.7.1 Relationship of Data, Information, and Intelligence

JP 2-0 makes a clear distinction between the nature of data, information, and intelligence: "Raw data by itself has relatively limited utility. However, when data is collected from a sensor and processed into an intelligible form, it becomes information and gains greater utility. Information on its own is a fact or a series of facts that may be of utility to the commander, but when related to other information about the operational environment and considered in the light of past experience, it gives rise to a new understanding of those facts, which may be termed "intelligence."

The distinctions among the three terms are important to understand. The most important type of intelligence to a commander is often the estimative intelligence product derived from a deep understanding of the adversary fused with data and information, from multiple sources, concerning the current situation. In this construct, intelligence is a synthesized product that is much more than the sum of its constituent parts. Data, which are comprised of discreet pieces of information that must be processed and placed in context to produce meaning, clearly does not meet the definition of intelligence. Providing a commander with raw data will overwhelm the commander with details that are contradictory, incomplete, or irrelevant. Information—while not intelligence—does share many of the features of intelligence. It is often perishable; like data, it can also be contradictory, incomplete, or irrelevant; and it may be inaccurate or misleading. While information may inform tactical actions under certain conditions, which are elaborated later in this publication, it is not the same as intelligence, which distills information from many sources and uses analytic methodologies and experience to predict future activity.

Figure 1-1 depicts the evolution of data-to-information-to-intelligence.

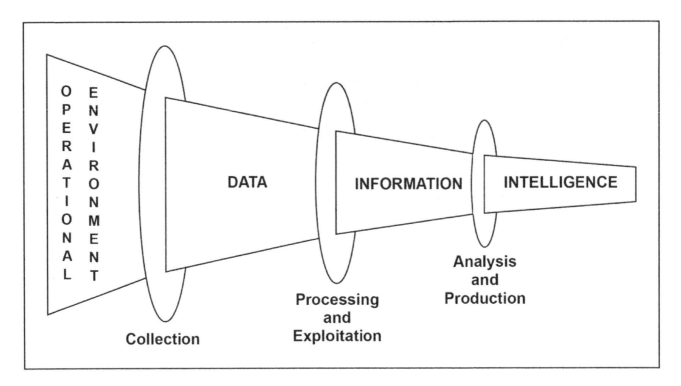

Figure 1-1. Relationship of Data, Information, and Intelligence

1.7.2 The Intelligence Process

JP 2-0 describes intelligence operations as "wide-ranging activities conducted by intelligence staffs and organizations for the purpose of providing commanders and . . . decision makers with relevant, accurate, and timely intelligence. The six categories of intelligence operations are: planning and direction; collection; processing and exploitation; analysis and production; dissemination and integration; and evaluation and feedback. In many situations, the various intelligence operations occur nearly simultaneous with one another or may be bypassed altogether."

The conceptual model used to describe the process that transforms raw data into information and information into intelligence is known as the intelligence process. Although it is an iterative process that, in practice, is not linear in nature, it is useful to view the process as cyclical progression of related activities that start with the planning and direction of intelligence operations and ends with the dissemination and integration of intelligence.

Figure 1-2 depicts the six interrelated operations that comprise the intelligence process. Note that the category of evaluation and feedback is depicted as a continuous feature of the process.

In addition to the key considerations listed here, it is useful to list additional aspects of the intelligence process that aid in understanding the application of the model in support of operations. First, good requirements are foundational to the success of the intelligence process. Ideally, requirements provided by the decision maker are clear, well defined, specific, and prioritized. Second, the data and information that feeds the intelligence process is typically derived from four main sources: (1) dedicated intelligence collection assets; (2) sensor data from surveillance or targeting systems; (3) data obtained from operational units in the field, and (4) data obtained from open-sources. In developing a collection plan, an experienced intelligence professional looks to leverage data from all sources to address intelligence gaps. Third, the value of a source is not necessarily related to the sophistication or cost of that source. In some situations, open-source press reports may provide more valuable information than dedicated assets and systems.

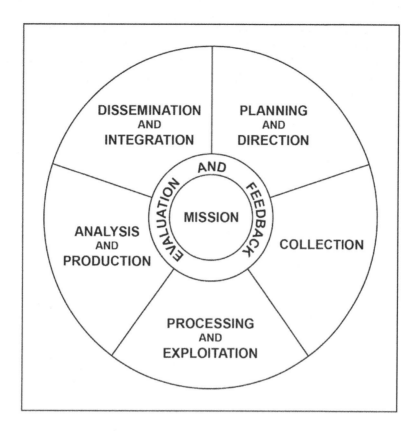

Figure 1-2. The Intelligence Process

1.7.3 Developing Intelligence

What is missing from the discussion above is the theoretical application of the intelligence process to developing a deep understanding of the adversary and the operational environment that the commander can use to make effective decisions. Developing the deep understanding of the adversary needed to estimate intentions is typically a lengthy process that requires the progressive development of detailed intelligence that is enabled by increasingly effective collection efforts and the analytic insights that come from focusing on a specific intelligence problem. The development of this understanding, particularly with regard to intent, is made even more difficult to achieve given active adversary deception efforts; the Clausewitzian "process of interaction" which requires the adversary to react to the activities of friendly forces; and the simple fact that the adversary may not have formulated their intentions and is still assessing options.

Although theoretical, there is a progression of intelligence development that ultimately leads to deep understanding of the adversary and an enhanced ability to predict adversary intentions. First is the foundational data that comprises basic intelligence or GMI. This includes measurable things such as: the numbers of personnel, ships, submarines, and aircraft that make up the adversary's Navy; the disposition, organization, and location of the adversary's forces; and the technical specifications of their weapons systems (range, effectiveness, performance, etc.). Second, intelligence must seek to develop an understanding of less quantifiable and more intangible factors such as: readiness, training, leadership quality, and morale; the adversary's methods, doctrine, and tactics, techniques, and procedures (TTP); and the performance of the adversary in past and current combat operations and exercises. Developing intelligence on the combination of factors above, while significant, is typically insufficient for understanding intent. When taken in combination, though, these factors can provide an excellent appreciation of adversary capabilities, which are those things that the adversary can do. Third, intelligence must develop knowledge of the deeper moral and cultural forces that drive the adversary's thinking to include: understanding institutions, preferences, and habits on a cultural level and, in the case of leaders, at an individual level; determining the values, goals, and past experiences that influence thinking at a societal level; and defining why the adversary fights and what they value. While these factors are intangibles and often the most difficult to collect and develop intelligence about, they are critical to understanding the adversary and determining not only what they can do (capabilities) but also what they will do (intentions). Developing this deep understanding of the adversary is critical for learning to think like the adversary and moving beyond a mere understanding of capabilities since the adversary will do what he thinks is possible, not what we think he can do.

This is particularly important in assessing adversaries from cultures that are significantly different from ours as mirror-imaging our values or ways of conducting operations to make sense of the seemingly irrational activities of an adversary may well lead to erroneous assessments of adversary intentions.

1.7.4 Developing Understanding

Intelligence is one piece of a continuum of information that is provided to the commander to build understanding. Figure 1-4 provides an overview of the hierarchy of information and makes it clear that the theory of intelligence would be incomplete without a discussion of how intelligence contributes to the development of the commander's understanding of the adversary and the operational environment. To develop understanding, the commander assimilates information from many sources to include intelligence and friendly force information. The commander then applies experience, judgment, and intuition to interpreting these disparate sources of information to achieve an understanding of the threat and the operational environment that is crucial to effective decisionmaking. The commander must also be mindful of their unique position in that many external and internal streams of data, information, and intelligence flow to the commander. Given this, the commander must take the responsibility to share information gained, particularly from external sources such as contacts with foreign leadership, with the intelligence team and other staff elements to ensure a fuller synthesis of available information occurs.

Figure 1-5 provides a conceptual view of the discussion thus far, showing the relationship between the information hierarchy and the intelligence process in developing intelligence that supports command decisionmaking. The depiction here is merely conceptual and is not meant to imply that the intelligence process follows a rigid, linear flow from planning and direction through dissemination and integration. In reality, the intelligence process is made up of highly interrelated but near simultaneously occurring intelligence operations, which is discussed in greater detail in chapter 3.

Intelligence Process Phase	Key Considerations
Planning and Direction	• Identifying intelligence requirements • Planning intelligence operations and activities • Supporting development of the commander's estimate of the situation
Collection	• Matching intelligence requirements to intelligence capability that best address the need • Using organic, attached, and supporting intelligence sources to collect intelligence
Processing and Exploitation	• Converting raw data into information and disseminating it in a form suitable to feed intelligence analysis
Analysis and Production	• Transforming processed and evaluated information into intelligence through the integration, analysis, and interpretation of all source data and the preparation of intelligence products in support of known or anticipated user requirements
Dissemination and Integration	• Providing timely intelligence, in the appropriate form, to those who need it
Evaluation and Feedback	• Conducted within every stage to ensure efficiency and effectiveness of intelligence operations in meeting requirements

Figure 1-3. Key Considerations of the Intelligence Process

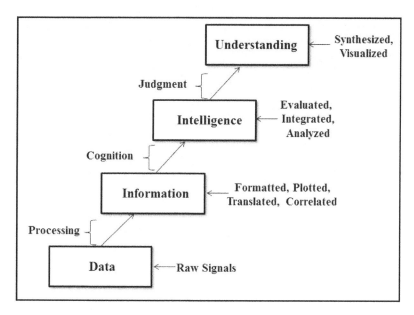

Figure 1-4. The Information Hierarchy

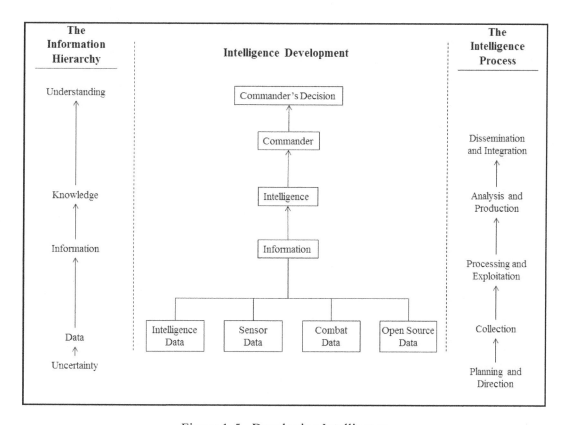

Figure 1-5. Developing Intelligence

1.7.5 Characteristics of Intelligence Excellence

While the role of intelligence in aiding the commander to develop understanding regarding adversary capabilities and intent has been discussed above, it is important to also consider the characteristic of good intelligence that ensure the commander can best assimilate the output of the intelligence process. Figure 1-6 provides a listing of the characteristics of intelligence excellence as defined by joint and other Service doctrine.

These characteristics are not meant to constitute a checklist which every intelligence product must meet but are merely guidelines that reflect characteristics that all intelligence exhibits to one degree or another. In some cases, there is a natural tension between some characteristics, such as timeliness and completeness, whereas other characteristics, such as timeliness, usability, and availability are mutually supportive.

Characteristic	Key Considerations
Anticipatory	Intelligence must anticipate the informational needs of commanders and their staffs in order to provide a solid foundation for operational planning and decisionmaking. This requires fully understanding the command's current and potential missions, the commander's intent, all relevant aspects of the operational environment, and all possible friendly and adversary COAs. Intelligence integration into operational planning must occur as early as possible.
Objective	Intelligence must be as free as possible of human bias or distortion and it has not been skewed to meet a preconceived judgment.
Complete	The intelligence meets the needs of the commander. Complete intelligence does not produce certainty; it merely means the intelligence provides sufficient depth of information upon which to plan or make a decision. Typically, the intelligence is provided with a confidence factor and addresses knowns and unknowns.
Accurate	The intelligence is factually correct. This does not imply certainty but merely means the facts and data used have not been distorted.
Timely	Intelligence must be delivered within required timeframes or it is of no use to the commander. Time considerations drive the type, format, dissemination path, and other factors in the intelligence process.
Useable	Intelligence products must be clear, concise, logical, and in a form that rapidly imparts meaning to the commander. Where possible, images and charts should be used since information is typically better understood when it arrives in the form of images.
Relevant	The intelligence must meet the commander's planning and decision-making requirements. The intelligence must be tailored to meet the needs of the commander.
Available	The intelligence must be readily accessible to the commander. This characteristic is a function of timeliness and usability but is also a function of information and knowledge management systems and communications architectures. Additionally, the use of security markings, which ensure sources of intelligence are protected while also ensuring commanders have reasonable access to intelligence based on their mission requirements, is a factor in making intelligence available.

Figure 1-6. Characteristics of Intelligence Excellence

1.7.6 Intelligence Practice

The intelligence process provides a model for a continuous and progressive process of intelligence development to meet the needs of the commander. Despite the technical sophistication and depth of intelligence capabilities employed by the United States, the practice of intelligence is a very human-centered process that requires close and active management throughout all its phases and the development and maintenance of individual and organizational relationships to ensure commanders are supported with the best possible intelligence.

The intelligence process is most easily understood when discussed as a cyclical process that starts with clear requirements in the planning and direction phase. It then involves a number of activities within each succeeding phase that ultimately result in the production, dissemination, and integration of intelligence products that meet decision maker needs. In reality, senior leaders do not always provide clear requirements for many reasons such as constraints on their time and lack of understanding about intelligence and what its capabilities can provide. Often it is up to the intelligence professional to develop a recommended set of prioritized intelligence requirements, based on their understanding of the mission and their commander's intent, and seek to have these requirements approved by the commander.[4]

In addition to the lack of clear requirements, other considerations that affect the practice of intelligence include:

1. Intelligence is dynamic. As stated in JP 2-0: "[i]n many situations, the various intelligence operations occur nearly simultaneous with one another or may be bypassed altogether." Simply stated, the intelligence process is, in actuality, neither linear nor cyclical. It requires continuous interactions between decision makers, planners, collectors, analysts, and others to ensure that requirements are understood and that resources are being effectively committed to address those requirements.

2. Intelligence is unlike other joint functions. Intelligence is one of the six joint functions, but there are more unknowns and less control in intelligence than in any other aspect of command. As stated in Marine Corps doctrinal publication (MCDP)-2, intelligence is unique because it deals directly with an independent, hostile will personified by the adversary that is purposefully attempting to frustrate the intelligence effort via denial and deception activities. Additionally, adversaries may not know their own intent so discerning the COAs they may choose is very problematic in those situations due to the lack of reliable indicators.

3. Commanders may lack understanding of intelligence. Intelligence processes are inherently opaque to the commander and the staff elements. For security reasons, intelligence is often classified and handled via special means, which isolates intelligence from other parts of the organization. Intelligence also employs specialized techniques and personnel and, unlike other organizational elements, these personnel and capabilities are often tied to theater or national-level capabilities. Additionally, it is often difficult to measure the impact that intelligence has on operations and, paradoxically, good intelligence may actually invalidate itself—such as in cases where warning provided by intelligence drive friendly force actions that cause an adversary to abandon their attack plans.

4. Quality is more important than quantity. Another unique feature of intelligence is that information obtained must be interpreted and what is most important is not the quantity of information but the quality of the information and the skill of the interpreters. Many intelligence failures have resulted from quality information that was incorrectly interpreted or from situations where analysts were overwhelmed by the amount of data requiring exploitation and analysis. Discerning the "signal" of relevant information from the "noise" of the vast amount of information available is a significant challenge for analysts and is another fundamental element of the operational art of intelligence.

5. Intelligence resources are finite. The intelligence process model does not discuss resource constraints, but there are resource capacity issues within every stage of the intelligence process that must be addressed by clear understanding of priorities and requirements and active management of the activities in each phase. Intelligence professionals must strive to strike a balance with the resources available and must make their

---

[4] Lowenthal, p. 7.

commanders aware of the fact that there are often tradeoffs required in the management of the intelligence effort. For example, the resources used to develop targets may be the same resources needed for bomb damage assessment (BDA) and so higher demands for one type of intelligence may result in sacrificing another. Intelligence must be treated like any other scarce resource. The intelligence process model assumes that collection assets are available and that they obtain the data needed to address intelligence gaps. This is often not the case. Weather, maintenance issues, or adversary activities may frustrate the collection effort or delay it to such an extent that the collection mission, if conducted, would no longer have any utility. Another important consideration is that intelligence systems collect what they can, which is quite different from collecting what is needed to address an intelligence gap. Additionally, many intelligence collection systems produce far more data and information than can be effectively managed, exploited, or disseminated. For these reasons, clear and prioritized requirements are essential for designing a flexible, adaptive, and focused collection, exploitation, and dissemination plans that mitigate the impacts of these factors.[5]

6. Intelligence cannot eliminate uncertainty. While it is a well-understood principle that intelligence can reduce but never eliminate the uncertainty found in warfare, the expectations placed on intelligence are often very high. Many commanders want more, better, and faster intelligence than the system can reasonably be expected to produce and both intelligence successes and failures may serve to increase the demand signal for intelligence. The intelligence professional must understand the capabilities of intelligence so that they can explain why intelligence products do not meet the level of detail or reliability desires of the commander. Also, intelligence professionals must understand that there are some questions that intelligence simply cannot answer either due to lack of capability or resource constraints and must work to make their commanders educated consumers of intelligence.

7. Cognitive biases are a threat to objective intelligence analysis. The intelligence process model also assumes that relevant information is recognized as such by the analyst and is assimilated into an all-source product that meets the commander's needs. In reality, the analytic process is a difficult and complex one given the fact that the analyst's mindset—the collection of assumptions, biases, and preconceptions they bring to the analytic process—may serve to filter out highly relevant data. Numerous studies have shown that people easily accept data and information that is aligned to their mindset and often question, discount, or devalue information that is inconsistent with their mindset. Chapter 3 and appendix F discuss the issue of cognitive biases and ways to mitigate their impacts on the intelligence process in greater detail but it is important to note that failure to objectively weigh and consider new data, especially data that is contradictory, is a major element of intelligence failures.

In summary, the intelligence process is exceedingly more complex than the intelligence process model indicates. Making the intelligence process work efficiently and effectively requires actively engaged and highly flexible intelligence professionals who understand the capabilities and limitations of intelligence systems at every level of war and are well versed at establishing the interpersonal and organizational networks required to leverage capabilities to address their commanders' requirements.

1.8 INTELLIGENCE OPERATIONS

Intelligence operations. The variety of intelligence and counterintelligence tasks that are carried out by various intelligence organizations and activities within the intelligence process. (JP 1-02. Source: JP 2-01)

Intelligence operations constitute the full range of activities that take place within the intelligence process. Intelligence operations are not limited to collection operations but occur within every phase and encompass the broad range of operational activities that are required to address intelligence gaps by transforming data into intelligence that imparts improved understanding of the adversary and the operational environment to the commander.

---

[5] Lowenthal, p. 7.

1.9 THE COMMANDER'S ROLE IN THE INTELLIGENCE PROCESS

"I want you to be the Admiral Nagumo of my staff," Nimitz had told me. "I want your every thought, every instinct as you believe Admiral Nagumo might have them. You are to see the war, their operations, their aims, from the Japanese viewpoint and keep me advised what you are thinking about, what you are doing, and what purpose, what strategy, motivates your operations. If you can do this, you will give me the kind of information needed to win this war."

Rear Admiral Edwin T. Layton, United States Navy (USN)
And I Was There

Finally, just like operations, intelligence is inseparable from command and is an inherent and essential responsibility of command. Intelligence failure is, ultimately, a command failure. To fulfill this command role, there are several principles that, if practiced, ensure that commanders are successful in meeting their intelligence responsibilities:

1. Understand intelligence doctrine, capabilities, and limitations. Understand the theory and practice of intelligence. They must understand intelligence doctrine and the capabilities and limitations of intelligence and from this understanding are able to recognize "good" intelligence from "bad" intelligence. Commanders should understand that intelligence provides estimates of adversary capabilities and intentions but it cannot predict the course of future events.

2. Provide planning guidance. Be personally involved in the intelligence effort by providing guidance, supervision, and judgment. Commander's intent, planning guidance, and identification of commander's critical information requirements (CCIRs) are key to developing relevant priority intelligence requirements (PIRs), a concept of intelligence operations, and coherent target development and nomination processes. Additionally, commanders must provide their Navy staff intelligence directorate (N-2) and the intelligence team with the authorities needed to plan and execute intelligence operations on the commander's behalf.

3. Define the area of interest. Define their areas of interest based on mission analysis, their concept of operations, and assessment of the relevant aspects of the operational environment.

4. Identify critical intelligence needs. Focus the intelligence effort by providing clear, specific, and prioritized requirements and establish limits and priorities for such things as the scope of the IPOE effort, dissemination formats and timelines, and bandwidth allocation. Commanders should not only specify what information is needed, but when it is needed. They should also understand that satisfying some intelligence requirements may require intelligence, surveillance, and reconnaissance (ISR) support from external organizations that require long lead times.

5. Integrate intelligence in plans and operations. Weigh and consider intelligence in their decisions since the primary purpose of intelligence is to support the commander's decision-making process. Commanders are ultimately responsible for ensuring intelligence is fully synchronized and integrated with plans and operations. It is the sole responsibility of the commander to make a personal analysis of the intelligence product and use it to arrive at an estimate of the situation that is the basis for the decision.

6. Proactively engage the intelligence staff. Actively engage their intelligence officers in discussions of adversaries, the operational environment, force protection, and future operations. Support the intelligence effort by being the advocate for the importance of intelligence and supporting requests for organic and external assets needed to address intelligence gaps.

7. Demand high quality, predictive intelligence. Hold the intelligence team accountable for providing predictive intelligence that meets the standards of intelligence excellence discussed in section 1.7.5. Evaluate the results of intelligence activities and provide feedback that leads to continuous improvement of the intelligence effort.

INTENTIONALLY BLANK

NWP 2-0

CHAPTER 2
Fundamentals of Naval Intelligence

Strategic intelligence does not differ from operational (combat) intelligence except in scope and point of view. In naval intelligence there is not, and can never be, any clear line of demarcation between strategic and operational intelligence. One flows into the other; distinction is possible only in terms of need and purpose. In what phase of a given operation is a particular item of intelligence needed? For what purpose is it needed? The viewpoint and needs of the commander about to engage the enemy will of necessity differ from those of the top strategic planner, yet a particular item of intelligence may be of value to both for different reasons, and may be used by each for a different purpose. Therefore the intelligence is operational when used by the commander immediately before or during battle and strategic when used by the top strategic planner.

United States (U.S.) Naval Intelligence School
Naval Intelligence, 1947

Nimitz's concept of intelligence was dynamic; facts were high grade ore to be sifted carefully, the pure metal of knowledge extracted and forged into a weapon to defeat the enemy.

Gordon W. Prange
Miracle at Midway, 1982

2.1 INTRODUCTION

Chapter 2 focuses first on defining the maritime domain, since it is the context within which fleet operations occur, and, then on defining naval intelligence, its enduring principles, attributes; and focus, and the qualities of the profession that are important to mission success. It also describes the functions of naval intelligence and the application of operational art to intelligence operations.

2.2 THE MARITIME CONTEXT

To enable an understanding of the principles of naval intelligence, it is first necessary to define the maritime operational environment, the unique characteristics of that environment and naval operations in general, and the major functions of the U.S. Navy, as these determine the scope and character of naval intelligence and its support to fleet operations. Three mutually supporting documents establish the context for the maritime operational environment:

1. A Cooperative Strategy for 21st Century Seapower

2. Naval Operations Concept 2010

3. Naval Doctrine Publication (NDP) 1, Naval Warfare.

Readers looking for more detailed explanations of the concepts provided in the overview that follows should refer to them. Additionally, understanding of these concepts is foundational to discerning the unique character of naval intelligence. Readers should also refer to Alfred Thayer Mahan's The Influence of Sea Power Upon History, 1660–1783, Julian S. Corbett's Some Principles of Maritime Strategy, and Wayne P. Hughes' Fleet Tactics and

Coastal Combat as these three works provide a solid foundation for the understanding of naval tactical, operational, and strategic concepts.

2.2.1 The Maritime Operational Environment

The maritime operational environment is complex and highly dynamic. Covering over 70 percent of the world's surface area, the oceans, rivers, and waterways of the world provide vital linkages between and among nations. Given that seaborne lift provides the most economical way to transport goods and commodities, roughly 90 percent of trade is carried on the seas and seaborne logistics provides the only feasible way to deploy and sustain a large joint force.

In addition to its value as medium for transport, much of the world's population lives within 100 miles of the sea. Advances in technology, globalization, and global prosperity over the last century has significantly expanded the use of the seas as a resource for food, minerals, oil, gas, and other commodities essential to modern societies. Maintaining freedom of the seas and protecting economic development in recognized territorial waters is an essential component of international stability and prosperity.

While the U.S. Navy has been the world's preeminent sea power since World War II, various trends will likely increase the complexity of the maritime operational environment and will pose significant challenges for naval forces. Competition for resources may cause nations to expand their territorial claims leading to increased friction between and among nations in a number of regions. Climate change will create new opportunities in previously inaccessible areas, like the arctic, but these same changes may bring new areas of competition and friction as well as lead to unanticipated changes in weather patterns that could cause humanitarian crises stemming from the effects of droughts, floods, destructive weather, and disease vectors. The rapid proliferation of technological and information capabilities available to state and nonstate actors will provide them with increasingly lethal and relatively easy to obtain capabilities that will be used to challenge the U.S. Navy's access to areas of interest and concern. Identifying, anticipating, and understanding how these changes in the maritime operational environment will potentially impact naval forces is a key focus of the naval intelligence professional.

Given the discussion above, it is apparent that the maritime operational environment consists of many components. First, there is the physical component, known as the maritime domain, which constitutes far more than just the "blue water" of the world's oceans. JP 3-32, Command and Control for Joint Maritime Operations, defines the maritime domain as "the oceans, seas, bays, estuaries, islands, coastal areas, and the airspace above these, including the littorals." The littorals, which constitute the boundary areas between the sea and the land, are defined as having two components. These include a seaward component, which constitutes the area from the open ocean to the shore, and a landward component, which is the area inland from the shore that the Navy can support and defend from the sea. The maritime domain thus also includes the seabed, subsurface, surface, air, and adjacent land. Operations in the maritime domain influence, and are influenced by, operations in the other physical domains as well as the information environment (to include cyberspace).

These physical and informational dimensions constitute different parts of the overall maritime operational environment but there are other significant aspects of that environment that the naval force must understand to effectively plan, execute, and assess operations. These other aspects of the maritime operational environment include weather and climate; history and sociocultural factors; and how neutral, friendly, and adversary forces employ the elements of their national power (their diplomatic, informational, military, and economic (DIME) resources) within each of the subdomains and within the integrated whole that forms the maritime operational environment. In addition to the physical environment and time, these other aspects of the operational environment are also commonly referred to as operational variables, which are described as the interrelated network of political, military, economic, social, information, and infrastructure (PMESII) factors.

Many Navy operations take place within the littorals and experience demonstrates that the closer operations occur to the littorals, the more complex and multidimensional they become. While the majority of the world's navies are small, regionally or territorially focused forces, these navies are only one concern that the Navy must contend with in planning and conducting operations at sea. Naval forces must also contend with adversaries who employ highly mobile, multimission platforms and an array of sea and land-based capabilities—such as mines, missile batteries, air forces, irregular forces, and electronic warfare forces—which an adversary may employ in an attempt to deter, degrade, disrupt, or destroy naval forces they perceive as a threat. The province of naval intelligence is to understand the physical characteristics of the multidimensional maritime operational environment and how they impact operations; penetrate the veil of secrecy regarding adversary capabilities, decisionmaking, and intentions; predict adversary courses of actions to include how, when, where, and to what degree the adversary will employ the elements of their national power to produce effects; and assess the potential or actual effectiveness of blue force naval force operations. Chapter 4 provides a fuller discussion of the maritime operational environment and its characteristics. Figure 2-1 provides an overview of the components of the maritime operational environment and how understanding of that environment supports the ability of the commander to visualize the current state of the maritime operational environment and how that environment should look at the conclusion of operations (the desired end state).

2.2.2 U.S. Navy Mission and Naval Force Operational Advantages

In accordance with Title 10 of the U.S. Code, it is the U.S. Navy's mission to be "organized, trained, and equipped primarily for prompt and sustained combat incident to operations at sea" and within the joint force the job of gaining and maintaining maritime superiority or supremacy—of engaging in and winning battles in the maritime domain and preventing conflict through presence offshore—falls almost exclusively to the naval Service. Given the need to counter a vast array of threats while simultaneously operating in and producing effects in a multidimensional environment, the Navy brings a unique and impressive array of capabilities to the joint force. The Navy presents the joint force commander with a highly adaptive force that is characterized by its lethality, speed, flexibility, scalability, mobility, and self-sustainability.

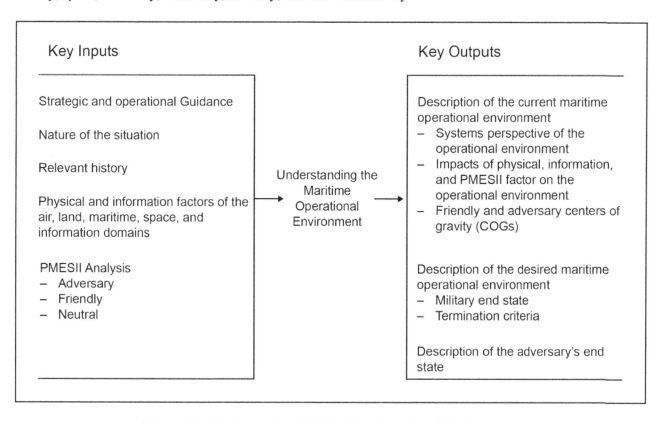

Figure 2-1. Understanding the Maritime Operational Environment

These characteristics provide two significant advantages, particularly when coupled with a model that ensures the forward deployment of naval forces to areas of concern across the globe. First, naval forces uniquely contribute to overcoming diplomatic, military, and geographic impediments to access, while respecting the sovereignty of nations. With their ability to use the seas as one vast maneuver space, naval forces can establish their presence in an area and use that presence to flexibly deter and, if necessary, engage with adversaries. This presence also positions naval forces so that they can rapidly respond to and address humanitarian and other crises. Use of the sea as maneuver space allows naval forces the ability to significantly complicate the decisionmaking of an adversary as naval forces can threaten large littoral areas by standing outside the detection range of adversary collection capabilities or far enough off shore that it is unclear to the adversary where to defend their coastline. The significant power projection capabilities provided by naval forces, such as those provided by tactical air, missile, and amphibious forces, enables the Navy to strike and, if necessary, capture targets far inland from the adversary's coast. Naval forces, through their maneuverability and scalability, can also rapidly escalate or deescalate the threat they pose to an adversary. Second, with their ability to bridge the seams between the air, land, sea, space, and information domains, the Navy provides an enabling force that can rapidly project combat power to hold adversary tactical, operational, and strategic targets at risk and secure the air and land access ports required by the joint force to generate and sustain combat power from landward bases.

2.2.3 Core Capabilities of the U.S. Navy

In addition to defining the complexities of the maritime domain, the mission of the Navy, and the unique role the Navy plays within the joint force, it is also important to define the core capabilities of the Navy as they provide the broad operational mission areas that naval intelligence professionals are required to support with collection capabilities and basic, current, and estimative intelligence analysis. The Navy's core capabilities are: forward presence, deterrence, sea control, power projection, and maritime security. Additional naval capabilities include FHA, strategic sealift, seabasing, and homeland security support.[1] The main features of these capabilities are:

1. Forward Presence. Naval Operations Concept 2010 defines forward presence as "[m]aintaining forward-deployed or stationed forces overseas to demonstrate national resolve, strengthen alliances, dissuade potential adversaries, and enhance the ability to respond quickly to contingencies." The Navy has a long history of being deployed throughout the globe—transiting and patrolling the world's vital sea lines of communication and positioning forces in areas of potential or actual conflict or disaster. Presence in international and foreign waters allows the Navy to protect U.S. shipping and other vital interests, promote trade and stability, enable influence, support diplomacy, provide indications and warning (I&W) of potential threats, and rapidly respond to crises. It facilitates all naval missions and underpins the ability to deter aggression, deliver decisive combat power, and set the conditions for follow-on operations by other components of the joint force. Presence is also instrumental to developing an understanding of the maritime operational environment as it enables observation of the behavior, culture, and capabilities of actors and familiarization with environmental conditions and geospatial/hydrologic features within the maritime domain. Naval intelligence supports forward presence by providing naval forces with analysis of the maritime operational environment and warning of potential threats. Additionally, naval intelligence uses presence operations to collect intelligence needed to support tactical, theater, and national decisionmaking and joint force operations. Naval intelligence also supports the assessment of forward presence operations effectiveness.

2. Deterrence. Deterrence is defined in joint doctrine (JP 1-02) as the "prevention of action by the existence of a credible threat of unacceptable counteraction and/or belief that the cost of action outweighs the perceived benefits." For deterrence to be effective there are two necessary components which must be in place. First, naval forces must be prepared to inflict unacceptable damage on the adversary and, second, the adversary must be aware of the potential threat and understand the risk so that they have the opportunity to refrain from hostile action. Naval forces have a number of credible threats with which to deter potential adversaries to include strategic nuclear capabilities, the forward posturing of conventional and unconventional combat power in areas of concern, and the ability to rapidly scale that combat power to increase or decrease pressure as needed. Given deterrent effects can be achieved by the application of all

---

[1] JP 3-32, Command and Control for Joint Maritime Operations.

elements of national power, effective theater security cooperation activities, partnerships, and even humanitarian operations are considered an extended form of deterrence as they can lead to increased stability and ameliorate potential sources of conflict. Naval intelligence supports deterrence activities through IPOE and analyses of adversary capabilities and decisionmaking that is instrumental for developing deterrent options and communicating them to the adversary. Assessment of the effectiveness of deterrent operations is also a key task for naval intelligence professionals.

3. Sea Control. Sea control operations is defined as the "employment of naval forces, supported by land and air forces as appropriate, in order to achieve military objectives in vital sea areas. Such operations include destruction of enemy naval forces, suppression of enemy sea commerce, protection of vital sea lanes, and establishment of local military superiority in areas of naval operations."[2] Also commonly referred to as the "essence" of seapower, the value of sea control is that it enables friendly use of the maritime domain in a specific area while denying adversaries the advantages inherent in maritime operations. Control of the seas provides not only a vast maneuver space for operating forces but also sanctuary and the ability to project power over the land and air of the littorals. In includes such operations as the identification, tracking, localizing, and destruction or neutralization of opposing adversary naval and air forces; the suppression of adversary seaborne commerce; protection of sea lines of communication; and establishment of local military superiority. Sea control is multidimensional, including all the subdomains of the maritime domain and may also include the seizure of land as required to protect the force and act as an enabler for other types of operations. While sea control remains a fundamental operational capability of the Navy, current trends demonstrate that increasingly capable blue water threats, theater antiaccess weapons, and area denial weapons employed in the littorals, such as mobile antiship cruise missile batteries and mines, create significant challenges to gaining and maintaining sea control. Countering these threats, which are likely proliferate among both state and nonstate actors, requires highly detailed understanding of the maritime operational environment. Naval intelligence supports sea control operations through IPOE, threat warning, and support to the entire detect-to-engage kill chain. Naval intelligence also aids in the assessment of the effectiveness of operations supporting sea control operations.

4. Power Projection. NDP-1 defines the naval context of power projection as "a broad spectrum of offensive military operations to destroy enemy forces or logistic support or to prevent enemy forces from approaching within enemy weapons range of friendly forces. Power projection may be accomplished by amphibious assault operations, attack of targets ashore, or support of sea control operations." In the context of joint operations, power projection is considered to be "the ability of a nation to apply all or some of its elements of national power—political, economic, informational, or military—to rapidly and effectively deploy and sustain forces in and from multiple dispersed locations to respond to crises, to contribute to deterrence, and to enhance regional stability. (JP 1-02) As a sea-based force, the naval force is not constrained by diplomatic, military, and geographic challenges to access and can project power ashore without access to ports or airfields. The naval force's power projection capabilities are lethal, scalable, flexible, and sustainable and can range from precision strikes on discreet targets using air/surface/subsurface fires, nonlethal effects, special warfare forces, and amphibious operations. These operations may support the degradation or destruction of specific adversary objectives and capabilities or include forcible entry operations required to establish the beachheads, ports, and airfield accesses needed to support further combat activities of the joint force. Naval intelligence plays a key role in power projection operations through such activities as identifying targets for strike operations; developing target intelligence packages; assisting with mission planning and weapon-target pairing decisions; producing intelligence products to support amphibious and special warfare operations; and assessing the effects of operations.

5. Maritime Security. Maritime security includes a collection of tasks that are derived from agreed-upon international law. Maritime security operations (MSO) are defined in NDP-1 as "operations to protect maritime sovereignty and resources and to counter maritime-related terrorism, weapons proliferation, transnational crime, piracy, environmental destruction, and illegal seaborne immigration." As discussed above, the economic strength of the United States and the stability of the world economy rest on secure sea lines of communication to conduct trade and on protected zones used for the economic development of the

---

[2] JP 1-02.

seas. There has been increasing use of the seas by various nonstate actors, such as criminal groups and pirates, which has the potential to create areas of regional instability and threaten legitimate trade activities. Additionally, the potential use of the seas to enable terrorist actions or weapons proliferation activities poses a threat to the United States and its allies. Maritime security, which includes operations such as maritime interception operations (MIO) and law enforcement detachments, can be performed individually, by one nation-state, or collectively, by a group of nations. Naval intelligence supports maritime security through intelligence collection and analysis efforts that support the planning, conduct, and assessment of MSO and provides specialized intelligence capabilities, such as the maritime interception operations–intelligence exploitation team (MIO-IET), to support the unique requirements of some MSO missions.

6. Strategic Sealift. Successful response to regional contingencies depends on sufficient strategic mobility assets in order to deploy combat forces rapidly and sustain them in an operational area as long as necessary to meet U.S. military objectives. Strategic sealift delivers the heavy combat units and their support equipment as well as the vital sustainment for deployed forces. The application of strategic sealift divides into three broad categories: pre-positioning, surge shipping during initial mobilization, and resupply sustainment shipping. A related mission, in combination with maritime security, is the protection of other nonstrategic sealift merchant shipping. The primary role of intelligence in strategic sealift is twofold. First, accurate intelligence assessments are critical to commanders who must decide when to employ strategic sealift capabilities and to determine where and how best to deploy these assets. Second, intelligence provides threat warning capabilities that are instrumental in ensuring these valuable assets are protected from adversary actions or operational environment impacts.

7. Seabasing. Seabasing is the deployment, assembly, command, projection, reconstitution, and reemployment of joint combat power from the sea without the reliance on land bases. It allows operational maneuver, and assured access to the joint force, while significantly reducing the footprint ashore, and minimizing permissions required to operate from host nations. As with strategic sealift, intelligence supports seabasing by providing the commander the insights needed to know when to deploy seabasing assets and the optimal positioning of those assets to best support operations while ensuring the protection of the seabase through threat warning activities.

8. Homeland Security Support. Maritime forces may also support the United States Coast Guard (USCG) in achieving its homeland security missions. USCG competencies and resources in support of the national military strategy and other national-level defense and security strategies include: maritime interception and interdiction operations; military environmental response; port operations, security, and defense; theater security cooperation; coastal sea control; rotary-wing air intercept; combating terrorism; and maritime operational threat response plan support. Naval intelligence would support the planning, execution, and assessment of these operations as they would for other naval missions but may need to be mindful of specific intelligence oversight or other considerations that must be addressed in supporting homeland security missions.

9. Foreign Humanitarian Assistance. FHA is defined in joint doctrine as "[p]rograms conducted to relieve or reduce the results of natural or man-made disasters or other endemic conditions such as human pain, disease, hunger, or privation that might present a serious threat to life or that can result in great damage to or loss of property. Humanitarian assistance (HA) provided by U.S. forces is limited in scope and duration. The assistance provided is designed to supplement or complement the efforts of the host nation civil authorities or agencies that may have the primary responsibility for providing humanitarian assistance. These activities include missions such as disaster relief (DR), support to civil authorities, and developmental assistance. Naval forces are uniquely suited to conduct these missions given their communications, lift, medical, basing, collection, and planning capabilities. Given the percentage of the world's population that lives close to the littorals and the Navy's forward presence activities, naval forces are often able to rapidly respond to FHA requirements. FHA can be proactive, when conducting operations such as support for humanitarian and civil development, and reactive, such as response to man-made or natural disasters. Naval intelligence supports FHA operations through IPOE efforts, collection and analysis efforts to support the planning and conduct of FHA missions, and collection and analysis efforts to assess the effectiveness of FHA in alleviating suffering or accomplishing development goals.

NWP 2-0

2.3 DEFINING NAVAL INTELLIGENCE

naval intelligence. Intelligence used to support naval and maritime decisionmaking, the planning, execution, and assessment of naval operations, and the naval personnel and associated organizations engaged in such activity.

2.3.1 Naval Intelligence—What it Is

The Navy has a long history of using its resources to collect and analyze information about foreign lands and the maritime environment. The concept that every sailor is both a potential collector and user of information is a time honored principle of the naval service. By the latter part of the 19th century, however, the Navy's leadership saw a need to adopt a more formal model for the collection of information and the analysis of that and other disparate information to produce intelligence estimates in support of policy, weapons development, acquisition, and fleet operations. In 1882, the Navy Department issued a general order which created the Office of Naval Intelligence (ONI), the nation's first permanent intelligence service. The order chartered ONI with "collecting and recording such naval information as may be useful to the Department in time of war and peace." Although ONI began humbly, with just one attaché in London and a few stateside analysts, the attaché system and analytic cadre grew throughout the period leading up to World War I. During that war, there was a significant expansion of intelligence capabilities in the Navy, not just at ONI but also with the designation of fleet personnel to formally perform intelligence collection and analysis functions in support of fleet planning and operations. While the interwar period saw a decline in naval intelligence capabilities and a shift in focus to stateside counterintelligence activities, this trend was reversed in the period just prior to America's entry into World War II. After Pearl Harbor, there was a significant expansion of naval intelligence across all levels of war. Although the size of the naval intelligence force has ebbed and flowed since World War II, it is clear from numerous policy and doctrinal documents since that period that naval intelligence plays a vital and important role in supporting senior level decisionmaking and the full range of Navy missions at each level of war.

While support provided by naval intelligence has evolved over time, so too has its definition. A training publication on naval intelligence written just after World War II states that naval intelligence "[w]hen used as an abstract noun to indicate material collected, processed, and disseminated for the use of naval authority may be defined as: 'The product of the subjection of information of naval interest to evaluation, analysis, and synthesis for the purpose of revealing its meaning and significance.'" This same publication also states that naval intelligence is a specialization within the larger field of specialization, known as intelligence, which requires professionals and organizations that "keep abreast of all military matters—technical developments, tactical doctrine, strategic theory, with special reference to navies, of course, but without neglecting land and air forces." The last capstone publication on naval intelligence described various aspects of the definition that can be summarized as: a system of personnel, procedures, equipment, and facilities both afloat and ashore, embedded at all echelons as either as dedicated or collateral duty assets, which follows a process designed to collect and analyze information that results in evaluated intelligence on an adversary's capabilities and intentions as a means to support planning and operations at all levels of warfare.[3]

The definitions above are consistent with the joint definition of intelligence which, as discussed in section 1.3.1, has a threefold meaning. Intelligence can refer to the product resulting from the intelligence process, the activity used to create the product, or the personnel/organizations engaged in such activity. Given the multidimensional nature of the maritime domain and the need for naval intelligence professionals to understand the PMESII aspects of the maritime operational environment, it is evident that while foreign naval forces are an enduring focus of naval intelligence, the range of topics and "products" that are "of naval interest" extend well beyond the capabilities and intentions of foreign navies and can vary significantly depending on operational requirements. For this reason, the broader definition of intelligence, as used by the joint community, should be adopted and modified to reflect these distinguishing characteristics of naval intelligence. With regard to the concept of naval intelligence as a product, it consists of intelligence products that are produced by intelligence professionals throughout the intelligence community (IC) that support naval decision makers and operational forces. While Navy intelligence organizations have well-defined collection and production responsibilities within a larger

---
[3] This definition was derived from various statements in NDP-2, Naval Intelligence, (30 Sep 1994) which was superseded in Mar 2010.

defense and national intelligence community framework, the nature of joint operations is such that intelligence produced from any organization within that community can potentially support the conduct of naval operations just as intelligence produced by naval intelligence organizations and professionals can support other Services in their operations. What distinguishes naval intelligence products from joint intelligence products are not the products, activities, or organizations involved in producing the intelligence but rather the consumer of the product. With regard to the activities used to create the product, there is no need to modify the joint definition as the intelligence process is used throughout the intelligence community. With regard to organizations engaged in the intelligence process, the joint definition must be modified to reflect the fact that naval intelligence, when used in that context, refers to the naval intelligence professionals who perform their duties within a system of afloat and ashore organizations (tactical, theater, and national). These organizations provide basic, current, and estimative intelligence on foreign navies and a broad range of issues pertinent to understanding all aspects of the complex maritime operational environment.

2.3.2 Naval Intelligence—What it is Not

Given that intelligence is a tool to reduce the commander's uncertainty about the adversary and the operational environment, naval intelligence professionals must educate consumers of intelligence on the strengths and limitations of their capabilities. First, commanders and supported activities must understand that intelligence can never eliminate uncertainty nor is it a "crystal ball" with which to foretell the future. Expectations concerning what intelligence can and cannot provide must be carefully managed to ensure the commander and supported activities understand where there are gaps in understanding of the adversary and the operational environment. Understanding what intelligence cannot address in a timely fashion will allow these consumers to either accept this uncertainty or advocate for more resources in an attempt to reduce the uncertainty further. There are simply some questions which intelligence cannot answer either because they are unknowable or because developing the human or technical accesses to the data and information needed to make an assessment would take years or a significant investment in resources. In these cases, the naval intelligence professional must be forthright in explaining why intelligence cannot address a specific question but must work with their commander to define other requirements that intelligence can potentially answer which may address the commander's concerns in a different way.

Second, intelligence is a finite resource and cannot instantly provide an answer to every question that the commander may have. Collecting, processing, exploiting, analyzing, and producing intelligence are all resource intensive activities that require individuals with highly specialized training and experience. Intelligence collection and analytic resources often operate at close to maximum capacity so reprioritizing and shifting the intelligence effort takes time, even in cases of significant crisis. For areas and regions that are not or have not been of great concern to the United States or the Navy, it takes time to develop and implement collection accesses and to build understanding of the operational environment and potential adversaries. Even for countries and issues of long standing concern, there may be a dearth of intelligence in certain subject areas and significant gaps in understanding due to lack of access, strong adversary operational security or denial activities, or the low priority of an issue relative to other requirements.

Finally, intelligence activities cannot violate the U.S. Constitution or U.S. law. Intelligence activities are constrained by laws, Executive Orders, and regulations that are designed to limit the scope of these activities to protect the rights of U.S. citizens. All intelligence operations are subject to extensive and rigorous oversight by the Executive Branch and commanders must ensure that intelligence operations are conducted in accordance with all applicable laws and orders.

It is also important to address here any potential confusion that may stem from the current organizational alignment of naval intelligence capabilities within the Navy. Naval intelligence is part of the Navy's Information Dominance Corps and is key component of the information dominance concept. Information dominance is defined as "the operational advantage gained from fully integrating the Navy's information, capabilities, systems, and resources to optimize decisionmaking and maximize warfighter effects in the complex maritime domain of the 21st century."[4] The fundamental capabilities of information dominance are assured command and control

---

[4] Office of the Chief of Naval Operations (OPNAV) N2/N6, Information Dominance Roadmap, 2013.

(C2), integrated fires, and battlespace awareness. Battlespace awareness is a capability focused on developing a penetrating knowledge of adversary capabilities and intent. The Information Dominance Corps provides an organizational construct for synchronizing and leveraging the Navy's core information capabilities—intelligence, cryptology, information operations, communications, meteorology, and space—to produce synergistic warfighting effects. The Navy is evolving the information dominance concept to address significant changes in the understanding and use of the information environment in warfare. These changes include:

1. Information as an enabler. Information capabilities have become a critical and vital enabler of naval operations. The volume of data that can be captured, stored, and transmitted along with the speed with which that information can be searched and analyzed has created significant operational advantages for U.S. naval forces.

2. Information in warfare. The information environment is an enabler of operations in other domains.

3. Information as a weapon. With the development of capabilities in the electromagnetic spectrum and cyberspace, platforms now exist that allow information professionals to produce warfighting effects.

4. Information as a threat. Adversaries are well aware of the operational advantages inherent in information capabilities and are also using those capabilities to counter U.S. forces.

5. Information as warfighting discipline. Application of all the elements of information dominance in a synergistic way requires a cadre of professionals who understand how to employ the full range of information capabilities to produce warfighting effects for the commander.

While naval intelligence is a key component in the information dominance construct, the construct does not fundamentally change the nature of naval intelligence. Intelligence, when used in coordination with other naval information capabilities, can produce synergistic effects within the information environment but support to information dominance activities is the same as intelligence support to other warfare areas. As the information dominance concept evolves and matures, naval intelligence is responsible, as it is for other warfare areas, to provide support for the planning, execution, and assessment of operations in the information environment by providing penetrating intelligence of adversary capabilities and intent and characterizing the information environment and its intersections with the physical domains.

Another potential source of confusion about naval intelligence stems from the organizational alignment within the command, control, communications, computers, combat systems, and intelligence (C5I) construct. The term C5I is nondoctrinal but is used by the Navy as an organizing construct to collectively refer to fleet information and combat systems capabilities. While this construct, from a fleet type commander perspective, has significant overlaps with the Information Dominance Corps construct, they are not synonymous. Further confusion arises from the use of another nondoctrinal term—cyber—by various fleet entities. While the term, at a very basic level, refers to information technologies, it also has many other meanings within the fleet depending on context. For example, Navy Cyber Forces is tasked as the global C5I type commander and so has a significant role in ensuring that the intelligence-related intelligence manning, training, and equipping needs of deploying forces are met. Fleet Cyber Command (FCC), however, is the Navy's component command assigned to U.S. Strategic Command and U.S. Cyber Command. FCC is concerned with numerous warfighting responsibilities associated with information operations and also acts as the Navy's service cryptologic component for the National Security Agency (NSA). Within the FCC construct, naval intelligence acts in a traditional N-2 role, providing intelligence support to FCC and its operational component, the U.S. Tenth Fleet. The purpose of this discussion is to highlight the fact that naval intelligence retains its unique identity regardless of the organizational constructs through which naval intelligence capabilities are aligned, employed, or provided to the fleet. While naval intelligence is heavily reliant on a vast array of "cyber" capabilities to perform its core tasks, these cyber capabilities do not define or circumscribe naval intelligence, which can still perform its core tasks without reliance on cyber capabilities.

2.4 FUNCTIONS OF NAVAL INTELLIGENCE

Naval intelligence has evolved a core set of functions that form an organizing construct for the development and delivery of analytic products designed to reduce uncertainty about the adversary, the operational environment, and blue force vulnerabilities. While there are similarities among all of these functions, each function is discreet and has unique features that drive tailored collection, exploitation, analysis, and dissemination approaches. Core functions are performed at all levels of naval intelligence. They are complementary and mutually supportive and, to the extent possible, occur concurrently and continuously based on resources available.

For example, a maritime intelligence operations center (MIOC), carrier intelligence center, or large deck amphibious ship joint intelligence center (JIC) has sufficient capacity to perform all of these functions simultaneously. To the greatest extent possible, collection capabilities, raw and exploited information, analytic products, and dissemination paths and methods are shared among intelligence divisions and work centers to produce synergistic effects. For example, IPOE may have been used to determine that adversary mine laying platforms are always preceded by a combatant vessel that sweeps the proposed minefield for threats and then stands off to provide force protection. I&W capabilities may detect this type of activity which is then passed to intelligence personnel who support the development of additional IPOE materials and targeting packages designed to strike at adversary mine laying capabilities. After the strike, intelligence personnel would assist with the assessment of the strike's effectiveness and use intelligence capabilities to monitor adversary reactions as part of force protection efforts.

By managing the intelligence team and the information management processes it uses, intelligence leadership ensures that all collection, processing, exploitation, analytic, dissemination, and evaluation resources are efficiently and effectively used to support the commander. While independent duty intelligence specialists or collateral duty intelligence personnel do not have the resources available to perform these functions concurrently, they are still responsible for these functions and typically perform them, to a greater or lesser degree, throughout their deployments.

2.4.1 Intelligence Preparation of the Operational Environment

Intelligence preparation of the operational environment is a foundational function that supports all other naval intelligence functions. JP 2-01.3, Joint Intelligence Preparation of the Operational Environment, defines IPOE as an "analytical methodology employed to reduce uncertainties concerning the enemy, environment, and terrain for all types of operations. Intelligence preparation of the operational environment builds an extensive database for each potential area in which a unit may be required to operate. The database is then analyzed in detail to determine the impact of the enemy, environment, and terrain on operations and presents it in graphic form. Intelligence preparation of the operational environment is a continuing process."

Note

Use of the term "database" in the definition provided is not a reference to a specific database or system. It refers to the collection of information on the adversary and the operational environment that is available to the naval intelligence professional in whatever format is available—electronic, hardcopy, etc.

Goals of IPOE include developing understanding of the adversary's forces, doctrine, tactics, and operational methods; the physical and environmental characteristics of the area; and the influence and interrelationships among operational variables (e.g., PMESII). In addition to building a comprehensive picture of the operational environment, IPOE is instrumental in identifying gaps in knowledge that are essential for driving intelligence collection, exploitation, analysis, and production efforts. Given the importance of IPOE to understanding the maritime operational environment, a more detailed discussion of the methodology is provided in chapter 4.

Joint doctrine (JP 2-01.3) emphasizes a distinction between joint intelligence preparation of the operational environment (JIPOE) (IPOE conducted at the joint task force (JTF) or combatant command level) and intelligence preparation of the battlespace (IPB) which "is specifically designed to support the individual operations of the component commands." While joint doctrine states that JIPOE and IPB products "generally differ in terms of their relative purpose, focus, and level of detail," it makes no distinction concerning the analytic processes used to create those products. JIPOE and IPB are complementary and mutually supporting processes and, while JIPOE typically takes a macro-analytic approach focused on identifying adversary vulnerabilities and COGs at the operational level; both processes may require the same high levels of detail during situations such as military engagements, security cooperation, deterrence operations, and crisis response. Figure 2-2 provides an overview of the relationship between JIPOE and IPB as discussed in joint doctrine along with the four process steps associated with both processes and the three major components of the maritime operational environment.

Navy tactics, techniques, and procedures (NTTP) 3-32.1, Maritime Operations Center, uses the term IPOE to describe the application of the JIPOE process at the operational level of war for the Navy. Given there is no distinction between the processes used for JIPOE, IPOE, or IPB and that each process may have the same focus and purpose depending on the circumstance, the term IPOE is used throughout this publication to describe the Navy's application of the JIPOE/IPB process at the operational and tactical levels of war and the products derived from that process. Since the term "battlespace" is no longer defined in joint or naval doctrine, use of the term IPB has become problematic and it is not used in this publication to describe the Navy's IPOE process or products.

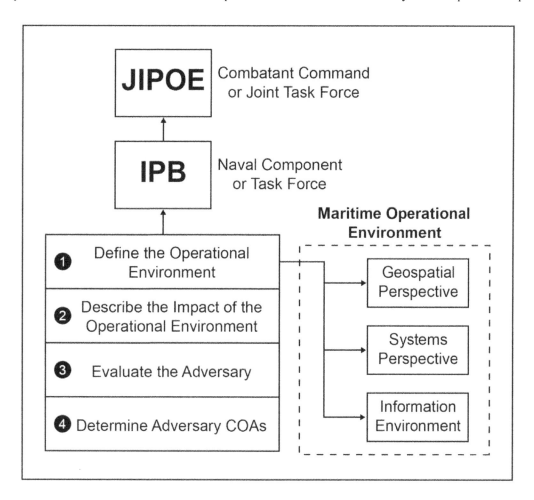

Figure 2-2. Intelligence Preparation of the Battlespace and Aspects of the Operational Environment

2.4.2 Indications and Warning

The goal of I&W is to provide early warning of potential hostile action. IPOE supports I&W by enabling development of indicators that may signal potentially hostile actions are imminent and allows for the interpretation of adversary activities within a broader operational context. I&W is conducted at all levels of war. At the strategic level, I&W is focused on detecting and discerning an adversary's preparations for war, policy shifts, and advances in military capabilities. With regard to time, space, and force considerations, strategic warning is concerned with issues that typically take months or even years to develop, deals with potential threats that can be many thousands of miles distant, and is focused on national-level capabilities, such strategic nuclear forces or signs of national mobilization efforts. At the operational level, the focus of I&W is on threats posed by regional actors and on changes in the operational environment, such as an impending natural disaster, that may require employment of naval forces. Time, space, and force considerations on the operational level typically are measured in weeks or months, concern threats with a given regional area, and are concerned with operational level capabilities, such as fleets. On a tactical level, I&W is critical for naval forces as they are frequently deployed to littoral regions where reaction times to threats are significantly decreased and typically consist of activities to detect, track, and derive the intent of adversary forces within the commander's area of interest. For tactical forces, warning problems may be measured in minutes or hours, concern threats within areas of operations or interest that may span tens or hundreds of miles, and involve the employment of tactical capabilities by the adversary, such as submarines or strike aircraft. While IPOE and training provide the foundation for the I&W function, naval intelligence professionals must work with their commanders to determine what collection resources are used to support I&W and must swiftly address any degradations in I&W capabilities given the criticality of this function to force protection.

Note

The term "indications and warning" is replaced in joint doctrine (JP 1-02) by the term "warning intelligence." For U.S. Navy purposes, and throughout this publication, the term "indications and warning" is used.

2.4.3 Current Intelligence

Current intelligence provides updated support for ongoing operations. It involves the integration of time-sensitive, all-source intelligence and information into concise, objective reporting on the current situation in a particular area. The term "current" is relative to the time sensitivities of the decision maker and the context of the type of operation being supported. For example, in some contexts intelligence may be considered "current" that is days or weeks old, whereas other circumstances may require current intelligence in near real time. Current intelligence provides the commander and supported activities with the situational awareness needed to support the conduct of current operations and to provide information on developing situations and potential contingencies that may require greater intelligence focus or the need to initiate planning activities in anticipation of future tasking or response actions. Examples of activities that support the current intelligence function are the production of a daily intelligence brief, intelligence update briefs on specific topics of interest to the commander or supported activities, and maintenance of the intelligence portion of the common operational picture (COP). IPOE and current intelligence are mutually supporting functions as IPOE provides the basic intelligence products that form the background material or context for current intelligence analysis and production efforts. The results of current intelligence analysis and production are used to update IPOE products and associated databases. Similarly, I&W and current intelligence are complementary functions that typically exploit and analyze the same raw data and information, but I&W is typically focused on a narrower and very specific range of issues since the time-sensitive nature of I&W threat warning requires more rapid analysis and dissemination of intelligence than does current intelligence.

2.4.4 Targeting

Joint doctrine (JP 1-02) defines targeting as the "process of selecting and prioritizing targets and matching the appropriate response to them, considering operational requirements and capabilities" and target intelligence as intelligence "that portrays and locates the components of a target or target complex and indicates its vulnerability and relative importance." Targeting is an analytic process that is focused on determining how to most efficiently and effectively disrupt, degrade, neutralize, or destroy adversary capabilities and determining the effect these actions will have on the adversary. Targets are developed and selected in accordance with the commander's guidance and objectives and may be initially identified through the IPOE process. Targets can be physical, such as bridges or missile batteries, or functional, such as an adversary's C2 capability, and can be interdicted or attacked using a variety of lethal and nonlethal capabilities such as strike aircraft, special warfare forces, or cyberspace operations. Intelligence supports the targeting process through the development of target intelligence packages, which are a compilation of essential information about a specific target that includes enemy order of battle (air, land, and sea), target description, vulnerabilities, lines of communication, general and localized photography, and other all-source intelligence information. Intelligence also supports the weaponeering process, which involves matching the appropriate offensive capability to the desired effect and support retargeting decisions through the BDA process.

2.4.5 Intelligence Assessment

JP 1-02 defines assessment as a term which refers to "(1) [a] continuous process that measures the overall effectiveness of employing joint force capabilities during military operations; (2) Determination of the progress toward accomplishing a task, creating a condition, or achieving an objective; (3) Analysis of the security, effectiveness, and potential of an existing or planned intelligence activity." Naval intelligence has a role to play with regard to each of these definitions. It supports the commander in determining the effectiveness of operations not only against adversary forces but also within the context of MSO and HA/DR missions. Additionally, naval intelligence, through the process of monitoring the operational environment, can assess the effects that naval force actions are having on neutral actors and coalition partners. Like intelligence support to planning and executing operations, naval intelligence professionals must plan for the collection, exploitation, analysis, and dissemination requirements associated with assessment and balance the needs of assessment against other intelligence priorities in accordance with the commander's guidance. Naval intelligence professionals must also continuously assess the effectiveness of each aspect of the intelligence process in supporting the commander's objectives and must address suboptimal or ineffective elements by developing and implementing new approaches for the planning, collection, exploitation, analysis, dissemination, and evaluation of intelligence as required. Evaluation and feedback consideration for each component of the intelligence process are discussed in chapter 3.

Note

While BDA, which JP 1-02 defines as the "estimate of damage composed of physical and functional damage assessment, as well as target system assessment, resulting from the application of lethal or nonlethal military force" is a structured and well known assessment methodology for evaluating the effectiveness of targeting operations, naval intelligence also has a broader assessment function that goes beyond BDA.

2.4.6 Security

Joint doctrine (JP 1-02) defines security as "(1) Measures taken by a military unit, activity, or installation to protect itself against all acts designed to, or which may, impair its effectiveness; (2) A condition that results from the establishment and maintenance of protective measures that ensure a state of inviolability from hostile acts or influences; (3) With respect to classified matter, the condition that prevents unauthorized persons from having access to official information that is safeguarded in the interests of national security." From a naval intelligence perspective, the security function is both offensive and defensive. While it is supported by all intelligence capabilities and incorporates analysis of information from all intelligence disciplines, it is primarily a counterintelligence mission that is focused on neutralizing, disrupting, or destroying hostile intelligence networks; supporting the operational force protection mission with intelligence resources; and supporting the planning, execution, and assessment of friendly operational security and deception activities. Counterintelligence and security measures are essential to identifying friendly force vulnerabilities, reducing operational risk, and achieving surprise in naval operations. Security must be considered in every aspect of planning and the execution of security measures must be tailored to the intended operations and the adversary's capabilities. Decisions regarding releasability of intelligence to naval forces and coalition partners must be made early in the planning process and reevaluated on a periodic basis to ensure that all activities are receiving necessary intelligence while also ensuring that there are no intelligence compromises. In all cases, the need to balance operational requirements with the need to protect sources and methods must be the paramount consideration in making decisions concerning information sharing.

2.4.7 Lead and Manage the Intelligence Process

Intelligence professionals have vast areas of concern. Naval intelligence professionals must take ownership of the intelligence process and manage each element of the process to efficiently and effectively meet operational needs. The key to focusing the intelligence effort is a clear understanding of the commander's priorities and the needs of supported activities. While the naval intelligence professional can support their commander by leveraging the considerable capabilities and resources of the greater intelligence community, coordinating that support and ensuring it is available in a timely fashion requires planning, anticipation of the commander's requirements, and the development and fostering of organizational and personal relationships. Effective leadership of the intelligence process requires clear articulation of goals and objectives for each phase of the process and the development of realistic plans made in coordination with subordinates, the commander, supported activities, and external organizations. Effective management of the intelligence process results from clearly stating priorities and standards as well as the willingness to delegate responsibilities to the lowest levels possible. It also requires flexibility and adaptability, which are needed to change or develop TTP in response to new situations or to address ineffective or inefficient outcomes. The commander expects the naval intelligence professional to have the masterful knowledge of the adversary and the operational environment needed to reduce uncertainty by predicting how events will unfold but just as importantly the commander expects the naval intelligence professional to be the expert in how to lead and manage the intelligence effort to support his or her objectives.

2.5 PRINCIPLES OF NAVAL INTELLIGENCE

Naval intelligence has a long and honored history of providing superior intelligence support to policymakers and the fleet. These years of experience in supporting operational forces has influenced the development of proven and enduring principles which, if followed, result in effective and successful intelligence operations.

2.5.1 Know the Adversary and the Operational Environment

The primary focus of the naval intelligence professional is the adversary. Achieving a deep understanding of adversary capabilities, intentions, and decisionmaking remains the primary objective of the intelligence effort. The two primary tools used to produce this understanding are the development and implementation of innovative intelligence collection strategies to address gaps in knowledge, and objective, relevant, accurate, and complete intelligence analysis. The naval intelligence professional must learn not only how the adversary thinks but must be able to adopt the adversary's way of thinking so that they can discern the strategies, goals, objectives, and intentions of the adversary. Deep understanding of adversary weapons systems, infrastructure, and other technical capabilities must be coupled with insights into adversary doctrine, training, actual methods of operations, readiness, and morale as these factors determine the boundaries of how the adversary will employ their capabilities.

Additionally, the naval intelligence professional should study the character, culture, social mores, customs, traditions, and history of the adversary as these factors have significant importance in determining adversary courses of action and methods of operation. Building knowledge of the adversary is a continuous process that also includes developing an understanding of the other aspects of the maritime operational environment—such as weather, oceanography, geography, hydrography, and informational factors—that impact adversary decisionmaking and operations. While the United States can never know adversary capabilities and intentions with absolute certainty, each new fact about the adversary can be integrated and fused into an overall assessment that leads to better understanding. Naval intelligence professionals must strive to develop this knowledge of adversary and the maritime operational environment to the greatest extent possible in times of peace.

Since naval forces are often the first to respond to a crisis, there is often little or no time to develop the deep understanding needed to support the planning, execution, and assessment of fleet operations once those operations have commenced. Since intelligence resources are finite, naval intelligence professionals must focus on the most likely scenarios they are expected to face and then develop and implement a plan that directs their collection and analytic efforts on those likely scenarios. Other potential adversaries and crisis situations should be monitored and contingency plans made to leverage the knowledge and understanding of the adversary and the maritime operational environment possessed by other tactical, operational, and national level resources in the event a potential or unanticipated crisis erupts.

2.5.2 Commander's Needs are Paramount

Naval intelligence professionals must strive to develop a deep understanding of their commander and his or her needs. It is the commander's requirements that provide focus for the intelligence effort, which ultimately concerns an attempt to reduce the commander's uncertainty about the adversary and the maritime operational environment so that they and their staffs can effectively plan and conduct operations. The naval intelligence professional must work closely with their commander, operations, and other directorates and participate in planning at the earliest possible stage. Assembling the intelligence products needed to support the mission; building a collection strategy and plan; conducting coordination to leverage other organizations' collection and analytic resources; and developing priority intelligence requirements with the commander are activities that all require substantial time and are needed to support the overall planning effort and effective decisionmaking. Ideally, the intelligence professional must learn to anticipate their commander's needs so that the focus of the intelligence effort can shift to meet the commander's requirements in a timely manner. Additionally, the naval intelligence professional must understand how their commander best receives information as this impacts how intelligence is produced, disseminated, and integrated.

Adherence to this principle requires that naval intelligence professionals become conversant in blue force (friendly force) mission, capabilities, operations, and tactics. Expertise on the adversary and the operational environment, while important, is insufficient if the naval intelligence professional cannot apply that knowledge to understand the impacts those factors may have on blue forces. The understanding of blue forces is also critical to successful red teaming. Adversaries will apply themselves to studying U.S. military capabilities, operations, and tactics in great detail and will base their calculations of how and when to commit their forces based on this understanding. Naval intelligence professionals must understand blue forces to the same level of depth to ensure they are delivering intelligence that meets the needs of the commander.

2.5.3 Ensure Unity of Intelligence Effort

Unity of intelligence effort provides for the efficient and effective use of intelligence resources and ensures that the intelligence process is focused on meeting the needs of the commander. The two main drivers for unity of effort are clear statements of requirements and the prioritization of those requirements. Since naval intelligence is one component of a far larger intelligence community team, clear and prioritized requirements are key to leveraging and focusing the support that may be provided by other tactical, operational, or national level intelligence capabilities. Achieving unity of the intelligence effort is not a simple matter and it requires leadership and the cultivation and maintenance of personal and organizational relationships to ensure that both internal and external intelligence resources clearly understand the commander's needs and are focused on meeting those needs.

2.5.4 Plan for Combat and Crisis Response

Crises often erupt with little warning. Preparation for combat and crisis response is a continuous process that requires close monitoring of the most likely crisis scenarios to determine as early as possible when the crisis may occur or when hostilities are likely to erupt. Effective planning for combat and crisis response requires knowledge of the adversary and the maritime operational environment, the understanding of the commander's needs, and the unity of effort that focuses the intelligence team. In times of peace, naval intelligence professionals must plan for combat and crisis response scenarios and conduct realistic training to support the full range of naval operations that may be needed to engage and defeat the adversary or respond to natural or man-made disasters. This planning should include identifying personnel, system, communication, collection, and analytic resource requirements as well as developing fallback options if these resources are unavailable due to issues such as resource constraints, system failures, or adversary action. Additionally, the plan should address how internal assets are realigned and surge capabilities integrated to meet combat and crisis response requirements and how plans for collection, processing and exploitation, analysis and production, and dissemination and integration will change to meet those requirements.

2.5.5 Use an All-Source Approach

The use of an all-source approach to the development of naval intelligence assessments is key to achieving relevant, accurate, objective, complete, and anticipatory intelligence for the commander. It is rare that a single source or sensor provides that key piece of intelligence needed to understand adversary capabilities and intentions. For example, some attribute the collection of a single communications intercept as the single piece of intelligence that led to the downing of Japanese Admiral Isoroku Yamamoto's plane in World War II. While the intercept was a significant and timely piece of information, it did not exist in isolation. The information from that intercept was fused with other data, such as IPOE on adversary airfields and the likely course, speed, and range of the aircraft, along with other aspects of the current intelligence picture to develop the plan for the shootdown. Fusion of data and information from multiple sources typically provides the most useful and complete intelligence assessments possible. All-source analysis also has a number of inherent benefits. First, it helps to define gaps in knowledge that are useful to developing collection requirements and an effective collection plan. Second, it allows for corroboration of data, which is needed to detect adversary denial and deception efforts and to guard against errors in analytic judgment. Finally, the all-source approach reveals conflicting data which forces the analyst to review the intelligence estimate and determine if this new data causes a reassessment of that estimate and leads to a more accurate understanding of the adversary and the operational environment.

Note

> JP 1-02 defines "all-source intelligence" as 1. Intelligence products and/or organizations and activities that incorporate all sources of information in the production of finished intelligence. 2. In intelligence collection, a phrase that indicates that in the satisfaction of intelligence requirements, all collection, processing, exploitation, and reporting systems and resources are identified for possible use and those most capable are tasked. The term "all-source" is used in this publication to mean "all available sources" of information since access to all sources of information is rare. Another way to view the term is that intelligence professionals fuse and analyze data and information from as many sources as are available in developing intelligence products.

2.6 CHARACTERISTICS OF NAVAL INTELLIGENCE EXCELLENCE

Naval intelligence has the same characteristics of excellence as the joint characteristics found in chapter 1. Intelligence that is anticipatory, timely, objective, relevant, thorough, accurate, available, and useable meets the standard of excellence expected from naval intelligence professionals in the practice of their craft. It must be remembered that these characteristics provide a guideline for the intelligence effort since attaining all these characteristics to the same high degree is not possible due to the fog of war and finite resources. For example, the most complete and accurate intelligence product is useless to the commander if it is not provided in time to support a decision. The naval intelligence professional must always be mindful of the needs of their commander and others they support to ensure that these characteristics are balanced in such a way as to meet their specific needs, which will likely change with each new situation or as an ongoing situation evolves. This requires consistent and continuous dialog with the commander and supported activities so that they are made aware when tradeoffs are needed and can weigh in with their perspectives and desires. The Battle of Midway demonstrates an example of naval intelligence excellence to illustrate how these characteristics are applied in supporting operations.

2.7 NAVAL INTELLIGENCE PROFESSIONAL CHARACTERISTICS

We have developed machines with uncanny powers of gathering and sorting information, but we have not thus far developed a machine with the power to think, and, by thinking, to convert information into intelligence.

U.S. Navy Intelligence School
Naval Intelligence, 1947

Despite the technical sophistication of the intelligence collection, processing, analysis, and dissemination capabilities available to naval intelligence professionals, the practice of intelligence remains a very human endeavor. Therefore, the characteristics of those who practice the profession of intelligence have a significant impact on the quality of intelligence. Like the characteristics of good intelligence, the characteristics of naval intelligence professionals provided below is meant to highlight personal attributes that contribute to the development and delivery of successful intelligence. Naval intelligence professionals may possess these characteristics to a greater or lesser degree depending on their education and experience, but cultivating these characteristics is one of the keys to intelligence excellence.

The Battle of Midway and the Characteristics of Naval Intelligence Excellence

Commander Edwin Layton, the Commander in Chief, United States Pacific Fleet (CINCPACFLT) intelligence officer, was tasked by Admiral Chester Nimitz to determine the time and method of an anticipated Japanese attack on Midway. Nimitz's PIR was clear and allowed Layton to focus collection and analytic resources on addressing this specific intelligence problem. Layton and his analysts received a decoded message from a Japanese aircraft carrier that contained the phrase "we plan to make attacks from a general northwest direction." Although the message did not designate a specific target, Layton and his team assessed the message referred to Midway and concluded that the Japanese would approach the island from the northwest on an approximate bearing of 315 degrees. Using the analysis of the Pearl Harbor attack as a guide to Japanese tactics, Layton estimated the Japanese would make their approach to Midway under cover of darkness and launch their attack aircraft at first light, which is when U.S. forces would launch their search planes.

Layton's close relationship to Nimitz and his staff counterparts enabled him to anticipate his intelligence requirements in support of the defense of Midway. The CINCPACFLT intelligence team combined new information with available intelligence to develop an all-source assessment that the Japanese would attack on June 4th. Layton and his team delivered an accurate and objective assessment, predicting that the Japanese force would approach Midway on a bearing of 315 degrees and that they would be sighted 175 miles from Midway at 0600. (Shortly after 0600 on June 4th, the main Japanese force was found on an approach of 320 degrees at a distance of 180 miles, causing Nimitz to later remark that Layton was "only five miles, five degrees, and five minutes off.") The estimate was delivered in a timely fashion, ensuring that operational planners had useable intelligence that allowed them to position search aircraft and combat forces in ways that negated the Japanese advantage of surprise and provided U.S. naval forces an operational advantage. The thoroughness of the assessment was evident in its level of detail and its specificity ensured its relevance to Nimitz and his staff.

While not an all-inclusive list, the characteristics of naval intelligence professionals include:

1. Adaptability. Early publications on naval intelligence used the phrase "flexibility of mind" to describe this characteristic. It is defined as "the ability to effectively meet new situations as they arise, and to predict the course events may take." As is stressed throughout this publication, naval intelligence professionals must know a great deal about many topics but must focus their resources on the most likely crises and contingencies they will face. Given the rapid pace of change in the world, especially during combat, naval intelligence professionals must have the ability to rapidly respond to changes in mission, the adversary situation, and the operational environment. These changes require the naval intelligence professional to

swiftly alter collection, processing, exploitation, analysis, and dissemination plans; clearly communicate changes in direction to subordinate personnel and those providing reachback support; and articulate resource gaps and tradeoffs to the commander and supported activities. Redirecting the intelligence effort on short notice requires bold and original thought to ensure that the intelligence process informs the planning, conduct, and assessment of operations in the face of rapidly changing and fluid circumstances.

2. Intellectual Honesty. Also referred to as "objectivity," this characteristic concerns the need for naval intelligence professionals to produce intelligence estimates that are based solely on analysis of data and information available and are not unduly influenced by biases or expectations of the analyst or the consumer of the intelligence. The naval intelligence professional must provide the analytic interpretation that is supported by the data and reflects objective judgment, vice providing an analysis that is what others want to hear or which supports a given agenda. Providing an assessment that may cause a complete reworking of the blue force operational plan and scheme of maneuver or providing an assessment that shows a perceived "successful" strike was, in fact, ineffective are difficult tasks but are necessary to ensure the commander has the best opportunity to defeat the adversary. Downplaying potential threats or tailoring intelligence to meet an agenda does not serve the commander and ultimately results in degrading confidence in the intelligence process and the integrity of its practitioners. Naval intelligence professionals must also consider how best to deliver intelligence that may be received negatively. Knowing the commander and others supported by the intelligence process ensures that products are developed that address their concerns in a way that enables them to best receive the intelligence and minimize negative reactions to it. In addition to ensuring intelligence is not skewed to meet a particular agenda, naval intelligence professionals must also be mindful of their own biases when analyzing data and information. A full discussion of these cognitive biases and techniques to mitigate their impacts occurs in appendix D.

3. Teamwork. While much is written about the solitary nature of "the analyst" and the intellectual nature of the intelligence profession, naval intelligence is most successful when teamwork is employed. The development and maintenance of internal and external relationships is critical to success in all phases of the intelligence process. To effectively plan and direct intelligence operations, the naval intelligence professional must understand the needs of the commander and activities that their intelligence organization supports. This requires continuous dialog and sharing of information about requirements, capabilities, and limitations. In addition to forging relationships outside of the intelligence organization, there must be leadership within the intelligence organization to ensure the efficient and effective use of intelligence resources in support of the commander. Collection is another element where teamwork is essential. All naval intelligence professionals, even those whose intelligence role is a collateral duty, are part of a much larger intelligence community that manages a tremendous number of human and technical collection resources. Leveraging these assets in the support of mission is a key function of the naval intelligence professional as is contributing collected intelligence to the broader community. In addition to leveraging tactical, theater, and national level capabilities from throughout the intelligence community, naval intelligence professionals also coordinate the use of operational assets to collect information needed to support the intelligence assessment. Similarly, processing and exploitation as well as analysis and production require significant teamwork as the Navy leverages processed, exploited, and analyzed information from a variety of intelligence community organizations to inform its basic, current, and estimative intelligence products. Additionally, teamwork is needed to build intelligence products which are objective, accurate, complete, relevant, and objective. Review of assessments by other intelligence experts typically leads to a stronger product as other intelligence professionals can add their experience to the assessment and critique the analytic methodology used and conclusions reached through that methodology. Finally, dissemination and integration of intelligence requires a team that understands how and in what forms intelligence must be disseminated and knows how to coordinate with combat systems, information technology, and communications professionals to resolve dissemination issues.

4. Ability to Speak and Write Clearly, Concisely, and Accurately. Naval intelligence professionals must be effective communicators. The commanders and activities they support have vastly different requirements with regard to how they receive and consume intelligence and naval intelligence professionals must accommodate these different needs to ensure effective intelligence dissemination and integration. Some commanders may prefer a brief oral report or graphic that provides the bottom line assessment while some

may wish to see a detailed report providing the sources and processes that were used to develop the assessment. In either case, close coordination with the commander and supported activities is needed to determine how to most effectively deliver intelligence. Additionally, these same skills are critical to every other phase of the intelligence process as well. Planning and direction of the intelligence effort requires the development of clear tasking to subordinates and to organizations across and up echelon that may support the overall intelligence effort. Clarity in spoken and written communication is also essential in driving and, if needed, redirecting the collection effort and coordinating changes to the processing and exploitation, analysis and production, and dissemination plans.

5. Curiosity. A natural curiosity about the world and a variety of subjects relevant to the practice of intelligence is essential to success as a naval intelligence professional. Some of these subjects include warfare (particularly naval warfare), history, politics, economics, geography, technology, sociology, psychology, and cultural studies. The naval intelligence professional must have sufficient knowledge of all fields related to their work to recognize the existence and general nature of any problem. Developing a strong foundation of knowledge about a wide variety of topics also enables the naval intelligence professional to assimilate quickly a large and perhaps diverse number of facts into a comprehensive whole, from which significant meanings may be drawn. Curiosity is also essential to determining where there are gaps in knowledge that will drive collection planning and is instrumental in the development of new accesses and innovative collection techniques that are needed to penetrate adversary secrets.

6. Resourcefulness. The naval intelligence professional must strive to swiftly attain complete familiarity with their duties and the capabilities available to them. By being cognizant of the body of work that is on hand and that exists in databases or at various organizations throughout the operational and intelligence communities, the naval intelligence professional can resourcefully leverage existing information and analyses to support their commander. Greater knowledge also leads to increased discernment about gaps in intelligence. Understanding of what intelligence and other information is available is critical to the efficient and resourceful use of intelligence resources. For example, the commander may require data about a potential antiship cruise missile battery in a country that has, heretofore, been of minor concern to the United States. The naval intelligence professional knows obtaining this information requires the use of a collection asset to verify the presence of the missile site since their experience shows there is little likelihood any intelligence databases will contain information on this potential site given its age and the relative low priority this country has as intelligence collection target. Knowledge of national to tactical intelligence collection plans indicates a theater asset with electro-optical capabilities will operate in the vicinity of the potential site for a separate purpose so the naval intelligence professional coordinates a retasking of this asset to meet their requirements and is thus able to use organic assets to meet other priority intelligence requirements.

2.8 NAVAL INTELLIGENCE AND OPERATIONAL ART

Joint doctrine defines operational art as the "cognitive approach by commanders and staffs—supported by their skill, knowledge, experience, creativity, and judgment—to develop strategies, campaigns, and operations to organize and employ military forces by integrating ends, ways, and means" and which also factors in the risks associated with courses of action. Operational art considers the arrangement and employment of friendly and adversary forces and other capabilities in time, space, and purpose. It possesses the characteristics of both a science and an art. The science of operational art derives from the use of factual or analytically derived data such as numbers and dispositions of forces; time, space, and distance calculations; and weapons capabilities. The art derives from the use of imaginative creativity to conceptually employ forces to achieve an objective with full consideration of the adversary and the operational environment. This requires activities such as determining and defining the objective; developing a concept of operations; and application of the principles of war to the development of a plan.

Clausewitz referred to this component of operational art as *coup d'oeil* which refers to the ability of the commander to intuitively comprehend the adversary and the advantages and disadvantages inherent in the operational environment. This aspect of operational art is "honed through operational experience and the study of military theory and history." Operational art is used to translate the operational design, which JP 1-02 defines as the "conception and construction of the framework that underpins a campaign or major operation plan and its subsequent execution," into the actions that sequence, synchronize, and integrate forces into accomplishing tactical and operational objectives known as the operational approach. Operational art also requires a broad vision of a future end state and the path to achieve it and the ability to anticipate how numerous variables concerning the adversary, friendly forces, and the operational environment will interact and promote or inhibit successful mission accomplishment.

Given intelligence is a form of operations, the use of the term operational art to describe the naval intelligence professional's application of judgment, perception, boldness, and character to planning, executing, and assessing the intelligence effort is apt. Like their counterparts in operations, naval intelligence professionals must consider the arrangement and employment of adversary and friendly forces and other capabilities in time, space, and purpose but the naval intelligence professional must do this on two levels. First, they must adopt the perspective of the adversary and determine the adversary's operational design and how the adversary applies operational art to deploy their forces in the accomplishment of their objectives. Second, the naval intelligence professional is also called upon to synchronize the activities of internal and external collection, processing, exploitation, analysis, and dissemination activities in ways that support their commander's operational design and intent. The application of operational art to the intelligence process results in efficient and effective use of resources and promotes unity of action as it draws upon the experience of the naval intelligence professional to integrate organic, theater, national, coalition, and blue force operational capabilities to address tactical and operational intelligence needs.

Since operational art concerns the balance of ends, ways, means, and risks, the naval intelligence professional can use that construct to ensure they are employing their resources in ways that meet the commander's need for insight into adversary capabilities and intentions and a deeper understanding of the operational environment. This construct can be applied to each phase of the intelligence process and to the process as a whole since each phase requires the application of judgment and experience to the coordination, sequencing, and synchronization of activities at some level. For example, when applied to developing a collection plan, the following questions should be asked during the plan's development:

1. What information is required to satisfy the commander's PIRs? (ends)

2. What intelligence collection discipline or combination of disciplines is most likely to address those information gaps and how should those collection activities be sequenced? (ways)

3. What resources are required and which are available to accomplish the collection plan? (means)

4. What is the likely cost or risk in executing the collection plan? Are there intelligence gain/loss considerations? What are the risks to the collection asset? (risks)

Figure 2-3 provides an overview of the operational art concept in graphical form. For further reading on the concept of operational art, readers are encouraged to review the following references:

1. JP 5-0, Joint Operation Planning

2. NWP 5-01, Navy Planning

3. NDP 1, Naval Warfare

4. Maritime Component Commander's Guidebook (published by Naval War College)

5. Operational Art Primer (published by Naval War College).

Additionally, appendix B, contains a case study of naval intelligence support to the U.S. invasion of Okinawa (Operation ICEBERG) that offers excellent examples of applying operational art to the practice of naval intelligence and highlights many of the concepts and themes discussed throughout this publication.

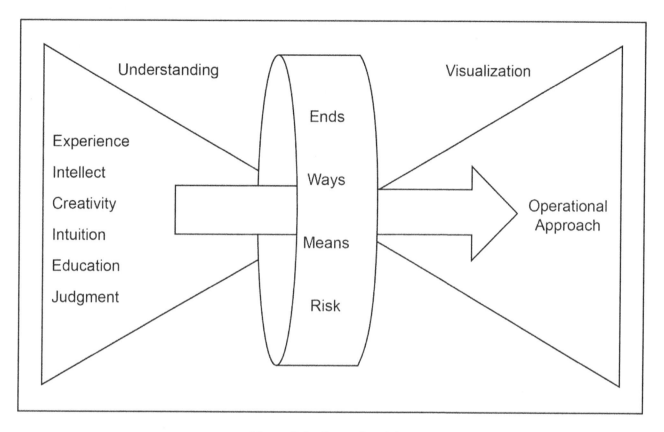

Figure 2-3. Operational Art

INTENTIONALLY BLANK

CHAPTER 3
Naval Intelligence Operations

> If you know the enemy and know yourself, you need not fear the result of a hundred battles. If you know yourself but not the enemy, for every victory gained you will also suffer a defeat. If you know neither the enemy nor yourself, you will succumb in every battle.
>
> *Sun Tzu*
> *The Art of War*

3.1 INTRODUCTION

Naval intelligence consists of a diverse mix of organizations and personnel that perform a wide variety of activities. These activities are aligned to the intelligence process and are collectively referred to as naval intelligence operations. The focus of these operations is on providing decision makers at all levels of war with the intelligence needed to reduce uncertainty on the battlefield and thus enable the successful planning and execution of naval operations.

This chapter describes the intelligence disciplines and elaborates on each component of the intelligence process—planning and direction, collection, processing and exploitation, analysis and production, dissemination and integration, and evaluation and feedback—as discreet yet highly interrelated naval intelligence operations. Effective employment of the intelligence process and the intelligence operations that constitute it requires naval intelligence professionals to apply their experience and operational art to ensure that intelligence resources are synchronized to efficiently and effectively meet command intelligence needs. Readers desiring a deeper understanding of the concepts discussed in this chapter should refer to JP 2-0, Joint Intelligence, and JP 2-01, Joint and National Intelligence Support to Military Operations.

3.2 THE INTELLIGENCE DISCIPLINES

Familiarity with each of the disciplines is foundational to supporting decision makers throughout the Navy with the intelligence needed to support their operations. This familiarity must include developing, through study and practice, an intuitive awareness of how each intelligence discipline contributes to understanding of the adversary and the maritime operational environment and how each can be leveraged and integrated to address intelligence requirements. The overview provided by figure 3-1 and the following subsections provide the JP 1-02 definition of each intelligence discipline and information about their tasking authorities. Readers desiring a more complete discussion of the intelligence disciplines should refer to appendix D as it provides a more complete overview of each discipline.

Fleet Intelligence is equipped with a wide inventory of ISR capabilities that enable intelligence collection operations that are executed according to the priorities and within the timelines directed by the commander—but there is also a vast array of resources within the U.S. IC that is not under the control of the commander. The inability to directly task such resources should not preclude naval intelligence personnel from requesting resources that may be uniquely suited to satisfy PIRs or essential elements of information (EEIs).

INTELLIGENCE DISCIPLINES, SUBCATEGORIES, AND SOURCES

CI—Counterintelligence

GEOINT—Geospatial Intelligence
- Imagery
- IMINT—Imagery Intelligence
- Geospatial Information

HUMINT—Human Intelligence
- Debriefings
- Interrogation Operations
- Source Operations
- Document and Media Exploitation

MASINT—Measurement and Signature Intelligence
- Electromagnetic Data
- Geophysical Data
- Materials Data
- Radio Frequency Data
- Radar Data
- Nuclear Radiation Data

OSINT—Open-Source Intelligence
- Academia
- Interagency
- Newspapers/periodicals
- Due Diligence
- Media Broadcasts
- Internet
- Alternative Collections

SIGINT—Signals Intelligence
- COMINT—Communications Intelligence
- ELINT—Electronic Intelligence
  - TECHELINT—Technical Electronic Intelligence
  - OPELINT—Operational Electronic Intelligence
- FISINT—Foreign Instrumentation Signals Intelligence

TECHINT—Technical Intelligence
- Weapon System Intelligence
- Scientific Intelligence

Figure 3-1. Intelligence Disciplines, Subcategories, and Sources

NWP 2-0

3.2.1 Counterintelligence

CI is information gathered and activities conducted to identify, deceive, exploit, disrupt, or protect against espionage, other intelligence activities, sabotage, or assassinations conducted for or on behalf of foreign powers, organizations or persons or their agents, or international terrorist organizations or activities. CI capabilities extending to Fleet Intelligence at the operational and tactical levels of war are components of the national CI enterprise under the cognizance of the National Counterintelligence Executive (NCIX). The Department of Defense (DOD) executive agent for CI is the Director, Defense Counterintelligence and Human Intelligence Center (DCHC). The Department of the Navy authority over naval elements of this enterprise resides with the Director, Naval Criminal Investigative Service (NCIS), and the theater authority with the combatant commander (CCDR) intelligence directorate of a joint staff (J-2) CI/joint force counterintelligence and human intelligence staff element (J-2X).

3.2.2 Geospatial Intelligence

Geospatial intelligence (GEOINT) is the exploitation and analysis of imagery and geospatial information to describe, assess, and visually depict physical features and geographically referenced activities on the Earth. Geospatial intelligence consists of imagery, imagery intelligence, and geospatial information. At a national level, National Geospatial-Intelligence Agency (NGA) is the functional manager for GEOINT. The Navy possesses a wide array of GEOINT collection capabilities but dedicated naval imagery analysts are concentrated at ONI in the fleet imagery support team (FIST). Navy GEOINT collection capabilities include airborne ISR capabilities, such as P-3 and Triton; tactical and theater-level unmanned aerial system (UAS) and other unmanned systems; weapons systems video from air and surface units; and handheld cameras employed by unit ship's nautical or otherwise photographic interpretation and examination (SNOOPIE) teams. Fleet requirements for GEOINT analytic support are met with augmentees from ONI fleet intelligence detachments (FIDs) or via reachback exploitation and analytic production at the FIST. Maritime operations centers (MOCs) and force-level ships (carrier, large deck amphibious ships, and command ships) have significant GEOINT processing and exploitation capabilities through the Distributed Common Ground System-Navy (DCGS-N). Other members of the intelligence community, such as the Air Force, also have considerable GEOINT capabilities that can be leveraged through a formal support request to higher headquarters.

3.2.3 Human Intelligence

Human intelligence (HUMINT) is a category of intelligence derived from information collected and provided by human sources. While Director of Naval Intelligence is responsible for Navy HUMINT programs, HUMINT is collected, processed, analyzed, and disseminated throughout the Navy. For example, SNOOPIE team collection opportunities and aircrew debriefings are examples of low level, overt HUMINT operations. Designated HUMINT capabilities extending to Fleet Intelligence at the operational and tactical levels of war is a component of the national HUMINT architecture under the cognizance of the Director of National Intelligence (DNI) National HUMINT Manager. The Navy authority over maritime elements of this architecture resides with the Service defense human intelligence executor (DHE), the Director of Naval Intelligence (OPNAV N2/N6). Authority as the theater DHE resides with the CCDR J-2. Navy HUMINT operations at the tactical level of war are generally conducted by MIO-IETs, under the administrative auspices of the Navy Expeditionary Intelligence Command (NEIC).

3.2.4 Measurement and Signature Intelligence

Measurement and signature intelligence (MASINT) is intelligence obtained by quantitative and qualitative analysis of data (metric, angle, spatial, wavelength, time dependence, modulation, plasma, and hydromagnetic) derived from specific technical sensors for the purpose of identifying any distinctive features associated with the emitter or sender, and to facilitate subsequent identification and/or measurement of the same. The detected feature may be either reflected or emitted. Due to the multiple subcategories and manifestations of MASINT, there is no central IC authority that directs and controls MASINT activities. The National Measurement and Signature Intelligence Management Office (NMMO) was formed as a Defense Intelligence Agency (DIA) component to oversee the "U.S. MASINT System"—but that system operates in a diverse and decentralized manner.

Accordingly, there is no single Navy organization with principal authority over MASINT in the maritime domain. Services and agencies have significant MASINT capabilities, especially since MASINT often involves using existing sensors in unique and innovative ways. For example, the Navy, given its maritime focus, has considerable resources for the collection, processing, analysis, and dissemination of acoustic intelligence (ACINT) products.

3.2.5 Open-Source Intelligence

Open-source intelligence (OSINT) is information of potential intelligence value that is available to the general public. Given the rapid rise of the Internet, there has been a significant increase in the amount and variety of open-source information available to naval intelligence professionals. In some cases, open sources may be the best or only source of information on a given topic. As mentioned in chapter 1, the value of intelligence has nothing to do with its classification. As with other sources, though, analysts must evaluate the open sources they are exploiting to determine the accuracy of the information provided and to detect adversary deception and strategic communication efforts. For example, some open sources may serve as unofficial or unrecognized government agents and may be a means by which certain governments communicate disinformation or are used to augment official strategic messaging capabilities. Additionally, naval intelligence professionals must carefully analyze open sources for organizational and author biases and factor these biases into their evaluation of information derived from open sources.

3.2.6 Signals Intelligence

Signals intelligence (SIGINT) is defined as: (1) a category of intelligence comprising either individually or in combination all communications intelligence, electronic intelligence, and foreign instrumentation signals intelligence, however transmitted; (2) intelligence derived from communications, electronic, and foreign instrumentation signals. NSA is the functional manager for SIGINT. The SIGINT enterprise extending to Fleet Intelligence at the operational and tactical levels of war is a component of the national SIGINT architecture under the cognizance of the Director, NSA/Central Security Service (CSS) as the functional manager for the United States Signals Intelligence System (USSS). The Navy authority over maritime elements of this architecture resides with FCC. The Commander, FCC is dual-hatted as Commander, Tenth Fleet (COMTENTHFLT). The theater SIGINT authority resides with the CCDR J-2.

FCC manages the functions of Navy information operations commands (NIOCs) and fleet information operations centers (FIOCs) at NSA locations worldwide. NIOCs are force providers to both NSA sites and FIOCs under COMTENTHFLT. Personnel assigned to NSA Cryptologic Centers are focused on national and theater-specific missions. The missions of the FIOCs varies, but typically includes fleet and unit-level cryptologic and/or information operations (IO) and intelligence analysis support, as well as providing fleet direct support personnel to deploying naval units.

3.2.7 Technical Intelligence

Technical intelligence (TECHINT) is intelligence derived from the collection, processing, analysis, and exploitation of data and information pertaining to foreign equipment and materiel for the purposes of preventing technological surprise, assessing foreign scientific and technical capabilities, and developing countermeasures designed to neutralize an adversary's technological advantages. There is no national TECHINT architecture under the cognizance of a single IC agency. The TECHINT enterprise extending to Fleet Intelligence at the operational and tactical levels of war is a component of the DOD foreign materiel program (FMP) administered by the DIA. DIA manages joint and service S&TI through the Joint Foreign Material Program Office (JFMPO). The Navy authority over maritime TECHINT and S&TI is the ONI, with the Navy FMP being managed by the ONI Collections Office.

3.3 THE INTELLIGENCE PROCESS

As illustrated by figure 3-2, the intelligence process is represented as a logical, circular flow—in which requirements generated via planning and direction activities drive collection of data that is transformed into information through processing and exploitation activities and then provided to analysts who produce finished intelligence products that are then disseminated to users of intelligence and integrated into policy formulation, planning, and operational activities. In reality, while the intelligence process model is useful for understanding how the various intelligence operations interrelate, no firm boundaries delineate where each operation within the intelligence process begins or ends. Each operation runs continuously to provide the commander, staff, and supported organizations with needed intelligence, and evaluation and feedback applies throughout the process. For example, a UAS mission may be tasked to collect on a particular information requirement. While the data it collects is sent to intelligence analysts for exploitation and follow-on all-source analysis, the data provided by the sensor may be directly downlinked to an operational unit where it is viewed by the commander and his staff without going through an exploitation or analysis node.

3.4 TASKING, COLLECTION, PROCESSING, EXPLOITATION, AND DISSEMINATION AND THE INTELLIGENCE PROCESS

actionable intelligence. Intelligence information that is directly useful to customers for immediate exploitation without having to go through the full intelligence production process. (JP 1-02. Source: JP 2-01.2)

The term tasking, collection, processing, exploitation, and dissemination (TCPED), is a term that is frequently used in discussions about the intelligence process. Despite its common usage in military intelligence lexicon, TCPED does not have an official definition in joint or naval doctrine. This doctrinal ambiguity results in the term having different meanings. In a systems acquisition setting (national-strategic), it may be used to describe the end-to-end architecture supporting a specific ISR program. In a war planning context (theater-strategic), it may be used to describe ISR-related force flows, communications paths, and federated exploitation plans. In a theater deployment setting (operational level), it may be used to describe the management of ISR sensors and data volume vis-à-vis in situ processing capacities. In a unit setting (tactical level), it may be used to describe the TTP of an organic ISR system. The lack of a definition for TCPED has also caused some to view the term as a description of the intelligence process, which is an inaccurate use of the term. Appendix E provides a deeper understanding of the operational and tactical level considerations associated with the TCPED, and their relationship to intelligence disciplines and the intelligence process.

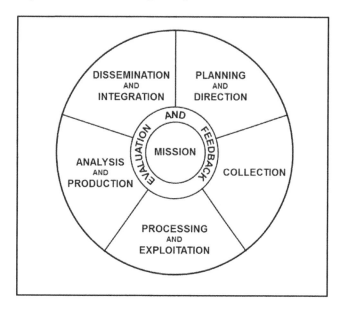

Figure 3-2. The Intelligence Process

3.5 PLANNING AND DIRECTION

planning and direction. In intelligence usage, the determination of intelligence requirements, development of appropriate intelligence architecture, preparation of a collection plan, and issuance of orders and requests to information collection agencies. (JP 1-02. Source: JP 2-01)

Planning and direction of intelligence operations occurs continuously and at all levels of war to ensure that intelligence resources are efficiently and effectively applied to meeting the needs of the commander, staff, and supported organizations. While a significant output of planning and direction is the intelligence collection plan, the planning and direction of intelligence operations is also applied to synchronize and focus all the activities occurring throughout the intelligence process. Successful execution of planning and direction activities requires naval intelligence professionals to have mastery in the application of operational art to their profession. Through deep understanding of the capabilities and limitations of organic intelligence resources and the broader intelligence community, naval intelligence professionals are able to visualize the entire chain of interrelated process steps that need to be executed to meet command intelligence requirements. For example, it is not sufficient that a collection asset is available to address a particular requirement. The end-to-end process of developing clear requirements; tasking the asset; determining where processing and exploitation of the data it collects will occur; what resources will be applied to incorporating the exploited information into a fused intelligence product; and how the resulting intelligence will be disseminated, integrated, and evaluated must all be considered and addressed in planning and directing the intelligence effort. Such an analysis of the end-to-end process may reveal gaps, such as insufficient analytic resources to provide a timely intelligence estimate or the failure of a critical communications circuit that requires a workaround. In the event such gaps are discovered, it is the responsibility of the naval intelligence professional to inform the commander of these gaps and propose solutions to address shortfalls. An overview of the major activities that occur in planning and direction is found in figure 3-3.

3.5.1 Intelligence Requirements Overview

The naval intelligence professional fully participates in the command's planning and decision-making process. Embedding intelligence professionals into this process enables intelligence to inform the planning and execution of operations and provides the naval intelligence professional with the insights needed to anticipate the commander's needs and focus the intelligence effort on the commander's priorities.

As part of the planning process, all staff elements develop information requirements, which are the things that must be known about the adversary, the maritime operational environment, or friendly forces in order to successfully plan for and execute the mission. Figure 3-4 provides an overview of the relationship between intelligence requirements and the broader set of information requirements generated by the command. Information requirements that are critical to mission success are nominated to the commander by each staff element. For the purposes of this publication, further discussion focuses only on the intelligence-related requirements.

3.5.1.1 Commander's Critical Information Requirements

commander's critical information requirement (CCIR). An information requirement identified by the commander as being critical to facilitating timely decisionmaking. (JP 1-02. Source: JP 3-0)

Figure 3-4 highlights that the CCIRs are comprised of an operational component—the friendly force information requirements (FFIRs)—and an intelligence component—PIRs.

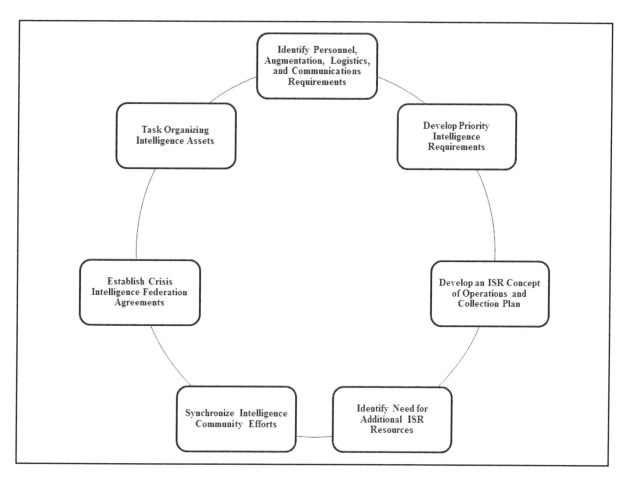

Figure 3-3. Intelligence Planning and Direction Activities

3.5.1.2 Priority Intelligence Requirements

priority intelligence requirement (PIR). An intelligence requirement, stated as a priority for intelligence support, that the commander and staff need to understand the adversary or other aspects of the operational environment. (JP 1-02. Source: JP 2-01)

PIRs are identified by the commander as the intelligence requirements (IRs) most important to mission accomplishment. PIRs are essentially "go/no go" degrees of knowledge, and should encompass the most urgent intelligence requirements of subordinate, adjacent, and supporting elements. They effectively constitute the commander's guidance for intelligence. During the execution of an operation, PIRs often equate to "decision points." Typically, resources are insufficient to address all of the commander's intelligence requirements, which is why PIRs are needed to focus the intelligence effort on the most critical requirements.

PIRs are the commander's priority intelligence requirements. It is the function of the naval intelligence professional to develop and nominate PIRs to the commander for approval. Once approved, PIRs are used to focus and synchronize all intelligence operations—collection, processing, exploitation, analysis, production, dissemination, integration, evaluation, and feedback. In addition to establishing priorities for organic intelligence operations, PIRs are also used as justification to advocate for external collection, analytic, communications, or other resources needed to meet command intelligence requirements.

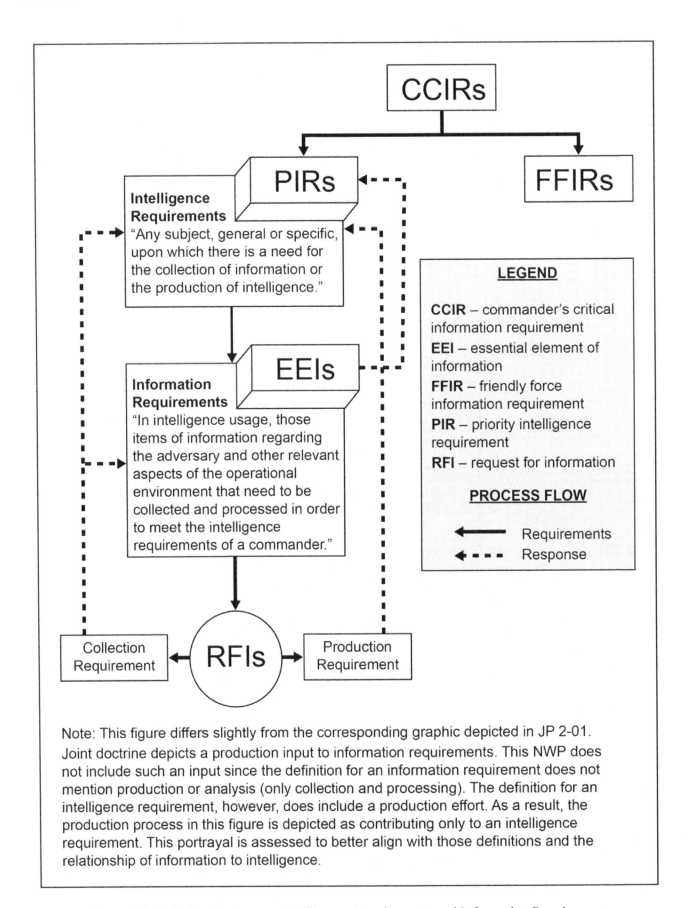

Figure 3-4. Relationship between Intelligence Requirements and Information Requirements

There is no standard list of PIRs nor is there a set rule for determining their number and content. The intelligence gaps for each situation are different and require distinct PIRs. It is up to the commander to determine the number of PIRs but naval intelligence professionals must advise the commander on the impacts to efficiency and effectiveness of the intelligence effort that result from the requirement to address a large number of PIRs. In general, though, PIRs are linked to the concept of operations and, for combat operations, they typically concern the adversary's most likely and most dangerous COAs and critical adversary vulnerabilities that the naval force can exploit. In designating a PIR, the commander establishes the following:

1. What he wants to know (intelligence required)

2. Why he wants it (linked to operational decisionmaking)

3. When he needs it (latest time the information is of value)

4. How he wants it (format and method of delivery).

PIRs are also not static—the naval intelligence professional must be in consistent dialog with the commander and other staff elements to determine if the list of PIRs is still valid and to cancel, modify, or create new PIRs as dictated by the commander's needs. As operations progress, particularly on the tactical level, PIRs will become more focused.

3.5.1.3 Intelligence Requirements

intelligence requirement. 1. Any subject, general or specific, upon which there is a need for the collection of information, or the production of intelligence. 2. A requirement for intelligence to fill a gap in the command's knowledge or understanding of the operational environment or threat forces. (JP 1-02. Source: JP 2-0)

Intelligence requirements serve as the foundation of PIRs. Intelligence requirements do not have the same criticality as PIRs, but still require intelligence assessments to support planning and COA considerations.

The nature of the term "intelligence" indicates that an intelligence requirement requires an analytical contribution to achieve the necessary insight. Therefore, satisfying an intelligence requirement involves some degree of analysis and production.

Intelligence requirements are decided by the senior naval intelligence professional within the command. Since addressing each intelligence requirement involves the application of intelligence resources, care must be given to limit the number of intelligence requirements and prioritize them to ensure that they do not impact efforts to address PIRs.

3.5.1.3.1 Priority Intelligence Requirement and Intelligence Requirement Characteristics

While there are exceptions, PIRs and intelligence requirements are generally broader in scope at the strategic and operational levels than they are at the tactical level. Without prioritized and specific requirements, the naval intelligence professional runs the risk of overburdening the commander, staff, and supported organizations with either too much intelligence or intelligence that does not address the commander's fundamental concerns. Either case complicates the commander's decision-making process and could undermine confidence in intelligence and have negative impacts on mission success. PIRs and intelligence requirements have the following characteristics:

1. Ask only one question

2. Focus on specific facts, events, or activities concerning the adversary and the operational environment

3. Are tied to mission planning, decisionmaking, and execution

4. Provide a clear, concise statement of what intelligence is required

5. Contain geographic and time elements to limit the scope of the requirement.

3.5.1.3.2 Priority Intelligence Requirement and Intelligence Requirement Management

Management of PIRs and intelligence requirements is a continuous and dynamic process. The following must be continuously assessed to ensure an accurate and relevant list of PIRs and intelligence requirements is maintained:

1. Importance of each requirement to mission success
2. Information and assets needed to satisfy each requirement
3. Resources presently committed toward fulfilling that requirement
4. Degree that the requirement has been satisfied by completed intelligence activities.

3.5.1.4 Essential Elements of Information

essential elements of information (EEI). The most critical information requirements regarding the adversary and the environment needed by the commander by a particular time to relate with other available information and intelligence in order to assist in reaching a logical decision. (JP 1-02. Source: JP 2-0)

Much as a PIR is a critical intelligence requirement, EEIs are the highest-priority information requirements.

3.5.1.5 Information Requirements

information requirements. In intelligence usage, those items of information regarding the adversary and other relevant aspects of the operational environment that need to be collected and processed in order to meet the intelligence requirements of a commander. (JP 1-02. Source: JP 2-0)

Information requirements are generally developed by the intelligence staff as detailed subsets of identified PIRs and intelligence requirements. The nature of the term "information," indicates that an information requirement does not require an analytical contribution, and could therefore be satisfied strictly through processing and exploitation.

3.5.1.6 Requests for Information

request for information (RFI). 1. Any specific time-sensitive ad hoc requirement for intelligence information or products to support an ongoing crisis or operation not necessarily related to standing requirements or scheduled intelligence production. A request for information can be initiated to respond to operational requirements and will be validated in accordance with the combatant command's procedures. 2. The National Security Agency/Central Security Service uses this term to state ad hoc signals intelligence requirements. (JP 1-02. Source: JP 2-0).

Note

In Navy usage, request for information (RFI) is also a general term for an information request that can be used to meet an information need associated with an operation.

Once intelligence and information requirements are established, existing intelligence databases and holdings are reviewed for answers to the requirements. If the intelligence does not already exist, RFIs are issued. An RFI can be initiated at any level of command, and leads to either a production requirement (if the request can be answered with information on hand) or a collection requirement (if the request requires new information be collected). Per JP 2-01, each echelon is responsible for validating, prioritizing, and, if possible, satisfying the RFI or collection requirement before forwarding it to the next level. RFIs should be satisfied at the lowest level possible. If the information required to satisfy an RFI does not exist, the requester is informed and a decision is made to initiate collection and/or production.

The generic process for submitting a production or other intelligence support requirement is discussed in the following section while the collection requirements process is discussed in section 3.6.

3.5.2 Procedures for Requesting Higher Echelon Intelligence Support

When the intelligence support requirements exceed the command's capacity, the senior naval intelligence professional at that command should coordinate with their commander and initiate a request for the needed support. The specific process for submitting and validating these requests varies by theater and depends on the organizational alignment of the command requesting support.

Key to the process of requesting support is validation of the requirement at each echelon. Per JP 2-01, validation is a process associated with the tasking, collection, and production of intelligence and confirms that an intelligence collection or production requirement is sufficiently important to justify the dedication of intelligence resources, does not duplicate an existing requirement, and may not be satisfied by previous collection or production. Tying requirements to specific mission needs and PIRs is critical to ensuring they are validated as they flow up echelon. Whenever possible, coordinating these requests as early as possible and providing information copies of the requirement to supporting intelligence organizations significantly increases the likelihood that the required support is provided in time to support operations.

Although the specific process of requesting support depends on the situation, joint doctrine does provide a general requirements flow for intelligence support requests for both crisis and noncrisis operations. Figure 3-5 shows a generic process flow for noncrisis requests. Crisis requests follow procedures established in each theater but may include direct coordination by the JTF J-2 staff with theater joint intelligence operations center (JIOC), national JIOC, joint and Service intelligence organizations, and national intelligence agency representatives to address the crisis requirement.

National intelligence production elements include: CIA, State Department, Federal Bureau of Investigation (FBI), Department of Homeland Security, Drug Enforcement Administration, Department of Energy, and Department of the Treasury.

Note

> Joint and Service intelligence production elements include: DIA, NSA, NGA, and Service intelligence elements, such as ONI.

3.5.3 Augmentation and Federation Requirements

Analysis of mission intelligence requirements may determine gaps that are not addressable using organic intelligence resources. In these cases, naval intelligence professionals at all levels should coordinate with their commander and other staff elements to request needed support through their operational chain of command. While shortfalls may exist in any intelligence operation, by far the most common gaps occur in the need for additional collection or analytic resources.

Federation enables one intelligence organization to task assets within another intelligence organization for a narrowly defined set of activities. For example, a fleet command headquarters may have imagery exploitation requirements that exceed their organic imagery analysis capabilities and so they establish federation agreements with other intelligence organizations to conduct exploitation on a defined set of targets. Given this is a resource drain for the supporting intelligence organization, federation agreements work best when they are negotiated as far in advance as possible and the task being performed by the supporting intelligence organization are of value to that organization as well. In a crisis situation, though, federation is often rapidly negotiated and agreements reached may be informal and more open-ended. While this situation may work in the short term, longer term shortfalls should be addressed through formal federation agreements or augmentation.

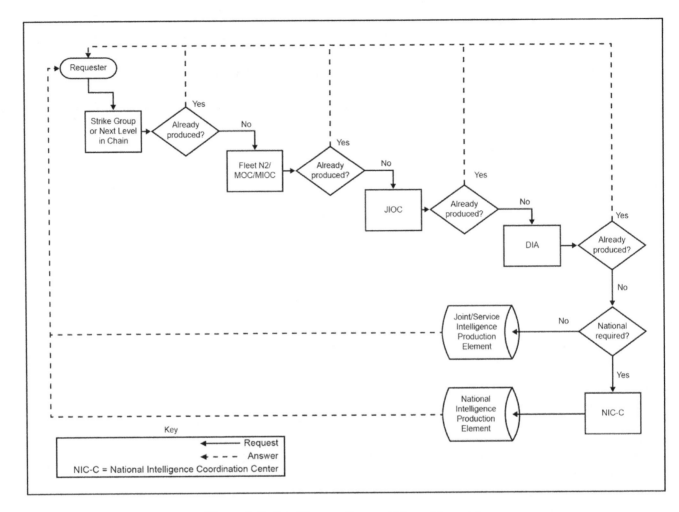

Figure 3-5. Intelligence Request Flow, Noncrisis

Augmentation is the process of requesting and receiving additional, dedicated support that is chopped directly to the supported intelligence organization. For example, all fleet imagery and strike analyst requirements are currently met via augmentation from the FID and are coordinated through Navy Cyber Forces as part of predeployment planning and workups.[1] Navy Cyber Forces also manages the fleet intelligence adaptive force, which is a ready pool of naval intelligence personnel located at various numbered MOCs to meet fleet and CCDR validated individual augmentation requirements, provide increased intelligence production support to the MOC, and provide greater focus on the operational commander's most critical, high-demand areas. Augmentation is not limited to fleet or ONI resources. For example, a theater JIC, another Service, national agency, or coalition partner may have specific technical or country expertise that is needed to support a specific operation and such support may be requested if needed for mission requirements.

Planners must consider that, with the exception of a significant crisis, coordinating intelligence federation and augmentation support outside of the Navy may take many weeks or months. Federation and augmentation requests must always be very specific about the requirement, which should always be tied to mission requirements and are approved by the commander. Nebulous or open-ended requests for support significantly slow the negotiations involved in establishing a formal support arrangement.

---

[1] FID augmentation requirements are coordinated with Navy Cyber Forces. Fleet imagery analysts are assigned to FID East, the imagery analyst center of excellence located at ONI. Fleet strike analysts are assigned to FID West, the strike center of excellence located at the Naval Strike and Air Warfare Center (NSAWC) in Fallon, NV.

3.5.4 Direction

Direction of the intelligence effort consists of the requirements development and management processes discussed above combined with the related functions of collection, production, and dissemination management and the establishment of the intelligence architecture needed to support operations. It requires the naval intelligence professional to apply operational art to the synchronization of limited intelligence resources to address the most important intelligence concerns of the commander, staff, and supported organizations.

3.5.4.1 Collection Management

Collection management is defined and discussed in detail in section 3.6.1 but is integrally related to the efficient and effective tasking of assets to collect on command intelligence requirements.

3.5.4.2 Production Management

Production management encompasses determining the scope, content, and format of each product; developing a plan and schedule for the development of each product; assigning priorities among the various intelligence production requirements; allocating processing, exploitation, and production resources; and integrating production efforts with collection and dissemination. Naval intelligence professionals must thoroughly understand the capabilities, limitations, and timelines associated with producing a wide variety of exploited information and finished intelligence products in determining how best to meet the needs of the commander, staff, and supporting organizations. For example, a national asset may collect imagery on a target of interest in the next 2 hours. Depending on the criticality of developing intelligence on that target, senior naval intelligence leadership may direct organic resources to exploit the raw imagery and give the results of that exploitation directly to planners. If the information is less time-critical, then the decision may be made to wait for a first phase exploitation product produced by another intelligence organization or to wait longer for a finished, all-source product concerning the target that may be produced by a Service or defense intelligence organization as part of a regular production cycle. The goal of production management is to make the most efficient and effective use of limited intelligence resources and to ensure those resources are focused on priority requirements.

3.5.4.3 Dissemination Management

Dissemination management involves establishing dissemination priorities, selecting dissemination means, and monitoring the flow of intelligence throughout the command. Its objective is to deliver the required intelligence to the appropriate user in the proper form at the right time while ensuring that individual consumers and the dissemination system are not overloaded by attempting to move unneeded or irrelevant information. Additionally, addressing the use of appropriate security controls is integral to the dissemination management effort. Writing for release and coalition information sharing requirements should be carefully considered when planning dissemination efforts. The naval intelligence team must always strive to strike a balance to ensure that needed intelligence is disseminated in a timely fashion to those who need it while still ensuring that the proper security controls are in place to protect intelligence and sources and methods from compromise.

3.5.4.4 Design, Establish, and Adapt the Intelligence Architecture

This activity is concerned with designing, establishing, and maintaining the intelligence architecture that is needed to meet the intelligence needs of the commander, staff, and supported organizations. When used in this sense, the term "intelligence architecture" is concerned with the people, organizational structures, processes, systems, and communication paths needed to effectively and efficiently provide intelligence support and is not limited to a systems or technological perspective. The naval intelligence professional looks not only at current requirements, as articulated in PIRs and intelligence requirements, but also anticipates the continuing intelligence needs of the naval force as it carries out its mission and designs an intelligence architecture that has the capability, flexibility, and redundancy to satisfy these needs. Naval intelligence professionals must continuously assess and reassess a number of factors associated with the intelligence architecture and must address inadequacies by requesting additional resources, realigning resources to meet emerging situational requirements, or developing innovative ways to use existing resources in a more effective and efficient manner. Some of the factors that are considered in designing, establishing, and adapting the intelligence architecture include:

1. Task organization of intelligence units

2. Identification of personnel, training, and equipment requirements

3. Requirement for liaison teams

4. Connectivity with national, theater, joint, and multinational force intelligence assets

5. Communications and information systems requirements

6. Logistic requirements

7. The need for specialized capabilities. For example, linguists, country, or technical specialists.

3.5.5 The Importance of Planning and Direction

Lord Nelson did not win at Trafalgar because he had a great plan, although his plan was great. He won because his subordinate commanders thoroughly understood that plan and their place in it well in advance of planned execution. You must be prepared to take action . . . when certain conditions are met; you cannot anticipate minute-by-minute guidance. . . .

Vice Admiral Henry C. Mustin III, U.S. Navy
Commander, Second Fleet
Fighting Instructions, 1986

In preparing for battle, I have always found that plans are useless but planning is indispensable.

General Dwight D. Eisenhower

These two quotes emphasize the fact that while there is no perfect plan, planning and direction in intelligence operations have ramifications for every other aspect of the intelligence process. Plans and orders should be detailed enough to ensure requirements are met but given intelligence operations are human-centric they must also allow naval intelligence professionals to exercise initiative and innovation in addressing the commander's needs for intelligence.

3.5.6 Evaluation and Feedback for Planning and Direction

Evaluation and feedback on planning and direction efforts are required to ensure those efforts are always in alignment with the mission and result in efficient and effective use of all assigned intelligence resources. Naval intelligence professionals must establish clearly defined policies and processes that are tailored to ensuring successful accomplishment of the range of intelligence operations to support naval force mission requirements. Ensuring these policies and processes produce the desired result is the responsibility of naval intelligence leaders at all levels. Assessing the efficacy of intelligence planning and direction efforts must be a continuous process and requires open communication within the intelligence team concerning ways to improve operational performance and the solicitation of feedback from intelligence users on the quality and utility of the intelligence they receive. Assessing planning and direction also requires naval intelligence leaders to review the outputs of all the intelligence operations under their cognizance, such as the collection plan or an analytic product, to ensure that these outputs contribute to or meet the standards of intelligence excellence described in chapter 1.

NWP 2-0

3.6 COLLECTION

> It will be vital to identify centers of gravity rapidly and determine the critical vulnerabilities that will be our pathways to them. We won't always have the luxury of a passive foe, and there's no natural law that says that every high-tech war must be fought in a desert with unlimited visibility and good weather.[2]
>
> *General Carl E. Mundy, United States Marine Corps (USMC)*

collection. In intelligence usage, the acquisition of information and the provision of this information to processing elements. (JP 1-02. Source: JP 2-01)

Given this definition, collection operations involve directing and managing the people, processes, and systems used to acquire information about the adversary and relevant aspects of the operational environment and then provide that information to processing and exploitation nodes.

Those who direct and manage intelligence collection must also understand the distinction between collection assets and collection resources.

collection asset. A collection system, platform, or capability that is supporting, assigned, or attached to a particular commander. (JP 1-02. Source: JP 2-01)

collection resource. A collection system, platform, or capability that is not assigned or attached to a specific unit or echelon which must be requested and coordinated through the chain of command. (JP 1-02. Source: JP 2-01.2)

For example, an EP-3 assigned to a theater would be a collection asset for the theater commander but to a carrier strike group (CSG), that same EP-3 would be considered a collection resource that could fly in direct support of their operations.

3.6.1 Collection Management

collection management. In intelligence usage, the process of converting intelligence requirements into collection requirements, establishing priorities, tasking or coordinating with appropriate collection sources or agencies, monitoring results, and retasking, as required. (JP 1-02. Source: JP 2-0)

The term collection management describes the overall process for turning intelligence requirements into collection requirements; prioritizing those requirements; tasking or coordinating collection on the requirements; gauging whether the requirements have been satisfied; and retasking collection, if necessary. Collection management is performed at all levels of war. Given the long lead times associated with the use of some collection assets and resources, collection management must be integrated at the earliest stages of planning so that collection operations can meet the commander's information needs within the timeframes imposed by operational needs. Figure 3-6 shows the collection management process, which comprises two distinct but interrelated functions, each of which is described in separate sections of this chapter:

1. Collection requirements management (CRM), which defines what intelligence systems must collect; is focused on intelligence user requirements; takes an all-source approach to address those requirements; and results in advocacy for the satisfaction of those requirements.

2. Collection operations management (COM), which specifies how the requirement is satisfied and is focused on how to use a single intelligence discipline (or specific systems within a single intelligence discipline) to address the intelligence requirement.

---

[2] General Carl E. Mundy, Jr., Reflections on the Corps: Some Thoughts on Expeditionary Warfare, Marine Corps Gazette, (March 1995), p. 27.

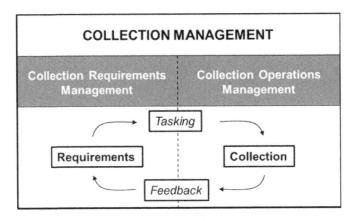

Figure 3-6. The Collection Management Cycle

All intelligence organizations, regardless of size, perform CRM and COM. While the functions are broken out separately because they are distinct, CRM and COM functions may be performed by a single organization or individual. In cases where separate organizations or individuals are performing the CRM and COM functions, there must be close coordination between the two to ensure efficient and effective collection operations.

CRM and COM are also performed at all levels of war, and each level interacts horizontally and vertically among various units, agencies, and organizations. Although formal processes exist for deliberate, ad hoc, and dynamic tasking of assets, the development of relationships with organizations that control collections resources and associated processing, exploitation, and dissemination (PED) capabilities is a key factor in the efficient employment of collection resources. These relationships enable an informal dialog that collection managers and collection resource operators can use to discuss requirements in greater specificity, determine acceptable trade-offs, or discuss the application of other collection resources to meet requirements.

As discussed, intelligence requirements flow up the operational chain of command and at each successive stage they are either satisfied, validated and sent to the next higher echelon, or rejected either because the intelligence is no longer needed or the determination is made that the requirement cannot be satisfied. Figure 3-7 shows a simplified flow of collection requirements from the unit to the national level.

For the Navy component commander (NCC), the MOC collection manager runs the Collection Management Board, which results in the NCC's intelligence collection plan for all domains and results in requests for theater and national collection resources to meet naval force collection requirements that cannot be satisfied at the NCC level.

NWP 2-0

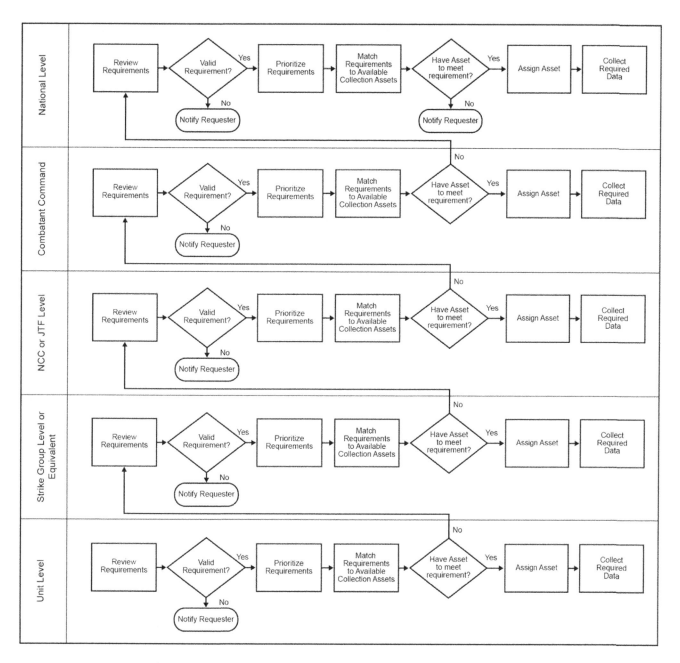

Figure 3-7. Simplified Collection Requirements Management Flow

At the theater level, the combatant command J-2 is kept apprised of all intelligence collection requirements levied against the collection assets and resources in theater. The theater J-2 has full management authority, which is the authority to validate, modify, or nonconcur with all intelligence requirements in that theater. Validated requirements for theater or national systems are forwarded to the theater intelligence collection management office for validation and, when validated, become part of the overall theater collection plan. To accomplish this validation process and plan for collection, a Joint Collection Management Board (JCMB) may be used at the theater level. The JCMB typically includes representatives from the J-2, operations directorate of a joint staff (J-3), Service components, and possibly national agency representatives or others as needed. The output of the JCMB is a joint integrated prioritized collection list. If resources beyond those allocated to the theater are required, the Joint Staff J-2 advocates for combatant command intelligence requirements to the Joint Staff, the intelligence combat support agencies, the Office of Secretary of Defense, and the Office of the Director of National Intelligence (ODNI). Naval intelligence professionals must understand this process to successfully leverage intelligence resources to meet their commander's requirements. Leveraging certain theater, national, or other Service intelligence collection capabilities must account for the time required to get requirements validated and approved. Knowing the timelines associated with getting collection approval for individual collection assets is key to winnowing out which assets may be available to support the naval force's requirements within the timeframes permitted by operational necessity.

3.6.2 Collection Management Responsibilities

Collection management occurs at all levels of war. Within this publication, the term collection manager is used to refer to an individual with responsibility for the timely and efficient tasking of organic collection resources and the development of requirements for collection resources that satisfy specific information needs in support of the mission.[3] For the Navy, dedicated collection manager positions are typically associated with the operational level of war (numbered fleet commander or higher) or as part of a Navy-led JTF. Regardless of whether a naval intelligence professional is formally designated as a collection manager, the processes and principles discussed below are performed by some member of the command's intelligence team since collection operations are a core component of the intelligence process.

3.6.2.1 N-2 Collection Management Responsibilities

N-2/Senior Naval Intelligence Officer collection operations responsibilities are to:

1. Develop and execute a concept of intelligence operations
2. Develop and train to standard operating procedures in support of the concept of intelligence operations
3. Assist the commander in the development of PIRs
4. Ensure that intelligence requirements from all staff sections are included in PIR recommendations to the commander
5. Develop a comprehensive list of intelligence requirements
6. Supervise the intelligence collection requirements development process
7. Develop and execute a concept of intelligence operations
8. Develop and train to standard operating procedures in support of the concept of intelligence operations
9. Assist the commander in the development of PIRs
10. Ensure that intelligence requirements from all staff sections are included in PIR recommendations to the commander

---

[3] The JP 1-02 definition is modified here to account for the fact that tactical units may request collection resources that are not "theater or national assets," which is a limiting qualifier in the joint definition of this term.

11. Develop a comprehensive list of intelligence requirements

12. Supervise the intelligence collection requirements development process

13. Ensure that intelligence is represented and engaged in all planning evolutions and focused on plan/decision supportability

14. Approve the collection manager's EEI and collection plan

15. Ensure that all intelligence watch standers fully understand the PIRs

16. Routinely engage the commander and supported organizations (i.e., composite warfare commanders (CWCs)) on the currency of PIRs and the development of new PIRs as changes in operations or the operational environment warrant.

3.6.2.2 Collection Manager Responsibilities

Collection manager responsibilities are to:

1. In conjunction with the N-2, develop intelligence collection requirements

2. Develop and manage the collection plan encompassing the N-2's IRs and always keep the PIRs in the forefront

3. Compare the collection plan to the capabilities and limitations of organic collection assets and inform the N-2 of gaps

4. Develop a collection strategy to optimize the use of collection assets and resources

5. Assist the N-2 in developing and maintaining the commander's PIRs and his/her IRs

6. Coordinate employment of organic sensors/collection platforms with the N-3 and CWCs

7. Integrate intelligence collection requirements into the collection plan

8. Align and synchronize collection processes with higher headquarters processes

9. Provide watch standers with collection templates so they can maintain situational awareness on the collection plan during execution and provide updates on key events.

3.6.3 Principles of Collection Management

Collection operations are resource intensive and as a fundamental aspect of intelligence operational art there has been considerable effort put into capturing best practices concerning how to lead and manage collections.

Figure 3-8 lists the four principles of collection management identified by JP 2-01.

Collection Management Principles
1. Identify Requirements Early
2. Prioritize Requirements
3. Take a Multidisciplinary Approach
4. Task Available Assets First

Figure 3-8. Collection Management Principles

1. Identify Requirements Early. Early engagement enhances the ability to respond in a timely manner, ensures thorough planning, increases flexibility in the choice of disciplines and systems, and allows the collection manager time to accomplish needed research, run prediction analysis tools if required, and establish and/or refine a point of contact list.

   a. As part of predeployment planning or in anticipation of a warning or operations order, the collection manager must consider the end-to-end processes associated with tasking collection resources and receiving the data or exploited information from those resources in time to support operational planning and execution requirements.

   b. A number of online tools and databases exist on SECRET Internet Protocol Router Network (SIPRNET) and Joint Worldwide Intelligence Communications System (JWICS) that provide specific information about the capabilities and limitations of various collection assets, their operating areas, and tasking. Additionally, tools such as the collection management workstation (CMWS) and online services, such as Keyhole Markup Language (KML) Google Earth COP feeds, enable collection managers to visualize planned and current activities of collection resources from the national to theater level and may include some tactical assets as well. Developing familiarity with these online resources and tools during workups enables the collection manager to swiftly ascertain what collection resources may be available to support their command's operational requirements.

   c. Figure 3-9 provides an overview of the data and questions a collection manager may ask before deploying to a particular theater or in preparing for a specific operation.

2. Prioritize Requirements. Given time constraints and the limited amount of available collection, processing, and exploitation assets, collection requirements must be prioritized to ensure that collection resources are tasked to address the most critical operational requirements. Nontime-sensitive requirements should initially be placed at a lower priority and periodically reviewed to determine whether they have been satisfied. If collection does not occur on a low priority requirement, then consideration should be given to raising the priority of the requirement or reevaluating the collection need.

3. Take a Multidisciplinary Approach.

   "The enemy is so well hidden that it takes multiple sources of intelligence to corroborate one another.

   —Signals intelligence (SIGINT), for example, can locate a target but may not be able to discern who it is.

   —Full-motion video (FMV) can track but not necessarily identify.

   —Human intelligence (HUMINT) can provide intent but may not be able to fix a target to a precise location.

   —Airborne intelligence, surveillance, and reconnaissance (ISR)'s effectiveness grows exponentially when it is cued to and driven by other sources of intelligence rather than operating alone.

Without a robust, collaborative intelligence network to guide it, sensors are often used in reactive modes that negate their true power and tend to minimize their full potential. These intelligence disciplines provide a start point into the enemy network that can be exploited through persistent and patient observation."

Flynn, M.T., Juergens, R., Cantrell, T.L.
Employing ISR: Special Operations Forces (SOF) Best Practices
Joint Force Quarterly, Third Quarter 2008

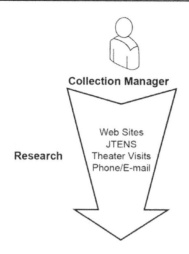

Figure 3-9. Collection Management Considerations

a. The various intelligence disciplines are complementary and collection managers must avoid becoming too reliant on a specific sensor, source, system, or method to address the command's requirements. Each discipline has limitations that can be mitigated by the capabilities of another discipline and often collection capabilities are used to "tip" or "cross-cue" (sometimes just referred to as "cueing") another collection asset to develop a deeper logical choice to accomplish a particular collection objective. Collection managers must be adaptable and determine other ways to meet collection needs using other assets or a combination of assets. Various factors such as weather or system maintenance (either for the asset itself or at its associated processing and exploitation node) could impact the availability of a given resource to satisfy a critical requirement.

b. Collection managers must coordinate closely with senior intelligence leadership, the commander, staff, and supported organizations to ensure the collection resource or resources used address collection needs. This requires synchronizing collection operations in a way that the complementary capabilities of all intelligence disciplines work to satisfy the commander's requirements. For example, HUMINT and SIGINT capabilities are suitable for discerning adversary intent, GEOINT is not.

c. Reliance on a single source of intelligence may lead to mission failure given the asset or resource used to collect on that source may become unavailable or the adversary may become aware of the collection effort and may take countermeasures to prevent the collection or deceive the collector. A multidiscipline collection approach is not only important in addressing command collection requirements but is also a key means of penetrating adversary deception efforts.

d. Synchronization of national to tactical collection assets and resources requires the application of operational art to visualize how multidiscipline resources and assets will come together in time and space to satisfy collection requirements. Despite the highly technical nature of many collection capabilities, the management of collection assets and resources is a very human endeavor. Collection managers should make every effort to assess what collection resources are available in a given theater and then build the relationships needed to facilitate the accomplishment of planned, ad hoc, and dynamic collection tasks using the best mix of collection resources possible.

4. Task Available Assets First. Using available collection assets allows a timely and tailored response to collection requirements and serves to lessen the burden on collection resources controlled by other units, agencies, and organizations. If available assets cannot satisfy a given collection requirement, then the collection manager should request collection from higher, adjacent, and subordinate units, agencies, and organizations. Use of organic collection assets provides two benefits. First, because the commander owns the assets, they can be more responsive to his/her needs, and the commander has greater understanding of the opportunity costs to employ assigned platforms. Second, full employment of organic sensors optimizes the broader collection system by not tasking theater or national resources to collect data that a command can collect itself.

3.6.4 Collection Requirements Management

collection requirements management (CRM). The authoritative development and control of collection, processing, exploitation, and/or reporting requirements that normally result in either the direct tasking of assets over which the collection manager has authority, or the generation of tasking requests to collection management authorities at a higher, lower, or lateral echelon to accomplish the collection mission. (JP 1-02. Source: JP 2-0)

Figure 3-10 lists the major elements of the CRM process described in this section.

3.6.4.1 Requirements Origination

Requirements flow from PIRs and intelligence requirements established by planning and direction activities. EEIs are used to determine the discreet information requirements that are collected to answer PIRs and intelligence requirements. All requirements should be logged for full traceability and may be tailored as needed by the organization. Figure 3-11 provides a template for a sample intelligence requirements management log. Figure 3-12 provides an example of a log tailored to support a CSG's collection requirements management tracking.

3.6.4.2 Collection Planning

Per JP 2-01, collection planning is a continuous process that coordinates and integrates the efforts of all collection units and agencies. PIRs and intelligence requirements are used to develop EEIs. These EEIs are then used to develop RFIs and are passed to naval intelligence personnel performing the CRM function. These personnel check to see if the required information already exists. For example, an RFI may ask whether a particular adversary destroyer has the latest variant of a particular missile system, a fact that may be determined by visually confirming a particular radar antenna configuration on that destroyer class. While satisfying this requirement would typically require tasking an asset or resource with an electro-optical sensor, the collection manager, who is familiar with the theater collection plan, looks back at collection activity from previous days and finds that a theater airborne ISR asset imaged the homeport of this particular unit 2 days prior. Notifying the analysts of this recent collection activity results in a subsequent search of imagery from that collection mission that reveals a high quality shot of the destroyer in question. Analysis of the image satisfies the collection requirement and analysts notify the collection manager who then marks the RFI as satisfied.

If the search for existing information fails to provide enough information to address the requirement, the collection manager checks the theater collection plan, which may also include the collection focus of certain national resources, to determine if any scheduled collection could satisfy the requirement. If so, the collection manager coordinates an ad hoc or dynamic collection requirement in accordance with strike group, Navy component, joint task force, or theater CRM policies.

Collection Requirements Management

- Requirements Origination
- Collection Planning
- Resource Availability and Capability
- Task Assets or Request Tasking of Resources
- Evaluate Reporting
- Collection Plan Update

Figure 3-10. Elements of Collection Requirements Management

PIR or Other Intelligence Requirements	Specific Information Requirements (SIRs)			Latest Time Intelligence is of Value (LTIOV)	Assets	Reporting	Action Required	Action Taken
	EEI	Indicators	Named Area of Interest (NAI)					
List PIR or other requirement. Leave enough space to have several indicators and SIRs in columns 2 and 3	List all indicators that satisfy each item in column 1	List all indicators that satisfy each item in column 1	Name or number of associated NAI	Time or event specific	Prioritize potential assets or resources by priority of effort or support against items in column 1	Include established communication requirements, reporting methods, formats, and report precedence	Examples Retask Update ISR plan Cue another asset	Record outcome of action taken and status of requirement

Figure 3-11. Sample Intelligence Requirements Management Log

IR #	Requestor	Date-Time Group Requested	Subject	Action	Date-Time Group Required	LTIOV	Closed
1	Carrier Air Wing	021100Z JUN	Condition of fuel farm at Base X	Electro-optical (EO) imagery taken and exploited by multispectral imagery (MSI)	061200Z JUN	081200Z JUN	Satisfied 052210Z JUN
2	Bravo Papa	021415Z JUN	Dispersal of strategic surface-to-air missiles (SAMs)	EO and radar imagery tasked twice daily; JIOC has exploitation mission	Daily	C +10	
3	CSG N00	021922Z JUN	Indications of loading sea mines at Base Y	Assigned to ISR orbit A	031200Z JUN	051200Z JUN	No longer required
4	Amphibious transport dock commanding officer	030500Z JUN	Indications of surface-to-surface missile activity in the vicinity of Obj B	Electromagnetic signatures added to Signals of Interest list	090001Z JUN	120001Z JUN	

Figure 3-12. Sample Tailored Intelligence Requirements Log for Carrier Strike Group

If previous or planned collection does not satisfy the information need, the collection manager must then develop collection requirements in collaboration with analysts as part of a deliberate planning process. The collection manager develops a collection plan. The collection plan may be an informal document that is generated via a hardcopy or automated worksheet and used only by the intelligence team or it may be a more formal document if required by the command. At a minimum, the collection plan should include PIRs, their associated EEIs and related indicators, and specific information requirements. These requirements are listed for complete traceability and they are arrayed against collection asset that will be tasked or requested, when the requirement needs to be satisfied by the LTIOV, and who requires the collected information. The collection plan is a tool that aids the collection manager by providing a method for transforming collection requirements into specific collection efforts that ensure optimal employment of collection assets and resources. Figure 3-13 provides an example of sample collection plan format. The process for populating the collection plan by matching requirements to available collection assets and resources is discussed below.

3.6.4.3 Resource Capability and Availability Determination

Once the requirements are defined, the collection manager must determine what collection assets and resources are available and capable of addressing the requirements. Availability and capability to collect depends on a number of factors. To determine suitability of particular collection assets to collect on a requirement, the collection manager compares characteristics of the collection target with the characteristics of available assets and determines which assets are best capable to address the requirement within the time frames required. Figure 3-14 provides an overview of the resource and availability factors. The process for determining availability and capability is described in greater detail below.

1. Determine the key elements that affect collection asset and/or resource nomination. Key elements are the parameters of the target's characteristics that can be compared with the characteristics of the available assets and/or resources and serve as discriminators in discipline and/or sensor selection. The key elements

typically considered are the characteristics of the collection target, the range to the target, and timeliness. These elements are described as follows:

a. Target Characteristics. These are the discernible physical, operational, and technical features of an object or event that are observable and collectible. They are derived from EEIs. Observables are unique descriptive features associated with the visible aspects of the target, such as the radar configuration of an antiship cruise missile site or the number of guns on a surface combatant. Observables are typically associated with GEOINT, MASINT, HUMINT, and CI. Collectibles concern the distinct descriptive features associated with a target's emanations. Typically, SIGINT is associated with collectibles as is MASINT. A target may have one or more observable or collectible characteristics and these can be compared against the capabilities of available assets and resources to determine which intelligence collection assets or resources are best postured to collect on the target.

b. Range. Determined by measuring the distance from a predetermined reference, such as an airborne ISR asset's current location, to the target, range can be used to quickly eliminate potential intelligence collection assets and resources from consideration. Both standoff sensors that cannot cover the target area and sensors on platforms that are unable to enter an area—due to environmental, political, legal, gain/loss adversary action, or other concerns—are not suitable to meet the collection objective and can be eliminated. For HUMINT or CI capabilities, inability to obtain source access is similar to the factor of range for other types of collection assets and resources.

Note

Intelligence gain and loss considerations must be factored into planning. Use of particular collection capabilities in particular ways may compromise that capability and result in a loss of intelligence from a particular source. For example, if the adversary discovers that SIGINT capabilities are being used to monitor a particular communications node, they may stop using that node or change frequencies, resulting in a loss of intelligence. A commander may determine that the intelligence gained in risking exposure of the collection asset is worth more than continued use of that asset to collect additional data and so may order the use of an asset even if the risk of asset compromise is high.

c. Timeliness. This factor concerns when the information or intelligence must arrive to be of value to the requestor, which is also known as the LTIOV. CRM planners must look at the entire process—from coordinating or tasking the asset, mission time, processing and exploitation time, and analysis, production, and dissemination timelines to determine if the information can be collected and provided in time to meet the requestor's need.

SAMPLE COLLECTION PLAN FORMAT					
Period Covered: From_____ To_____					
PIRs or Other Intelligence Requirements	Indications	Specific Information Sought	Assets to be Tasked/Resources Required	Place and Time to Report	Remarks

Figure 3-13. Sample Collection Plan Format

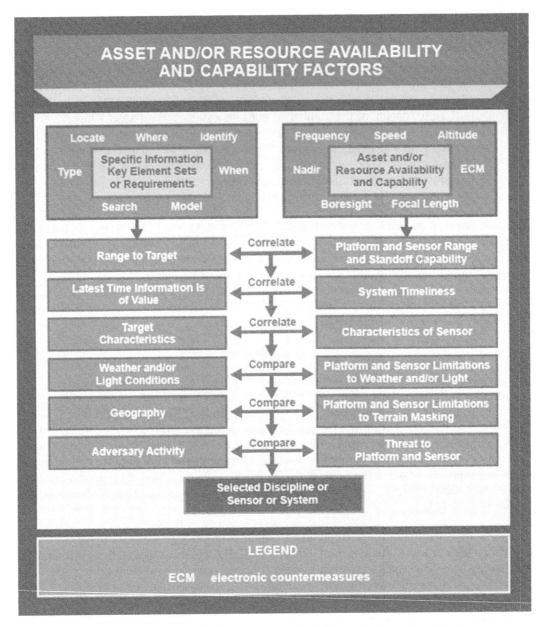

Figure 3-14. Asset and/or Resource Availability and Capability Factors

2. Develop collection capability factors. Collection capability factors are developed by taking the capabilities and limitations of available sensors, systems, or disciplines and creating a set of characteristics that can be compared against the key element sets discussed above. Capability factors should be considered with availability to determine what assets and resources should be tasked. Descriptions of these capability factors are provided below:

   a. Performance characteristics. Factors typically considered as performance characteristics include a given system's ability to collect required data, the quality of the data output, and the location accuracy of the sensor. Collection ability is a determination of whether a particular system in a given intelligence discipline can collect the required data. For example, GEOINT may be chosen as the best capability to collect against a particular target but, because the target is hidden in a jungle environment, there would be a low probability of detecting the target using an electro-optical sensor but a higher degree of probability using an infrared sensor.

Note

Data quality refers to the level of detail that the collection capability can provide. For example, a gyro-stabilized weapons systems camera will likely produce a higher quality image at a greater range than a handheld camera used by a member of the SNOOPIE team. Location accuracy refers to how detailed and accurate the coordinates provided by the sensor are, which is particularly important if the intelligence collection need is in support of a targeting scenario. In the case of a targeting scenario, the requester may require high quality images for precision point mensuration while an analyst confirming whether a particular unit was active in a particular area for an order of battle update product would just need to know if the unit being collected on was in the vicinity of the area of concern.

   b. Platform/sensor range. This factor deals with whether a sensor can cover the requested target. This may be a factor of the sensor's sensitivity or detection range alone or may concern the sensor's capability and the ability of the sensor's associated platform to reach an adequate detection range. Specific issues considered in making a determination concerning platform/sensor range suitability include the maritime operational environment (both environmental factors and threat); commander's guidance and rules of engagement; physical range of the platform; and receive ranges for the platform/sensor data transmission.

   c. Dwell time. Dwell time is defined as the amount of time that an asset or resource can maintain access to the collection target. This factor is an important consideration if the requestor has a requirement for persistent surveillance, tracking, threat-warning, and time-sensitive targeting operations.

   d. Revisit time. This is defined as the period of time after which the collection asset or resource can return and resume collection on a specific target or issue. This can be measured in hours or days and is helpful to know if designing a collection plan that requires multiple collections on the same target over a period of time to assess any changes to the target or the maritime operational environment.

   e. Timeliness. Timeliness is determined by considering the time it takes to complete the end-to-end process that starts with selecting an available and capable collection asset or resource and ends with dissemination of needed information or intelligence (see figure 3-15). Times vary based on mission priority; collection asset or resource availability; mission planning time; the collection mission; and the time to process, exploit, analyze, produce, and disseminate the intelligence. If the collection asset or resource cannot meet LTIOV requirements, it should be dropped from consideration or further coordination should be done to determine if some change can be made to timeliness factors, such as an increased priority for the collection effort or receiving a direct downlink of raw data for exploitation vice relying on another organization's exploitation node.

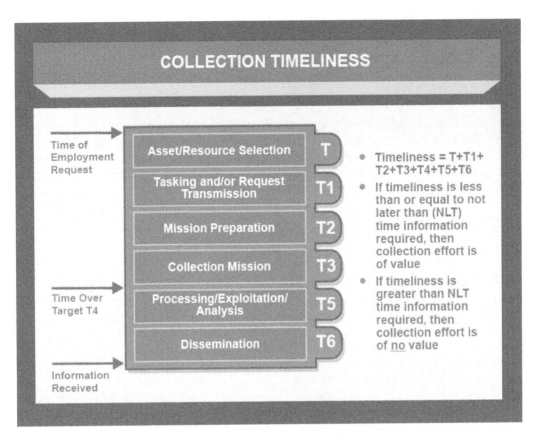

Figure 3-15. Collection Timeliness

3. Correlate key element sets (target characteristics) to collection capability factors to identify candidate sensors/platforms that are technically able to collect on the target within the required timeframe.

4. Compare candidate sensors/platforms with operational environment factors to support final sensor selection. Those operational environment factors include the threat, terrain, contamination, solar position, electromagnetic interference, and weather that might influence the particular discipline or sensor selection. Depending on the operational environment factors, a technically capable sensor may be dropped from consideration. One consideration that must be addressed is sensor vulnerability, which is the degree that adversary countermeasures may impact the collection platform and the sensor. Those platforms operating over adversary territory are typically the most vulnerable while stand-off sensors and overhead capabilities are usually less prone to adversary action. Assessing whether to use an asset in a threat environment is a decision made by the commander and must be a risk calculation that weighs the expected benefit of the collection against the military risk (loss of the asset) and political sensitivity associated with using the asset. Operations in "denied territories" may require their own protection packages and the employment of additional collection assets and resources to conduct threat warning for the mission. These types of missions may also require approval by a higher echelon.

5. Compare the list of viable collection assets and resources with current availability. This comparison should include estimates of downtime for assets and resources that are not currently available. For organic assets, collection managers work closely with other internal intelligence organizations, operations, or J-3 personnel to determine the availability of these assets for collection tasking. Availability of theater and national assets can be coordinated with adjacent or higher headquarters although resources do exist to determine the availability of many of these resources. Automated tools and collection management-related SIPRNET and JWICS Web sites run by Navy component N-2, theater J-2 collection management offices, and the intelligence agencies can be used to determine the availability of many intelligence resources.

Additionally, availability of assets from other Services may be determined through direct coordination with other Service units, if direct liaison authorized has been granted, or requested via the chain of command.

3.6.4.3 Collection Tasking

Figure 3-16 is an example of a collection tasking worksheet, which is used to translate requirements from CRM into tasking that is used in COM. The collection manager starts with the highest priority requirement and then takes each requirement in order to determine how best to satisfy each. The recommendations for planning, scheduling, and control are passed to COM and should include collection activities; target and area to be covered; time and date requirements; and PED tasking and instructions to ensure collected data is provided back to the requester in sufficient time to meet their requirement.

1. Integrate collection resources. Collection to satisfy a requirement may occur at any level—tactical, theater, or national—and may be conducted by a wide variety of organizations, to include coalition partners. Whenever possible, requirements should be integrated into an ongoing or planned mission. This process, called resource integration, is preferred as it makes efficient use of collection resources and may decrease overall costs and risk associated with certain types of collection activities. When preparing a request, collection managers must review existing tasking of assigned missions, which can be obtained through coordination with other units, higher headquarters, agencies, or multinational forces or by using automated systems, such as CMWS, Planning Tool for Resource Integration, Synchronization, and Management (PRISM), HUMINT Online Tasking and Reporting, and collection management-related SIPRNET and JWICS sites. The difficulty with resource integration is weighing the priority of new requirements against those in the planned collection deck since it is possible that adoption of a new collection requirement may bump a lower priority collection opportunity. Additionally, intelligence gain/loss considerations, which are carefully considered in the deliberate mission planning for collection assets, must also be weighed if redirecting an asset in response to ad hoc or dynamic tasking.

COLLECTION TASKING WORKSHEET

COLLECTION TASKING WORKSHEET

Organization: ____________ Registration Number: ____________
DTG: ____________ Collection Manager: ____________
Specific Information Requirements: ____________
Time: ____________ Priority: ____________ Target Range: ____________
Characteristics: ____________

Assets/ Resources	Range	Timeliness	Characteristics	Weather	Geography	Threat	Capability	Remarks
HUMINT								
CI								
IMINT								
COMINT								
ELINT								
MASINT								
TECHINT								

Assets/Resources Selected: HUMINT ____ COMINT ____ MASINT ____ CI ____
IMINT ____ ELINT ____ TECHINT ____

Figure 3-16. Collection Tasking Worksheet

2. Tasking new collection. Tasking a new collection mission should only occur if there is no other way to obtain the needed data in the timeframe required to support the mission. Each intelligence discipline and systems within that discipline have structured processes for requesting collection resources and these request procedures are typically augmented with additional theater and NCC guidance on request and validation procedures. Tools such as PRISM provide a single interface for entering requirements for various intelligence discipline capabilities but at the tactical level requests may be initiated with verbal, e-mail, chat, or record message requests to a higher echelon. If considering tasking a national capability to satisfy a requirement, collection managers should consult the guidelines found in Figure 3-17.

3. Consider multidiscipline and redundant collection approaches. In some circumstances, multiple collection assets and resources may be focused on a single target or set of targets for a variety of reasons. For example, multiple disciplines may be employed to counter adversary deception or counter-ISR capabilities; developing intelligence on the target may be critical to the operational plan and so redundancy is built into the collection plan; or multiple disciplines may be needed to fully characterize the target, such as may be needed to support an ONI S&TI study of a new adversary submarine class. Additionally, there may be a need to overlap collection resources to achieve a long dwell over a target area, to develop a pattern of life analysis, or to maintain track a particular target, such as high interest vessel. Redundancy and use of multiple disciplines to collect on a single area or target is a powerful collection strategy but, since it consumes significant collection resources, the gain achieved by adopting such a strategy must be justified and weighed against the cost of missed collection opportunities against other priority requirements.

4. Evaluate reporting. The collection manager or CRM staff must track the status of all outstanding requirements to their satisfaction or reject them due to inability to meet the requirement in a timely fashion or changes in the operational situation. The collection manager should confirm receipt of tasking and ensure that tasking is understood by the collecting activity. The collection manager should record when collection results are received and should review collected information for completeness. The collection manager tracks the progress of exploited information, ensuring these products are received by the requester. In coordination with the requester, the collection manager determines if the requirement has been satisfied. Information concerning the status of the requirement (i.e., collection tasked, products received, satisfaction status, etc.) should be annotated in the requirements tracking log discussed in section 3.6.4.1 (figures 3-16 and 3-17). Maintaining an accurate understanding of collection requirement satisfaction is crucial to ensuring that collection assets are freed to satisfy other priority requirements.

5. Update collection plan. The collection manager maintains the collection plan and continuously reviews it for currency. Based on the results of each mission, the collection plan is updated by deleting satisfied requirements; annotating retasking decisions; and recording new collection requirements.

3.6.5 Collection Operations Management

collection operations management (COM). The authoritative direction, scheduling, and control of specific collection operations and associated processing, exploitation, and reporting resources. (JP 1-02. Source: JP 2-0)

The major elements of the collection operations management process are:

1. Collection mission planning

2. Execution

3. Exploitation

4. Collection plan update.

Per JP 2-01, COM develops strategies for collection against requirements in cooperation with CRM; predicts how well a system can satisfy requirements; evaluates the performance of the collection systems; allocates and tasks collection assets and/or resources and processing and/or exploitation systems; and monitors and reports the operational status of collection systems. Figure 3-18 provides an overview of the COM management process.

GUIDELINES FOR REQUESTING NATIONAL RESOURCE COLLECTION

Areas of Interest	National systems are best employed against high-priority targets outside the range of theater sensors, beyond standoff collection range, and/or in high-threat areas.
Exploitation and/or Analysis Timeliness	Targets must be chosen such that, under applicable timeliness constraints, exploitation reports will reach the commander in time to react or influence decision making.
Justifications	Request justifications must fully explain the request for information, address why current information does not satisfy the requirement, and identify any required unique sensor capabilities that are unattainable from other assets.
Sensor Capabilities	Target descriptions must place minimum restrictions on systems' use, unless specific parameters are required.
Sensor Accessibility	The targets' accessibility must be determined when possible before a collection request is forwarded.
Exploitation and/or Analysis Requirements Clarity	Specific essential elements of information directly related to the target will add clarity in addition to concise, explicit exploitation statements.
Exploitation and/or Analysis Requirement Purpose	Exploitation and/or analysis requirements must state the purpose of the information desired and how it will benefit the interpreter and/or analyst.
Preplanned Collection	Preplanned target sets submitted in advance of an operation can relieve the workload and must be considered where the tactical situation permits.

Figure 3-17. Guidelines for Requesting National Resource Collection

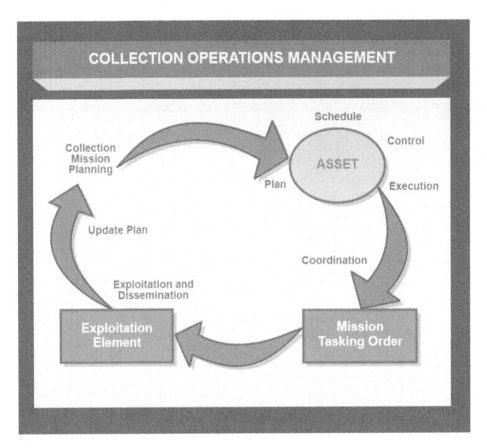

Figure 3-18. Collection Operations Management

3.6.5.1 Collection Mission Planning

Collection mission planning deals with the identification, scheduling, and control of collection assets and resources. Developing the collection plan requires the naval intelligence professional to apply operational art in coordinating and synchronizing the activities of collection assets and resources from the tactical to the national level in support of command requirements. The planner must review the requirements received from the CRM process concerning sensor or intelligence discipline needs; ranges; threats exploitation, analysis, and dissemination requirements; weather; and timeliness. This data is considered when reviewing technical, administrative, and logistical details of various collection systems that could potentially satisfy the requirements. Once a particular collection asset or resource is paired to a particular requirement, the planner may need to coordinate specific tasking and availability issues such as operations, logistics, communications, weather, maintenance, and exploitation nodes to ensure objectives and mission profile are clear, that the asset can meet requirements, and that the data collected is exploited or analyzed and provided back to the requestor. The COM mission planning process results in mission tasking orders for assigned collection assets.

Intelligence collection plans integrate and synchronize the application of the best collector for each collection requirement. Most collection plans are graphical in design and display collector operations against a time-phase matrix and may also include a geospatial plot of collection tracks or areas of interest assigned to specific collection assets (see figure 3-19). This visual format supports rapid understanding of the consequences of the loss of a collector, the rapid identification of what alternative collection capabilities might be available, and a high level assessment of the opportunity costs and ripple effects of retasking a given collection platform.

NWP 2-0

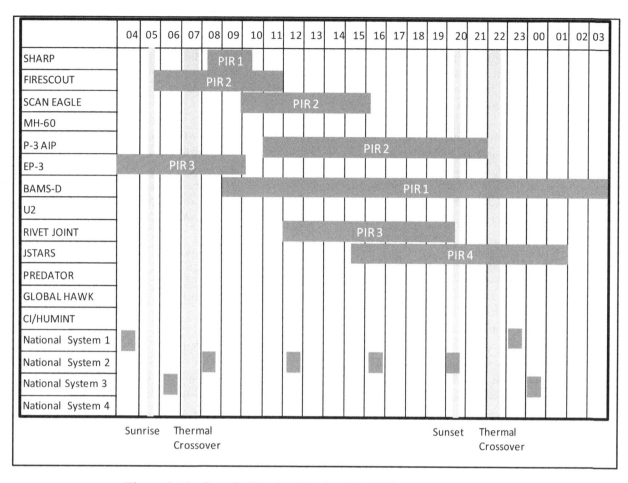

Figure 3-19. Sample Synchronization Matrix for a 24-Hour Period

3.6.5.2 Execution

Once a unit has been selected to conduct a specific collection operation, a mission tasking order or mission type order is sent to that unit. The unit typically decides on the mix of platform, sensor, equipment, and personnel it will employ to satisfy the requirement. In the case of well-established collection operations, mission tasking orders may be received on a periodic basis and provide collection windows over a defined period with specific mission tasking occurring closer to the actual mission date. For example, P-3 and EP-3 reconnaissance missions are typically scheduled well in advance of the mission date to support standing theater collection priorities and integration of these capabilities with other theater, Service, and national systems. Specific tactical-level intelligence collection tasks can be integrated as deliberate, ad hoc, or dynamic requirements in accordance with theater procedures.

3.6.5.3 Exploitation

Management of the exploitation, analysis, and dissemination of the collected intelligence is closely associated with the management of collection assets and resources. In general, the organization that controls or operates the collection capability also controls the sensor-unique PED resources associated with that collection asset or resource. Increasingly, though, systems are being designed to deliver raw intelligence data to the tactical edge, which has the advantage of getting needed data to naval units more expeditiously but also has the potential to overwhelm the limited exploitation, analysis, communications, and data processing and storage capabilities of those units. Striking a balance between the need for timeliness and the capacity of organic PED and analysis capabilities should be considered when developing the exploitation plan. Exploitation is discussed in more detail in section 3.5.

3.6.5.4 Collection Plan Update

Once exploitation is completed, the information or intelligence is provided to the requester, who makes a determination regarding the satisfaction of their requirement. If the requirement is not completely satisfied, the requester coordinates with the collection manager for retasking of an asset or resource to meet the need. At this stage, the requirement goes back to the CRM process.

Collection planning is a dynamic and continuous process so, while evaluation of the success of a recently executed collection plan drives new collection planning, it also causes changes to IRs and PIRs. Operational execution will perturb the operational environment and produce unexpected results that drive changes in the collection plan or collection strategy. The naval intelligence professional needs to balance all of the variables affecting collection operations and oversee processes that affect collection planning. He or she must also ensure that the planning process is transparent to affected commands and that changes are communicated in a timely fashion.

3.6.6 Collection Management Best Practices

Figure 3-20 provides a list of best practices for successful naval intelligence collections operations.

3.6.7 Evaluation and Feedback for Collection

Commanders should require a brief on the output of the collection operations evaluation process as a standard element of the intelligence update. Commanders who understand collection system performance over time are better able to anticipate the ability of that system to address their PIRs and can factor that knowledge into their planning efforts.

The evaluation process deals with whether specific collection requirements have been satisfied and how well the collection system is performing. The naval intelligence team performs both functions, but the determination of whether specific collection requirements have been satisfied is the timelier, tactical action; it drives collection planning updates that may reassign immediate retasking of a collection platform or sensor. System performance is also very important, but it drives collection tasking over longer timelines and also drives sensor and platform selection.

Collection Management Best Practices

1. PIRs are the foundation of everything—build a process that works and execute it.
2. Building relationships with collection managers at higher headquarters is critical to success.
3. Share PIRs with all stakeholders—especially with MOC/combined air operations center and other off ship ISR owners.
4. Designate a representative to each CWC and hold them accountable for capturing intelligence requirements and educating the commander and staff on intelligence processes.
5. Empower (develop standard operating procedures (SOPs), train, and qualify) watch officers to perform critical collection management functions as the 24/7 component of the collection management effort.
6. Remember that collection timelines and architectures need to account for PED.
7. Manage expectations; always be realistic and honest about capabilities and timelines.
8. Collection management is more than IMINT; all intelligence disciplines are key contributors.

Figure 3-20. Collection Management Best Practices

Depending on platform and connectivity, the evaluation process can begin while the platform is still operating. In most cases, commands engage in cyclical operations and the primary evaluation effort occurs once the collection platform has left station or returned to base. The goal is to identify whether scheduled collection mapped to PIRs or commander decision points has been successful, and if not, whether it can be retasked to provide information prior to LTIOV. This process requires that the naval intelligence professional evaluate collection satisfaction against the collection plan.

In the case of organic collection, debriefing the collector is an important element of the collection operation. Carrier air wing aircrews routinely debrief their flights, but when a crew has been engaged in a collection operation, the debriefer needs to be aware of that fact and what the specific collection tasking was. Naval intelligence professionals should develop debriefing procedures for all assigned operational platforms that might be employed in an ISR role and practice the process during workups. For surface ships, this might be providing independent duty intelligence specialists or collateral duty intelligence officers with debrief templates and training.

The output of this evaluation process is feedback on the success of the collection operation that is used to update the collection plan.

3.7 PROCESSING AND EXPLOITATION

processing and exploitation. In intelligence usage, the conversion of collected information into forms suitable to the production of intelligence. (JP 1-02. Source: JP 2-01)

Processing, the conversion of data into a form that is understandable by a user of intelligence, does not add meaning to data and is, in many cases, a technical activity. Converting an image into a format that an intelligence analyst can exploit and analyze and translating a document or communications intercept are examples of processing. For some forms of data, processing may be minimal or not required while in other cases collected data may require significant processing. The time required for processing and exploitation operations is a factor that must be considered when developing a collection plan to ensure that timeliness requirements are met.

Exploitation, on the other hand, is an analytic activity that takes the data received from a single intelligence discipline or system and transforms that data into information. For example, an imagery interpreter may exploit an image to determine the number and type of naval combatants in a given port. The activity is distinct from analysis and production given the fact the resulting information has not been subjected to a full analytic assessment. Processed and exploited information may be sent directly to the intelligence personnel and decision makers, depending on the situation. For example information to support threat warning, targeting, or personnel recovery is often passed directly to commanders, planners, tactical action officers (TAOs), and warfare commanders given the time-sensitive and perishable nature of the processed and exploited information and meets the definition of actionable intelligence discussed in paragraph 3.4.

3.7.1 Processing and Exploitation Considerations

Given that some processed and exploited information can be sent directly to those who require it, security classification and foreign disclosure requirements must also be considered when planning processing and exploitation operations. In the case of processed data, which may be transmitted via automated means, naval intelligence personnel should coordinate closely with systems personnel to ensure that multilevel security and automated guard solutions are properly configured to both protect sources and methods and transmit information to those who require it. Similarly, write for release, sanitization, and tear line reporting policies should be developed and implemented for time-sensitive information that comes from a highly classified source.

Planning for processing and exploitation operations typically takes place as part of collection operations because the collection operation element usually also controls the sensor-unique processing, exploitation, and analysis equipment. Advances in communications and information technologies are pushing more and more processing and exploitation capabilities to the tactical level, which significantly improves the speed with which numerous units can simultaneously receive direct feeds of relevant data and information. For example, P-3 aircraft can now directly downlink FMV and still images to properly equipped surface vessels to aid in building situational awareness and to support mission planning and execution activities. These same improvements in technology have also driven the development of effective reachback solutions, such as ONI's FIST, which enables a remote node to process and exploit needed information and rapidly disseminate exploitation results to the supported unit. Processing and exploitation requirements must be considered and planned for with the same level of fidelity given to collection planning. Close coordination with processing and exploitation nodes is also needed to ensure these nodes understand the requirements they are working and the priority of the requirements to the supported organization.

Figure 3-21 provides an overview of the activities associated with processing and exploitation operations.

3.7.2 Evaluation and Feedback for Processing and Exploitation

Evaluation and feedback on processing and exploitation operations require the establishment of formal or informal processes to assess factors such as timeliness, accuracy, usability, relevance, and availability of processed and exploited information. Naval intelligence professionals must clearly understand the TCPED chains associated with internal and external providers of processed and exploited information so that they can quickly assess what component of that chain (communications, unclear tasking, etc.) is broken or suboptimized and take action to correct it or develop workarounds. Processed and exploited information can be directly disseminated to decision makers, so naval intelligence professionals should solicit feedback from those intelligence users to determine whether the information provided was useful, delivered in a timely manner, and easily understood. Issues identified during the feedback process should be addressed with the provider of the information so that necessary improvements can be made. Since intelligence analysts also receive the output of processing and exploitation operations, they should also provide feedback on how timely and useful the information provided was in supporting intelligence analytic and production efforts.

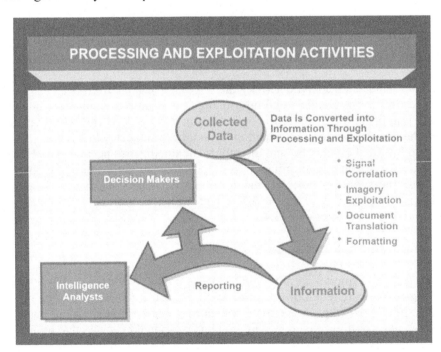

Figure 3-21. Processing and Exploitation Activities

3.8 ANALYSIS AND PRODUCTION

> Major intelligence failures are usually caused by failures of analysis, not failures of collection. Relevant information is discounted, misinterpreted, ignored, rejected, or overlooked because it fails to fit a prevailing model or mind-set. The "signals" are lost in the "noise."
>
> *Richards J. Heuer, Jr.*
> *Psychology of Intelligence Analysis*

> During this . . . phase . . . the trained intelligence [professional] and the skilled analyst work with quantities of raw information collected from a great variety of sources, selecting, verifying, comparing, and interpreting items of value, and acting upon the results to produce usable intelligence. The basic problem is one of determining the significance of information in the light of past experience, present circumstances, and possible future developments. To be utilized by the analyst in solving this problem are the principles of logic, reasoning, and exhaustive research. As is true of the social sciences, intelligence is based on observed phenomena or facts from which generalized conclusions are drawn. Therefore, . . . [analysis] . . . is much more than the orderly assembling of related facts and the determining of their significance; it involves the deriving of total meaning from these facts when related to other intelligence already available.
>
> *U.S. Naval Intelligence School*
> *Intelligence for Naval Officers*
> *1956 Edition*

analysis and production. In intelligence usage, the conversion of processed information into intelligence through the integration, evaluation, analysis, and interpretation of all source data and the preparation of intelligence products in support of known or anticipated user requirements. (JP 1-02. Source: JP 2-01)

Analysis, the cognitive process of integrating and interpreting diverse amounts of data, and production, which consists of communicating the results of analysis to intelligence users in a meaningful and timely way, are core competencies of the naval intelligence profession. Analysis and production provide the answer to the commander's questions about the adversary and the maritime operational environment. Analysis drives the activities of collection, processing, and exploitation—which provide the data and information needed to address intelligence gaps—and is bounded by planning and direction and dissemination and integration—which shape how and in what form the analysis is produced and provided to the requester. Sound analytic practice and strong communication skills are foundational to the profession of naval intelligence as they ensure commanders receive timely, objective, accurate, complete, and relevant intelligence that incorporates all available sources, uses methods tailored to the specific intelligence problem being addressed, and communicates results in an easily understood and persuasive form.

Intelligence analysis and production operations occur in response to tasked and anticipated intelligence requirements. These requirements may be standing, recurring requirements, such as the need for the red track database management (RDBM) to maintain and communicate the current threat picture in the Global Command and Control System–Maritime (GCCS-M) COP, or special requirements, such as the need to develop an intelligence estimate to support the planning of an amphibious operation. Regardless of the purpose, all analysis and production requirements should be approved by the senior naval intelligence professional assigned to a command or that senior person's designated representative. Since analysis and production are resource-intensive operations, leadership oversight is needed to ensure that production is prioritized, nonduplicative, and is conducted using the most efficient and effective combination of intelligence resources.

Satisfaction of intelligence analysis and production requirements may require requesting or tasking those requirements to intelligence organizations that have greater capacity or specialized expertise unavailable at the tactical or operational level. In these cases, naval intelligence leaders must ensure the supporting intelligence organization clearly understands the requirements for the intelligence product, to include the timeframe in which it needs to be completed and disseminated. Additionally, intelligence analysis often involves significant coordination and collaboration to ensure the commander is supported with the best possible intelligence assessment. This coordination and collaboration may occur solely within the command's intelligence team or may involve other analysts from throughout the intelligence community, defense, other government agencies and departments, academia, or other experts as required. Naval intelligence professionals should always seek collaboration with other experts, as time allows, to ensure the relevance, accuracy, completeness, and objectivity of their analysis.

The following subsections provide an overview of key concepts and processes related to intelligence analysis, discusses specific intelligence analytic products that are commonly used by naval intelligence professionals, and touches on a few key issues related to analytic tradecraft. The subject of intelligence analysis and production is complex and deserving of greater independent study given how critical the analytic skill set is to successful naval intelligence operations. Appendix F, which discusses analytic tradecraft, provides more information on specific analytic tools and warns of the dangers of cognitive biases and analytic pitfalls that are inimical to sound analytic practice. In addition to the joint publication series on intelligence and other Service intelligence doctrine, all of which provide useful perspectives and insights on intelligence analysis and production in a military context, readers are encouraged to seek out other sources on this subject. Suggested works include, but are not limited to, Edwin Layton's I Was There; Richards J. Heuer's Psychology of Intelligence Analysis; Richards J. Heuer and Randolph H. Pherson's Structured Analytic Techniques for Intelligence Analysis; James S. Major's Communicating with Intelligence; Timothy Walton's Challenges in Intelligence Analysis; and Roberta Wohlstetter's Pearl Harbor: Warning and Decision.

3.8.1 Sources and Methods

source. 1. A person, thing, or activity from which information is obtained. 2. In clandestine activities, a person (agent), normally a foreign national, in the employ of an intelligence activity for intelligence purposes. 3. In interrogation activities, any person who furnishes information, either with or without the knowledge that the information is being used for intelligence purposes. (JP 1-02. Source: JP 2-01)

The intelligence analyst must understand the concepts of sources and methods and the distinctions between them as it is instrumental in assessing intelligence gaps; driving collection operations to meet those gaps; and enabling an assessment of the overall quality of a finished intelligence product. Before discussing the concept, the terms source and method need to be defined as they are used in this publication. A source is defined as a "person, thing, or activity from which information is obtained." A method[4] is the means by which that information is obtained. Understanding sources and what they can provide to support the intelligence effort and the methods available to collect against these sources is key to naval intelligence operational art.

Note

A fundamental premise in assessing the relevance of various intelligence disciplines to address threat factors is the distinction between sources and methods. The customary conjunction of these two terms often leads to confusion that they are either synonymous or interchangeable; when, in fact, there is an important difference between them.

Any information or intelligence requirement presumes that an answer exists to the question posed. The unique nature of intelligence, however, assumes that an adversary is taking active concealment or deception measures to protect those answers. Therefore, any intelligence-based collection effort requires a two-fold process:

1. Determine where the answer resides. This determination is the source. In most cases more than one source can address a particular requirement, although credibility across various sources may vary. Considerations when determining possible sources include operational time constraints, existing accesses, assessed authenticity and risk.

---

[4] "Intelligence method" has no formal joint or Navy doctrine definition.

2. Determine how to gain access to the answer. This determination is the method (often the intelligence discipline). In most cases more than one method can exploit a particular source, although the relevance of a particular intelligence discipline may vary. Considerations when determining possible methods include asset availability, exploitation timeliness, program cost, and risk.

It is important to underscore that the classification of an intelligence collection effort can be determined by the sensitivity of either the source or the method. A widely known collection method (e.g., HUMINT) may generate a highly classified report when a well-placed source is used. Conversely, a routine source may be the subject of a highly classified report when a collection method involving leading-edge (e.g., SIGINT) technology is used. Either way, the need to protect either the source or the method from compromise is what delineates intelligence operations from other research and analytic efforts.

The relationship and difference between sources and methods is depicted in figure 3-22. This graphic portrays scenarios wherein various intelligence disciplines may be used to exploit a particular source.

Understanding of the difference between sources and methods keeps naval intelligence professionals focused on information needs as the foundation for collection planning. Most intelligence successes can be traced to developing innovative ways to exploit a source when established methods—or those readily available to fleet commanders—were not sufficiently addressing requirements.

3.8.2 Conversion of Data and Information into Intelligence

Figure 3-23 shows a notional example of the steps taken to transform collected data into the processed and exploited information that, when combined with other sources and previous intelligence assessments, results in a finished intelligence product. The operations involved with collection, processing, and exploitation were discussed in sections 3.4 and 3.5. This section discusses the process used to convert data and information into intelligence.

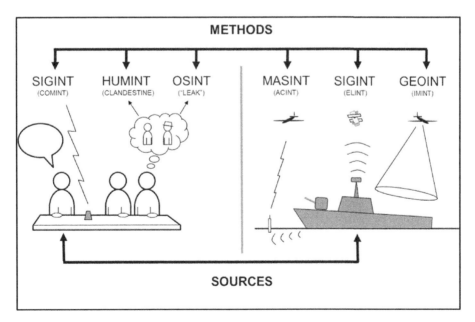

Figure 3-22. Distinction between Sources and Methods

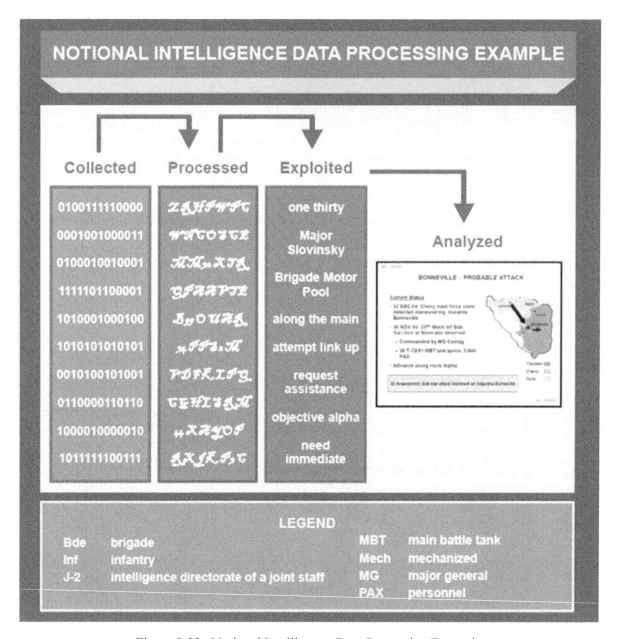

Figure 3-23. Notional Intelligence Data Processing Example

Information is converted into intelligence through a structured process shown in figure 3-24. While this process is shown as sequential, it is possible for the various steps of integration, evaluation, analysis, and interpretation to occur in parallel. The process may also be iterative. While it is useful to establish an information cutoff date to ensure that intelligence is produced to meet LTIOV requirements, it is often the case that the steps shown in this figure are revisited during the production cycle as new intelligence is received; new insights are developed from the analytic process; or different perspectives provide fresh interpretations of the existing data and information.

While a critical skill set, the discussion on analysis below is at a very high level. Appendix F provides a more detailed discussion of analytic tradecraft that is an overview of the subject taken from a similar appendix in JP 2-01, Joint and National Intelligence Support to Military Operations, and other sources. It discusses the analysis process, provides examples of various analytic products and tools, and deals with the subject of analytic pitfalls and cognitive biases that are inimical to good analytic tradecraft.

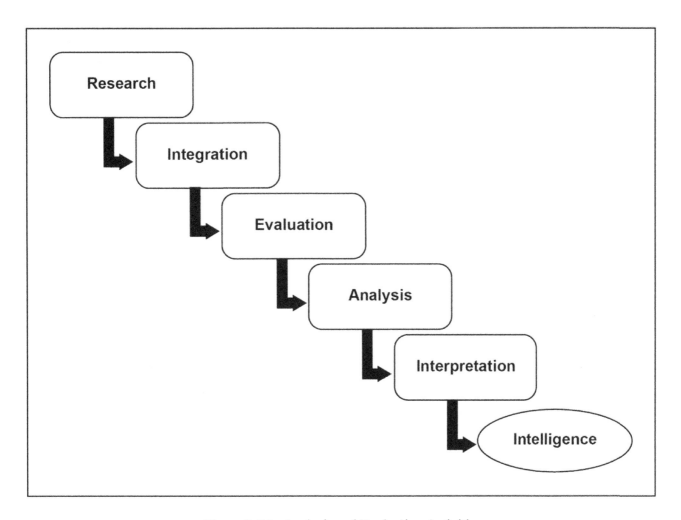

Figure 3-24. Analysis and Production Activities

3.8.2.1 Research

Research is a continuous process conducted by the naval intelligence analyst to determine the extent of finished intelligence and other information that exists on a given intelligence problem. While used during planning and direction to determine the intelligence gaps that drive production and collection requirements, research is also done while awaiting new data, information, or finished intelligence products needed to inform an assessment. Research is multisource, consulting the widest possible variety of classified and open sources. For a given intelligence problem an analyst should have a well-defined set of sources and information providers they consult on a daily basis as well as others that are checked on a periodic basis. This should only be a starting point, however, as the naval intelligence professional may need to absorb data and information on a given topic from a number of perspectives—such as historical, political, technical, economic, sociological, and cultural—to address the intelligence problem and identify gaps. Often the information needed by the analyst is either already available or is already being sought by collection assets so close collaboration with other analysts and collections personnel is essential to maintaining awareness of available or potentially available information that can be used to address the intelligence problem.

Note

> While research is not listed as a component of converting data and information to intelligence in joint doctrine, it is an implied task that is critical to establishing and maintaining a baseline of knowledge from which to make sound analytic judgments.

3.8.2.2 Integration

During integration, information from single or multiple sources is received, collated, and entered into appropriate databases by the analysis and production elements of intelligence community organizations. Using predetermined criteria, naval intelligence analysts integrate and group newly received data and information with related pieces of information. For example, a good analytic practice is to maintain electronic folders of relevant data on various topics such as a particular country's naval combatants or on a specific piracy ring. Archived intelligence products on these topics are kept in these electronic folders. As new, relevant information is received it is reviewed and the information it provides could be used to update a standing assessment. This new information is then added to the archive.

3.8.2.3 Evaluation

Intelligence produces an assessment or interpretation of adversary capabilities and intent and the maritime operational environment, which enables the development of predictive estimates in support of operational missions. The estimate is not a certainty and the validity of the estimate rests on a combination of factors. Two of the most significant factors are the analytic acumen of the naval intelligence professional or intelligence team conducting the analysis and the quality of the data and information used in the analytic process. While the first factor is the result of experience, training, and intuition, even the most seasoned analyst finds it difficult to make a relevant and accurate analytic judgment if the quality of data upon which to rest that judgment is poor.

Evaluation of sources and methods is a key factor in analytic tradecraft. General Colin Powell's famous instructions to his intelligence team—tell me what you know, tell me what you don't know, tell me what you think, and always distinguish which is which—is sound advice for any analyst. In the early stages of a crisis, there may be a dearth of finished intelligence products relevant to the crisis situation, few sources to exploit, and fewer methods with which to exploit them. The intelligence analyst must always be forthright about the quality of the sources upon which their assessment is made so that the commander, staff, and supported organizations can factor the analyst's level of confidence in their intelligence assessment into the planning and execution of operations.

Each new item of information received must be evaluated in terms of the reliability or the source and the credibility of the information. Both must be assessed independently to ensure the evaluation of one factor does not bias the other. While both are subjective judgments, they are typically rooted in the naval intelligence analyst's experience with the particular source and their subject matter. In assessing reliability, the question answered is how well or how often the particular source has produced accurate information and whether the source is a primary, secondary, or tertiary provider of the information. In the case of a technical collection means, like the electro-optical sensor of a P-3, reliability will probably be assessed as very high whereas a new HUMINT source, one without a reporting history, may be of questionable reliability. Credibility is assessed by how consistent the information provided is with known facts. For example, an OSINT report from a foreign news service claiming that a relatively small country with no known blue water naval capabilities is now in possession of three nuclear powered submarines would most likely be assessed as a low credibility report. Once reliability and credibility are assessed, the naval intelligence analysts should assign a confidence factor to the data or information. This confidence factor can take the form that works best for the analyst or the recipient of the intelligence but is typically done using an alphanumeric rating (i.e., B3, C2), a low-medium-high assessment, or a percentage rating. Figure 3-25 provides an example of reliability and accuracy scale based on an alphanumeric scale while Figure 3-26 provides an example of a low-moderate-high scale.

Source Confidence Scale—Example #1	
A = Completely reliable	1 = Confirmed by other sources
B = Usually reliable	2 = Probably true
C = Fairly reliable	3 = Possibly true
D = Not usually reliable	4 = Doubtful
E = Unreliable	5 = Improbable
F = Reliability cannot be judged	6 = Truth cannot be judged

Figure 3-25. Alphanumeric Confidence Scale[5]

Source Confidence Scale–Example #2

Low
- Uncorroborated information from good or marginal sources
- Many assumptions
- Mostly weak logical inferences, minimal methods application
- Glaring intelligence gaps exist

Terms/Expressions
- Possible
- Could, may, might
- Cannot judge, unclear

Moderate
- Partially corroborated information from good sources
- Several assumptions
- Mix of strong and weak inferences and methods
- Minimum intelligence gaps exist

Terms/Expressions
- Likely, unlikely
- Probable, improbable
- Anticipate, appear

High
- Well-corroborated information from proven sources
- Minimal assumptions
- Strong logical inferences and methods
- No or minor intelligence gaps exist

Terms/Expressions
- Will, will not
- Almost certainly, remote
- Highly likely, highly unlikely
- Expect, assert, affirm

Figure 3-26. Low-Moderate-High Confidence Scale

---

[5] Major, J., Communicating with Intelligence, p. 21.

The evaluation of sources and methods should also extend to understanding the quality of the sources used to create finished intelligence products employed by the analyst in building their assessment. The mere fact that an intelligence organization has produced and posted a finished intelligence product does not mean that product is free from error or represents the best analytic judgment now possible. For example, an intelligence organization produces a report on a new adversary naval gun system that is almost exclusively based on a single source HUMINT report. They use a single source to build the report because the information is deemed critical and efforts to use additional sources to confirm the information were unsuccessful. That report would represent the organization's best analytic judgment on the system at that time and would likely be caveated as a low confidence assessment. Months later, another analyst may be charged with producing a report on that same naval gun system. That analyst now has access to SIGINT and GEOINT data that contradicts the information about the gun system capabilities given by the HUMINT source in the first report. Further research on the initial finished intelligence report reveals that the source is actually a minor government functionary with an unreliable reporting record. The analyst would most likely discount much of the data contained in the initial intelligence report produced on this naval gun system. If there was high corroboration between the SIGINT and GEOINT data, the new assessment of the naval gun system's capabilities may be delivered with a moderate or high level of confidence. Figure 3-27 provides an overview of the main types of data an analyst may encounter.

3.8.2.4 Analysis

The analysis process consists of comparing integrated and evaluated information with known facts and predetermined assumptions. The purpose of this comparative process is to deduce potential patterns or to derive new insights about some aspect of the adversary or the maritime operational environment.

3.8.2.5 Interpretation

Interpretation is an objective mental process of comparison and deduction based on common sense, life experience, military knowledge covering both adversary and friendly forces, and existing information and intelligence. This mental process involves identifying new activity, recognizing the absence of activity, and postulating the significance of that activity.

Term	Definition	Example
Fact	Verified information; something known to exist or to have happened.	A confirmed inventory of a resource of one's own service.
Direct Information	Information relating to an intelligence issue under scrutiny the details of which can, as a rule, be considered factual, because of the nature of the source, the source's direct access to the information, and the concrete and readily verifiable character of the contents.	Foreign official report providing a specific piece of information within their purview or human intelligence from a U.S. diplomatic officer who directly observed an event.
Indirect Information	Information relating to an intelligence issue the details of which may or may not be factual, the doubt reflecting some combination of the source's questionable reliability, the source's lack of direct access, and the complex character of the contents.	Human intelligence from a reliable agent citing secondhand information from a source of undetermined reliability.
Direct Data	Organized information that provides context for evaluating the likelihood that a matter under scrutiny is factual.	Charts, graphs, or tables depicting organized data collected by U.S. personnel or trusted agents.
Indirect Data	Organized information that provides context for evaluating the likelihood that a matter under scrutiny is factual.	Charts, graphs, or tables depicting organized data collected by a liaison intelligence service.

Figure 3-27. Types of Intelligence Data

Note

All analysts must be mindful of analytic pitfalls and cognitive biases that may cause them to reject a given piece of data or information, even though that data and information is accurate. Naval intelligence analysts must carefully evaluate all data and information. Just because a particular source's information does not conform to the established analytic judgment on a particular intelligence problem it must still be considered in building the intelligence assessment.

3.8.3 Production

Figure 3-28 highlights standard types of intelligence products that are developed to support operational requirements. While not an all-inclusive list, it provides details on the major classes of intelligence products typically produced by naval intelligence analysts. This list is consistent with types of intelligence and functions of naval intelligence discussed in chapters 1 and 2.

3.8.3.1 Indications and Warning

The I&W process analyzes and integrates operations and intelligence information to assess the probability of hostile actions and provides sufficient warning to preempt, counter, or otherwise moderate their outcome. The nature of I&W varies at each echelon. For example, at the theater-strategic level, I&W products may focus on a wide variety of geopolitical and military factors that provide indications of potential mobilization or preparations for hostilities. At the tactical level, I&W is typically very specific. For example, indicators of fire control radars in the vicinity of strike aircraft is likely an adversary attempt to target those aircraft and must be reported in the most expeditious manner possible.

I&W analysis and products are rooted in scenarios. At the operational level of war, there may be multiple scenarios developed that cover a range of potential adversary courses of action or that describe alternative ways a particular situation will unfold. For each of these scenarios, a list of potential indicators is developed. These indicators must be:

1. Observable and collectable—need to have a reasonable chance of collecting and observing the activity

2. Valid—correlates to verifying the scenario

3. Unique—only measures one thing

4. Reliable—others will reasonably interpret the data the same way

5. Stable—should be able to monitor change over time.[6]

Analytic and collection assets are used to determine if there is a change to the indicators and, as change is noted, assessments are made as to the potential likelihood of the scenarios developing as predicted. While strategic and operational indications and warning problems follow a very well-defined scenario-indicator development process, this same process is at use, albeit in an abbreviated and more informal form, on the tactical level. Concern over fire control radar activations is predicated on a general understanding of military tactics coupled with an assessment of how a specific adversary employs their weapons systems. This is, in essence, part of a detect-to-engage scenario and the observation of a particular predetermined indicator (the fire control radar) is what triggers the warning report.

---

[6] Heuer, R. J. and Pherson, R. H., Structured Analytic Techniques for Intelligence Analysis, p. 119–134.

INTELLIGENCE PRODUCTS	
Indications and Warning	**Current intelligence reports** from theater assets, theater I&W support and correlation of force movements in the joint operations area (JOA). **National level** provides tipoff and warnings of imminent or hostile activity.
Current Intelligence	**Military and political events** of interest from joint intelligence operations center (JIOC), joint intelligence support element (JISE), and national sources. Counterintelligence on current foreign intelligence activities. **Reports** on joint force operations. **Summaries and briefings** by JIOC, JISE, and national organizations. **Open-source intelligence** in the JOA.
General Military Intelligence	**Tailored to specific mission:** Political, economic, and social aspects of countries in JOA. Information on organization, operations, and capabilities of foreign military forces in the JOA. Counterintelligence on foreign intelligence capabilities and activities, as well as terrorism, which impacts the force protection mission. **Formats:** Military Capabilities Assessment, Military-Related Subject Assessment, Adversary Course of Action Estimate, Foreign Intelligence Threat Assessment.
Target Intelligence	**Target systems analyses.** **Electronic target folders** containing target materials describing characteristics of selected targets. **Target lists.** **Combat assessment products.**
Scientific and Technical Intelligence	Examines foreign developments in basic and applied sciences and technologies with warfare potential, particularly enhancements to weapon systems. It includes S&TI characteristics, capabilities, vulnerabilities, and limitations of weapon systems, subsystems, and associated material, as well as related research and development. S&TI also addresses overall weapon systems, tactics analysis, and equipment effectiveness.
Counterintelligence	**Counterintelligence** analyzes the threats posed by foreign intelligence and security services and the intelligence activities of non-state actors.
Identity Intelligence	Results from the fusion of identity attributes (biologic, biographic, behavioral, and reputational information related to individuals) and other information and intelligence associated with those attributes collected across all intelligence disciplines. Identity intelligence utilizes enabling intelligence activities, like biometrics-enabled intelligence, forensics-enabled intelligence, and document and media exploitation, to discover the existence of unknown potential threat actors by connecting individuals to other persons, places, events, or materials, analyzing patterns of life, and characterizing their level of potential threats to U.S. interests.
Estimative Intelligence	**Estimates** provide forecasts on how a situation may develop and the implications for planning and executing military operations.

Figure 3-28. Intelligence Products

I&W relies on tip-offs of time-critical intelligence from sources at all levels. It also relies on well-established intelligence architectures to function correctly. Intelligence requirements for I&W may include:

1. Adversary intentions, capabilities, preparations, deployments and associated activities, and attack methods
2. Adversary motivations
3. Changes in adversary force dispositions, alert or readiness status, or mobilization activities
4. Information operations activities that may indicate hostilities.
5. Status of other military forces in the areas of concern
6. Civil activities that may be indicators of potential military action
7. Nonmilitary factors that could significantly alter political, economic, or social situations
8. The capability of a local or regional government to deal with a particular situation.

3.8.3.2 Current Intelligence

While similar to I&W, in that it requires close monitoring of world events and specific activities in areas of concern, current intelligence focuses on producing intelligence concerned with describing the existing situation. Current intelligence is done at every level of war. The information used to build current intelligence products may include but is not limited to the following:

1. Adversary capabilities, intentions, and will to use military force—where, when, in what strength, with what forces and weapons
2. Adversary operational plans
3. Adversary COGs
4. Adversary vulnerabilities
5. Analysis of the maritime operational environment
6. Impact of meteorology, oceanography, hydrography, and space effects
7. Military, political, social, and economic events
8. Status of transportation nodes
9. Adversary weapons of mass destruction (WMD) capabilities
10. Adversary foreign intelligence and security activities
11. Adversary cyberspace capabilities.

Current intelligence and general military intelligence (GMI, also referred to as basic intelligence) are interdependent.

Note

> Current intelligence products include such things as the command daily intelligence briefing, which typically covers conditions of the operational environment; tactical or strategic level adversary military force locations, disposition, readiness, activities and intentions; and updates on geopolitical issues of potential concern to the naval force. As the intelligence contributor to the COP, which is typically maintained using a system, such as GCCS-M, naval intelligence analysts provide all-source intelligence to detail adversary locations and dispositions in a near real time visualization environment accessible by the commander, staff, and supported organizations.

Joint Doctrine defines the term operational intelligence as intelligence support to the operational level of war but does not associate the acronym OPINTEL to the term. However, for naval intelligence professionals, the term OPINTEL refers to the products resulting from an all-source intelligence analysis process that can provide the commander, staff, and supported organizations the best intelligence estimate possible to support their planning efforts and operations.

OPINTEL is a term often used by naval intelligence professionals in discussing current intelligence concepts or processes yet it has no doctrinal definition. While the discussion that follows is not an attempt to firmly define the term, it does provide historical context for the evolution of the concept to enable a better understanding of the concept and an appreciation for the role of naval intelligence professionals in conceiving and maturing it.

The term OPINTEL was first coined by naval intelligence in World War II in response to a perception that the term "combat intelligence" was too limiting for the type of operationally focused current intelligence support that was being provided to naval forces. Initially, the term combat intelligence, which was defined as "information about enemy forces, their strength, disposition, and probable movements," was used by naval intelligence at the start of the war as it reflected accepted doctrine that the intelligence officer was the expert on the enemy who was, in a sense, the enemy's representative on the commander's staff. As the war evolved, naval authorities believed the term combat intelligence to be too limited in that it did not fully encompass the breadth and depth of information commanders actually desired and needed to plan and execute their operations. For this reason, naval intelligence professionals began using a new term, "operational intelligence," which was defined as "intelligence needed by naval commanders in planning and executing operations, including battle" and was described as: Intelligence needed by naval commanders charged with:

1. The operations and support of operations of naval forces in theaters of operations

2. The successful employment of forces under their command against enemy or hostile forces in the immediate zone of combat and in those areas from which the issue of battle can be materially influenced or affected before, during, and immediately after battle.

In the World War II period, the term was considered a more overarching concept that encompassed a range of intelligence support activities associated with what would today be called current operations or near-future operations. So, during that period, it would be technically correct to discuss "operational intelligence for air operations" or "operational intelligence for amphibious operations" if one was being specific about the specialized intelligence requirements and the methodologies needed to support those specific types of operations.

Over time, the term OPINTEL evolved and began to become more closely associated with particular aspects of naval intelligence support during World War II. In Christopher Ford and David Rosenberg's book The Admiral's Advantage, OPINTEL is defined as "the art of providing near-real-time information concerning the location, activity, and likely intentions of potential adversaries" and "that intelligence information needed . . . for planning and conducting operations . . . in the most immediate sense." Used in this context it describes both a conceptual framework and an all-source analytic methodology originally used by naval intelligence analysts to develop current intelligence on adversary naval forces. While consistent with the discussion above, this particular application of OPINTEL had its origins in the all-source analytic techniques explored by U.S. and British fleet intelligence centers during World War II to find, localize, and track German and Japanese naval units and supply convoys.

While not limited to the detection, localization, tracking, and analysis of adversary naval intentions as discussed above, the term OPINTEL became nearly synonymous with that type of naval intelligence activity during the Cold War. The need to find and track Soviet naval units and worldwide merchant ship activity became the major focus of naval intelligence during that period and both the OPINTEL concept and the associated all-source, multimethod intelligence analysis practices pioneered by U.S. and British naval intelligence analysts during the World War II period were adapted to the new challenge posed by the Soviet naval threat. With the evolution of automated systems for visualizing the maritime operational environment, OPINTEL methodologies were applied to developing and maintaining an intelligence picture that was integrated with operations in many ways, to include contributing red threat analysis to the COP. The techniques that evolved during the Cold War period provide the core methodologies used today for RDBM and associated analytic activities required to provide a near-real-time view of threat and the maritime operational environment.

OPINTEL is not and never has been specifically limited to developing and maintaining a current intelligence picture, despite the fact this was a major focus of the Navy's OPINTEL efforts in building the network of fleet ocean surveillance information facilities/Fleet Ocean Surveillance Information Centers (FOSICs) and the Ocean Surveillance Information System nodes that supported fleet operations in the Cold War. As Ford and Rosenburg document, the FOSICs evolved into some of the JICs that today support regional combatant commands. As has been shown by naval intelligence's support in responding to numerous regional conflicts, the global war on terrorism, and emerging threats, naval intelligence professionals have consistently been able to adapt OPINTEL concepts and methodologies to serving their commanders' needs for relevant intelligence to support the planning and conduct of current and near-term operations.

While the tools for the analysis and dissemination of OPINTEL have changed since World War II, the process for conducting OPINTEL remains one of rapidly evaluating each new piece of intelligence and integrating that into a rigorously maintained all-source product that can provide the commander, staff, and supported organizations the best intelligence estimate possible to support their planning efforts and operations. Readers desiring to know more about OPINTEL are encouraged to consult The Admiral's Advantage as well as Patrick Beesley's Very Special Intelligence, Donald McLachlan's Room 39, and Allen Harris Bath's Tracking the Axis Enemy.

3.8.3.3 General Military Intelligence

GMI is informed by the development of current intelligence products (and relies on the same sources of data and information that is used to build those products) and on analysis and estimates of future trends, alternate outcomes, and the likelihood of various future possibilities. GMI focuses on military matters and associated concerns, related to known and potential adversaries as well as other countries and nonstate actors deemed of interest by decision makers for their planning purposes. It is conducted at every level of war, from the tactical to the strategic. It can include information on the organization, operations, facilities, capabilities, plans, and estimated futures of foreign military forces and pertinent information concerning the environment (political, economic, topographic, geodetic, demographic, and sociological aspects of foreign countries).

Figure 3-29 provides a list of the major information elements used in GMI analysis. This list is not all-inclusive.

General Military Intelligence Elements

1. Foreign training, doctrine, leadership, experience, morale of forces, state of readiness, and will to fight
2. Foreign order of battle, also known as inventory, both current and projected through at least 20 years in the future
3. Foreign military strengths and weaknesses, force composition, location, and disposition, including command, control, communications, computers, and intelligence, logistics and sustainment, force readiness and mobilization capabilities
4. Foreign perception of threats, force mobilization plans, and war plans, and plans and triggers for military escalation
5. Foreign military operations and tactics
6. Foreign plans and capability to reconstitute forces during wartime, including reinforcements, replacements, and repair
7. War termination goals and vulnerabilities
8. Foreign denial and deception concepts and practices
9. Foreign alliances and other cooperation with third parties
10. Strategic warning of new adversaries that may arise over the next 20-plus years and why they may become adversaries, as well as of positive opportunities that may arise
11. Basic infrastructure (power, resources, health, population centers, public infrastructure, and production and storage facilities for toxic industrial materials)
12. Hydrographic and geographic intelligence, including urban areas, coasts and landing beaches, troop handling zones, and geological intelligence
13. Capability and availability of all transportation modes in the operational area
14. Military materiel production and support industries
15. Military economics, including foreign military assistance
16. Insurgency and terrorism
17. Military-political and sociological intelligence
18. Location, identification, and description of military-related installations
19. Survival, evasion, resistance, and escape
20. Government control

Figure 3-29. General Military Intelligence Elements

Some examples of GMI products include:

1. Military Capabilities Assessment. A fused, all-source intelligence product that assesses how capable a particular military force is. The six major components assessed in a military capabilities assessment include: Leadership and C2, Order of Battle (OB), Force Readiness and Mission, Force Sustainability, Technical Intelligence, and Cyberspace Threat Assessments.

2. Military-Related Subjects Assessment. This type of assessment is used to provide indicators of an adversary's capabilities and vulnerabilities, to include an evaluation of warfighting sustainability. The major components of a military-related subjects assessment include: Communications System and Cyberspace, Defense Industries, Energy, Military Geography, Demography, Transportation, Space Systems, Environmental Considerations, Medical, and Meteorological and Oceanographic (METOC).

3.8.3.4 Target Intelligence

Target intelligence portrays and locates the components of a target or target complex and indicates its vulnerability and relative importance. While the targeting process is most closely identified with lethal strike operations, it has wider applicability and can be applied to the development of targets for nonlethal fires and other types of operations. Targeting products developed by intelligence provide comprehensive vulnerability analysis of adversary fixed and mobile capabilities; play a key role in detecting, localizing, and tracking mobile targets; support effective weaponeering; and assess the effects of targeting operations. Consistency of targeting intelligence throughout the joint force and Service components ensures the efficient and effective use of offensive capabilities, force protection, and the minimization of collateral damage. Global Command and Control System (GCCS), MIDB, and standardized electronic target folder production processes are among the tools used by naval force targeteers to share information about targets and to ensure a common visualization of the operational environment within which targeting operations are conducted. Target intelligence efforts typically occur at the operational and tactical levels of war.

Target intelligence production requests may include the following types of products:

1. Target system analysis
2. Electronic target folders
3. Targeting materials
4. Target lists
5. Collateral damage estimates
6. Addition, modification, and validation of target data in MIDB
7. Combat assessments.

3.8.3.5 Scientific and Technical Intelligence

S&TI products focus on adversary weapons systems and other relevant military capabilities to determine the characteristics, vulnerabilities, limitations, effectiveness, research and development, and manufacturing efforts associated with those capabilities. S&TI products typically support the development and acquisition of friendly systems and countermeasures by providing information on the threats posed by adversary capabilities to friendly forces and by delineating the vulnerabilities of adversary systems. While an all-source product, much of the data for S&TI usually comes from foreign military exploitation, foreign material acquisition, and captured adversary equipment programs. S&TI is not generally done at the operational or tactical levels of war as it typically involves analysts with advanced engineering and technical degrees and specialized analysis equipment. For example, ONI has extensive capabilities to produce S&TI on foreign naval and maritime systems and is the primary producer of S&TI on these subjects within the intelligence community. Long-term S&TI focuses on developmental and future weapons, technologies, and threats.

3.8.3.6 Counterintelligence

CI products include multidiscipline CI threat analyses that provide intelligence on foreign intelligence and security services disciplines, terrorism, foreign-directed sabotage, and related security threats. For the Navy, NCIS produces standing and specialized products that are critical to naval force protection, deception, counter-ISR, and operational security efforts. Counterintelligence products support all levels of war.

3.8.3.7 Estimative Intelligence

While not limited to the IPOE process, estimative intelligence analysis and products are discussed at length throughout chapter 4. Estimative intelligence analysis and production occurs at all levels of war and focuses on predicting potential adversary courses of action or forecasting how a particular situation will unfold. Creating accurate, complete, and relevant estimative intelligence requires a firm grasp of adversary capabilities and understanding of the aspects of the maritime operational environment that are relevant to the specific mission. This in-depth understanding of the adversary and the operational environment is needed to visualize COAs from the adversary's perspective, assess their relative advantages and disadvantages to the adversary, and predict how the adversary will act. In the case of COA analysis, naval intelligence analysts should develop a range of potential ACOAs and, at a minimum, fully develop an assessment on the most likely ACOA and that ACOA determined to be most dangerous to the naval force.

3.8.3.8 Identity Intelligence

Identity intelligence results from the fusion of identity attributes (biologic, biographic, behavioral, and reputational information related to individuals) and other information and intelligence associated with those attributes collected across all intelligence disciplines. Identity intelligence utilizes enabling intelligence activities, like biometrics-enabled intelligence, forensics-enabled intelligence, and document and media exploitation, to discover the existence of unknown potential threat actors by connecting individuals to other persons, places, events, or materials, analyzing patterns of life, and characterizing their level of potential threats to U.S. interests.

3.8.4 Analysis and Production Considerations

All analysis and production is driven by intelligence requirements, which should be, but are not always, defined during planning and direction. Ad hoc requests for completed or tailored versions of intelligence products to meet emergent needs are common. Those involved with analysis and production must be mindful of the time requirements under which their efforts must be completed, which should include a consideration of the time it takes to package, transmit, and deliver the intelligence product to the intelligence user. Another key consideration, particularly when operating with coalition or multinational forces, is the time it takes for foreign disclosure officer review and approval to release products outside of U.S. channels.

Clearly understanding and, in some cases, anticipating the intended audience, the transmission path, and the desired format of the intelligence product from the beginning of the analysis and production effort is one of the keys to efficient intelligence operations. Without this understanding, naval intelligence analysts may waste considerable time developing a graphically rich intelligence product only to find that the intended audience does not have the bandwidth, system capabilities, or security clearance needed to receive such a product. The needs of coalition partners, multinational partners, and forces with limited communication capabilities should be considered in developing intelligence products. To the greatest extent possible, a "write for release" policy should be adopted to ensure naval intelligence analysts produce intelligence at the lowest level possible and incorporate tear line reporting, or similar measures, to ensure a wider sharing of intelligence to naval forces or multinational partners who do not have the clearance for more sensitive intelligence products.

While the format of the intelligence product is driven by requirements, there are general guidelines for intelligence production that should be followed, particularly in creating intelligence briefs and written reports. Intelligence products should have a beginning, middle, and an end. The beginning should provide the bottom line or relevance of the assessment up front. It is a summary of the entire product and may consist of a single analytic judgment or several. Even though the summary is provided at the front of the product, it is always written last and it should stand alone as often the intelligence user will read no further than the summary of the product. The middle section of the report recounts the sources of data, information, and intelligence that were used in the analysis and detail the analytic methodology that was used to create the assessment. Use of annexes and appendixes may also be necessary for an in-depth analytic product on a complex subject. To the greatest extent possible, charts, tables, graphs, geospatial overlays, and other tools should be used to convey detailed data sets and analyses. The end of the intelligence product should contain an overview of the analysis and then provide the predictive estimate based on the analysis.[7]

---

[7] Major, pp. 8–11.

Effective intelligence writing is an acquired skill that is obtained through reading good intelligence reporting, and good writing in general, and through practice in developing a wide range of intelligence products. The ability to accept honest and constructive feedback on the quality of the intelligence product is also a key to improving one's skills in this area. Intelligence products, like all good written works, should be clear, concise, and correct. Clarity is achieved through simplicity in writing, making careful word choices, avoiding jargon, and making clear transitions from one thought to another. Conciseness comes from using as few words as possible to convey the intended meaning. Correctness has two main elements. An intelligence product should be factually correct and the language in it should be as precise as possible. Use of vague language must be avoided. Known facts, gaps in knowledge, and assessments should always be identified and distinguished from one another. In addition to factual correctness, an intelligence product should be mechanically correct. There should be no misspellings, usage of words and phrases should conform to generally accepted definitions for those words and phrases, and correct punctuation must be used.[8]

Intelligence products should also be appropriate, complete, and coherent. Appropriateness concerns tailoring the intelligence product to the specific needs of the intelligence user. A complete intelligence product is one where everything that the analyst needs to say is in the product. Finally, a coherent product is one that has one voice—it provides a unified whole and is constructed around a central idea or theme. An intelligence product is often produced through collaborative effort but ultimately each product should have one main analyst or editor to ensure that the product reads as one consistent whole and not the work of a committee.[9]

Two techniques for ensuring effective intelligence production are outlining and proofreading. Outlining should be done prior to developing any new intelligence brief or report. Whether created collaboratively or by an individual analyst, the outline enables naval intelligence professionals to consider the user of their intelligence and the organization and flow of their analysis before writing. The outline thus provides a "roadmap" for the intelligence product and aids in identifying major themes as well as gaps and can be used for tasking specific components of a larger analytic product. While outlining should be done at the beginning of the production process, proofreading comes at the end. All intelligence products should be proofread by the analyst who created them and checked for the factors, such as clarity and completeness, discussed above. Additionally, the proofreader should find and correct any grammatical or spelling errors in the product. Whenever possible, the product should be reviewed by at least one other individual who can provide an objective and frank assessment of whether the intelligence product meets the standards of intelligence excellence before being passed to the intelligence user. If no other reviewer is available, reading of the product aloud by the analyst who produced it is a best practice for detecting grammatical issues with a product.

3.8.5 Evaluation and Feedback for Analysis and Production

Analysis and production operations should be evaluated against the standards of intelligence excellence discussed in chapters 1 and 2. Figure 3-30 provides an example of questions that may be asked about the analytic effort, and the products resulting from it, that would assist in evaluating analysis and production operations.

Feedback on intelligence analysis and production operations should be solicited from within the intelligence team and from the intelligence user. Naval intelligence professionals should consistently look at analytic and production processes and their outputs to look for quality defects that need to be addressed and to determine if more efficient and effective means of analysis and production can be implemented. Feedback must always be honest, clear, constructive, and delivered swiftly when standards are not met to ensure corrective action can be taken quickly. Changes to processes resulting from evaluation and feedback efforts should be based on frank discussions about the relative pros and cons on adopting new techniques, which may cause a short-term disruption or training period in exchange for more efficient and effective intelligence analysis and production operations over the long term.

Commanders, staffs, and supported organizations should be queried concerning the value of intelligence products and should comment on whether there were any quality or timeliness deficiencies. Positive feedback should be

---
[8] Major, pp. 29–51.
[9] Major, p. 29.

communicated to the naval intelligence team and used to reinforce sound analytic practices and production methods. Negative feedback must be closely assessed to determine whether the issue or issues highlighted involve analytic or production operations or a combination of the two. Naval intelligence leadership should immediately resolve identified analysis and production deficiencies by addressing the root causes of the issues involved, which may include the need to request and advocate for additional resources or training to meet intelligence analysis and production requirements.

3.9 DISSEMINATION AND INTEGRATION

Success in developing information superiority depends upon integrating information from a range of sensors, platforms, commands, and centers to produce all-source intelligence. This intelligence must be part of a portrayal of the operational environment characterized by accurate assessments and visual depiction of friendly and enemy operations which makes the operational environment considerably more transparent for a United States commander than for the adversary and forms the basis for superior decisionmaking. In short, intelligence must be displayable, digestible, and manageable. Interoperable will not be good enough. Integration into the common operational picture is required.

Rear Admiral L. E. Jacoby, U.S. Navy
Joint Staff J-2, 2001

dissemination and integration. In intelligence usage, the delivery of intelligence to users in a suitable form and the application of the intelligence to appropriate missions, tasks, and functions. (JP 1-02. Source: JP 2-01)

ATTRIBUTES OF GOOD INTELLIGENCE

- **Objectivity:** Does it avoid mirror imaging, cultural bias, and prejudicial judgments?
- **Relevance:** Does it relate directly to the end user's area of responsibility and mission?
- **Accuracy:** Did the producer clearly articulate the level of confidence in its accuracy?
- **Precision:** Does it have the required level of detail to satisfy the needs of the end user at his or her operational level?
- **Completeness:** Were all of the user's requirements addressed in the level of detail necessary to satisfy his or her needs?
- **Usability:** Did it arrive in a format that the end user can easily understand and assimilate into his or her decision-making process?
- **Availability:** Was it readily accessible to the commander at the appropriate security classification?
- **Anticipation:** Does it anticipate the intelligence needs of the commander and staff?

Figure 3-30. Attributes of Good Intelligence

Per JP 2-01, intelligence must be disseminated in such a manner that it is readily accessible by the user. Planning for dissemination and integration must be conducted as early as possible as the format and transmission path required by the intelligence user drives decisions made in processing, exploiting, analyzing, and producing intelligence. Without an executable plan for disseminating intelligence to the users who need it, when they need it, and in the form they need it, the entire intelligence effort is for naught. The senior naval intelligence professional at each command manages the dissemination effort although this responsibility may be delegated in larger intelligence organizations. In addition to ensuring timely delivery of intelligence, leaders must ensure that intelligence receives the widest possible release, meeting the needs of tactical forces and coalition partners, while still ensuring the security of sources and methods used to collect, process, exploit, and analyze intelligence.

3.9.1 Form

Determining the form used to disseminate intelligence is a function of several factors. It is driven by the purpose for which the intelligence was produced; the urgency with which the intelligence is needed to support operations; the type and volume of the intelligence; the user's capabilities and capacity to receive and integrate the intelligence; and the dissemination methods available. Briefings, verbal reports, video-teleconferences, telephone or Voice over Internet Protocol reports, facsimiles, textual reports, graphical reports, formatted man-to-machine or machine-to-machine messages, and e-mails are just some of the forms that can be used to disseminate intelligence. Graphical reports typically can convey a wealth of intelligence information in an easily digestible form but these reports are often large files that may be difficult to disseminate to intelligence users, such as a patrol craft or a deployed construction battalion unit, with limited communications capabilities. In those cases, a voice or text report with essential intelligence may suffice, especially if the intelligence need is time sensitive.

Regardless of the form, intelligence should be disseminated in standard formats whenever possible. This practice has three main benefits. First, it enables better integration of intelligence by the intelligence user. By employing a standard form, the intelligence user knows what to expect. They do not have to spend time trying to "figure out" how to interpret the form and they are able to focus on the intelligence that is most relevant to meeting their needs. Second, standard forms promote efficiency and effectiveness in intelligence production as intelligence analysts are able to quickly update intelligence products rather than developing new products for every situation. Finally, standard forms are sometimes required to facilitate man-to-machine intelligence integration. For example, over the horizon-Gold message formats are used to integrate intelligence information into the GCCS-M COP and intelligence information reports follow a fixed format to ensure that the intelligence they provide is readily searchable and accessible by the entire intelligence community.

3.9.2 Dissemination Methods

While most intelligence is transmitted via electronic means, the dissemination method must meet the needs of the intelligence user and should minimize, to the greatest extent possible, the burden on communications systems. Requirements should be the driving force in determining what intelligence to provide and how to provide it. For example, if a riverine unit with a limited communications capability requests to know if the upper reaches of a river are passable, a textual imagery interpretation report describing the conditions in the area of concern meets the requirement. Forwarding such a unit a large, annotated image of the area of concern would not only provide them more intelligence than requested but it may result in the user never receiving the required intelligence as bandwidth restrictions may preclude them from receiving such a large file. The discussion below addresses the concepts of "push" and "pull" as they relate to intelligence dissemination and further elaborate on electronic and hard copy dissemination methods.

3.9.2.1 Push and Pull of Intelligence

Per JP 2-01, dissemination consists of "push" and "pull" control principles. The push concept pertains to the dissemination of intelligence from one intelligence organization to another organization either in response to an RFI, to provide threat warning, or to provide other intelligence that the "pushing" organization deems relevant to the receiving organization. In many cases, the push of intelligence comes through automated means or in response to standing requirements. For example, NSA's TRIBUTARY network enables a push of threat warning information via a voice communications circuit to deployed units while the integrated broadcast service (IBS) provides a wealth of intelligence reporting that is fed directly into the GCCS-M COP and other systems. Other examples of push intelligence may include e-mail dissemination of a daily intelligence brief to a defined set of users; the use of chat capabilities to disseminate textual accounts of active ISR missions; or the use of the Global Broadcast System, which can disseminate a wide variety of intelligence, to include bandwidth-intensive feeds from FMV and imagery sensors, that users can access at all levels of war.

Many push arrangements are established as part of the planning and direction of intelligence operations. These standing arrangements are monitored to ensure that the intelligence received is timely, relevant, accurate, and usable. With the exception of extremely perishable intelligence, such as threat warning reporting, there is typically some level of coordination that is done prior to forwarding intelligence to another organization. Prior coordination ensures that the organization receiving the intelligence has an opportunity to judge its relevance to their operations and can resolve any issues, such as bandwidth or classification concerns, that may cause difficulties for the receiving organization.

The pull concept involves direct electronic access to databases, intelligence files, or other repositories by intelligence organizations at all levels. Almost all intelligence organizations have some presence on the SIPRNET and JWICS versions of Intelink, which provides naval intelligence professionals the capability to search for and retrieve a vast amount of raw, exploited, and finished intelligence products. Additionally, special enclaves, like STONEGHOST, which provides the ability to pull intelligence from coalition partner networks, may be established. The advantage of intelligence pull is it allows the intelligence user to expeditiously find the intelligence they require and retrieve it for their use without having to engage in a time-intensive RFI/production request submission process. Additionally, pull promotes efficiency in the intelligence process. For example, rather than spending time packaging an e-mail and managing a distribution list to disseminate the amphibious ready group's (ARG's) daily briefing product, a naval intelligence analyst can post the product to a shared site, thus ensuring dissemination to all cleared individuals who have access to that site. Password protections or public key infrastructure certificates can also be used on these sites to restrict access, if required, due to classification or sensitivity concerns.

While both the push and pull methods of intelligence dissemination have many advantages, naval intelligence professionals should be mindful of three considerations when using either method.

1. Products stored on file servers or posted to Web servers should always be at or below the classification level of the network on which that server operates to ensure that higher classification intelligence is not compromised.

2. Alternative dissemination methods should be planned for and practiced. In many cases there are alternate paths to receiving needed intelligence. Considering how to access those alternate paths should be completed well before the primary path fails to ensure an uninterrupted flow of intelligence, particularly in a crisis situation. Even if those alternative paths are less optimal than the primary path, they may still provide intelligence of value. For example, an equipment malfunction may cause a disruption that prevents a DCGS-N-equipped ship from accessing the Joint Concentrator Architecture (JCA) to receive requested imagery. To respond to that event, the ship's imagery analysts could search the NGA Web-based access and retrieval portal (WARP) or National Reconnaissance Office's OMAR site to "pull" the imagery they need or to obtain older imagery that, while less optimal, still meets their requirement. Alternatively, arrangements could be coordinated with ONI FIST to "push" the specific imagery files to the ship via a file transfer or e-mail.

3. Naval intelligence professionals must account for the limitations of supported units. Given smaller units may have bandwidth or personnel constraints, large intelligence centers must still accommodate those limitations and ensure that "challenged" intelligence users have access to the intelligence needed to support their operations. For example, it is insufficient for an analyst aboard a carrier or large deck amphibious ship to tell a combatant they should just download a product they need from their servers without considering the limitations that unit may have. What may take a matter of minutes to download on the carrier, especially if that product is held on a local server, may take hours to download for the combatant.

Naval intelligence professionals must first confirm the requirement and then determine how best to address that requirement given communications and other limitations.

3.9.2.2 Electronic Dissemination

Electronic dissemination is by far the most predominant method for disseminating intelligence. The advantages of electronic dissemination are many. It enables naval intelligence professionals and those they support to rapidly access needed intelligence from a central location and provides them access to current intelligence products and also archived intelligence that may still be of value for mission planning and execution. Electronic dissemination also alleviates the administrative burden and storage concerns associated with managing a large library of classified hard copy intelligence products. Finally, electronic dissemination enables swift and seamless integration of intelligence into systems such as GCCS-M, electronic target folders, and weapons systems to support a wide variety of command requirements such as the need for situational awareness and efficient targeting processes.

3.9.2.3 Hard Copy Dissemination

The dissemination method used is based on the needs of the user and, in some cases, a hard copy product may still be required to ensure the commander, staff, and supported organizations receive the intelligence they need. For example, operations involving multinational forces or nongovernmental organizations (NGO) may require naval intelligence professionals to produce a sanitized, hard copy version of an intelligence report since no viable electronic means exists to get the intelligence to the multinational or NGO partners. Another example is planning for amphibious operations, where a large format, hard copy, annotated map or chart may be the best means of conveying needed intelligence to planners and to aid in wargaming activities. While electronic dissemination is preferred, as it offers many advantages, naval intelligence professionals should plan for hard copy dissemination requirements and ensure they have adequate supplies and equipment available to support the hard copy dissemination requirements of those they support.

3.9.3 Intelligence Integration

The intelligence user defines how and in what form they will receive disseminated intelligence with a goal of easily integrating that intelligence to support their mission. Naval intelligence professionals must work closely with those they support to truly understand how they intend to integrate intelligence into their operations. In some cases, the naval intelligence professional, who has the advantage of serving a wide variety of intelligence users, can help shape and define how best to disseminate and integrate intelligence based on experience gained from previous intelligence support opportunities. Integration may take many forms. In some cases, integration may be a purely mental activity. For example, a TAO who receives a verbal threat warning report will mentally integrate that intelligence with their understanding of the maritime operational environment and then take appropriate actions to defend their ship. In other cases, intelligence may be integrated into the planning process and associated products, such as when the textual and graphical intelligence estimate product is used to enable wargaming. In any case, by understanding the user's goal for integration, the naval intelligence professional can determine the dissemination form and method that best meets the user's integration needs.

JP 2-01 emphasizes that the COP is the primary vehicle for integrating intelligence and operations as it enables a broad merging of inputs from a wide variety of tactical, operational, and national sources into a single picture that serves a broad set of users for multiple purposes. It is important to note here that the COP is a concept and is not dependent on a particular system. An annotated map or chart that is consistently updated may serve as a COP if so designated by the commander. That said, the GCCS family of systems has been specifically designed to enable the automated integration of operations and intelligence data and provides a common set of protocols. GCCS enables rapid sharing of information and enables an integrated view of the operational environment that is accessible to a wide variety of decision makers and users. This view can be easily shared with other Services and joint command and control elements and is tailorable by the individual user to support planning and operations. Additionally, through applications like Command and Control Personal Computer and Google Earth KML feeds, users can receive GCCS COP track data, which provides current information on friendly, threat, and neutral forces, using a standard workstation.

Intelligence is integrated into the GCCS COP through a variety of means. National systems, such as IBS provide data from a number of sensors that is automatically ingested and plotted in the GCCS COP via automated, machine-to-machine messaging. Additionally, GCCS-M is fully integrated with MIDB, which enables any GCCS user to access a vast amount of current intelligence and OB information contained in the MIDB data stores. Naval intelligence personnel at the MOCs/MIOCs and on force-level ships (carriers, large deck amphibious ships, and command ships) have extensive capabilities to integrate intelligence into the COP using the DCGS-N system. This system, which operates on both JWICS and SIPRNET, is interoperable with GCCS-M and enables intelligence watch standers, operating in the JWICS environment, to develop all-source intelligence on adversary forces (red tracks) and push sanitized intelligence to the SIPRNET through automated guard solutions like RADIANT MERCURY. Additionally, naval intelligence professionals equipped with DCGS-N can associated a rich amount of intelligence data—images, FMV, briefings, textual reports—to GCCS-M tracks that operations personnel can access through simple point and click interfaces. While the GCCS-M COP is just one example of how intelligence is integrated with operations, GCCS-M provides a powerful tool for ensuring that operations and naval intelligence professionals share a common picture of the maritime operational environment and enables a rapid means of updating and disseminating current intelligence to a wide variety of intelligence users.

3.9.4 Evaluation and Feedback for Dissemination and Integration

At the point in the intelligence process where the value of the intelligence produced is typically realized, naval intelligence professionals should establish processes for determining how well the intelligence produced satisfies the requirements given. The evaluation of dissemination and integration operations should consider every aspect—requirements, format, dissemination method, etc.—to determine if the intelligence was received in a timely manner and met the needs of the requester. Evaluation should also consider ways to make dissemination and integration operations more efficient and effective. Naval intelligence professionals should also seek open and honest feedback from those they support on the value of the intelligence provided and should, whenever possible; examine how intelligence was integrated into operations to better understand the needs of current and future intelligence users.

3.10 VALUATION AND FEEDBACK

Evaluation and feedback is considered an intelligence operation in its own right. Each section of this chapter included specific considerations for conducting evaluation and feedback as a continuous process. For all intelligence operations, the evaluation standards used should be derived from the qualities of naval intelligence excellence discussed in chapters 1 and 2. Naval intelligence professionals should carefully evaluate the full range of intelligence operations and the outputs of those operations to ensure they meet applicable standards for anticipatory action, timeliness, objectivity, relevance, completeness, accuracy, availability, and usability.

Feedback from the intelligence team and from intelligence users should be solicited at many points in the intelligence process. Intelligence team members, particularly junior members or those new to the team, may have fresh insights and perspectives that can be used to make intelligence operations more efficient and effective or they may provide innovative ways to use available intelligence resources. Intelligence users are the ultimate judge of intelligence excellence so every effort should be made to determine their level of satisfaction with received intelligence. In cases where deficiencies are noted, naval intelligence professionals must swiftly analyze the issue to determine the root cause of the problem, which could occur at any part of the intelligence process, and develop solutions to resolve or mitigate the impact of the issue. For example:

1. Were requirements unclearly defined or understood in planning and direction?

2. Was the collection asset used unable to collect required data due to sensor limitations or was it a maintenance issue?

3. Was there insufficient exploitation capacity at the exploitation node to exploit collected data in a timely fashion?

4. Did the analytic product contain too much irrelevant detail for the intelligence user?

5. Did intelligence dissemination fail because of receiving unit bandwidth limitations?

Evaluation and feedback processes may also uncover best practices that may be shared throughout the fleet and the intelligence community to improve the overall efficiency and effectiveness of a wide range of intelligence operations. For example, a naval intelligence analyst may discover a way to efficiently combine MASINT, GEOINT, and SIGINT data into a product that reveals significant intelligence on an adversary capability and which can be used to effectively monitor the development and employment of that capability over a long period. Such a best practice should be disseminated as other analysts studying the same intelligence analysts could use this technique or it may have broader applicability and can be applied to other intelligence problems.

Evaluation and feedback is also important as it enables naval intelligence professionals to develop lessons learned that can be shared with the fleet and throughout the defense community through the Navy's Lessons Learned Information System. Navy Warfare Development Command (NWDC) oversees the Navy Lessons Learned System (NLLS) in accordance with Chief of Naval Operations instruction (OPNAVINST) 3500.37C. Naval intelligence professionals should be familiar with the NLLS, which is located at NWDC's SIPRNET Web site (http://www.nwdc.navy.smil.mil). Researching the NLLS database enables naval intelligence professionals to benefit from the experience of other intelligence teams in the planning and conduct of intelligence operations. NWDC also provides instructions on how fleet users can submit their own lessons learned to the database so that others can benefit from best practices they have developed.

INTENTIONALLY BLANK

CHAPTER 4

Intelligence Preparation of the Maritime Operational Environment

joint intelligence preparation of the operational environment (JIPOE). The analytical process used by joint intelligence organizations to produce intelligence estimates and other intelligence products in support of the joint force commander's decision-making process. It is a continuous process that includes defining the operational environment; describing the impact of the operational environment; evaluating the adversary; and determining adversary courses of action. (JP 1-02. Source: JP 2-01.3)

4.1 INTELLIGENCE PREPARATION OF THE MARITIME OPERATIONAL ENVIRONMENT

Intelligence preparation of the operational environment is a continuous process that is used to produce intelligence estimates, assessments, and other products to support decisionmaking. A separate discussion of how the IPOE process is employed to develop understanding of the adversary and the maritime operational environment is needed as the products of the IPOE process are foundational to many intelligence support activities at all levels of war. IPOE is integral to Navy planning and execution efforts across the range of military operations.

The IPOE process generates understanding of adversary capability and intent by taking a holistic, systematic view of the adversary, the operational environment, and how that operational environment may favor or constrain the activities of adversary and friendly forces. By creating a baseline set of intelligence products, such as an order of battle analysis or a beach study, IPOE provides a foundation that other specialized analytic processes, such as targeting, can use as a starting point. These products also help to identify intelligence gaps that can be addressed by the application of additional collection or analytic resources.

The IPOE process can be tailored to meet numerous situational or level of war requirements. While the process has many components, each unique intelligence problem requires application of the steps to a greater or lesser degree. For example, a Navy component commander conducting IPOE to support a major operation plan (OPLAN) may spend many months conducting the IPOE process given the scope and scale of the OPLAN effort. On the tactical level, many of the substeps in the IPOE process may be completed rapidly given the tightly bound nature of tactical level problems and the short timelines typically experienced in planning at that level of war.

This chapter is not intended to be a full guide to the IPOE process. Readers looking for more detail on IPOE and the outputs of each IPOE step should refer to JP 2-01.3, Joint Intelligence Preparation of the Operational Environment. As discussed in that publication and as shown in figure 4-1, the four steps of the IPOE process as applied to naval warfare, are:

1. Define the maritime operational environment

2. Describe the impact of the maritime operational environment

3. Evaluate the adversary

4. Determine adversary courses of action.

Intelligence preparation of the maritime environment is derived from the IPOE principles and processes described by JP 2-01.3. Refer to the JP for more detail on IPOE and the outputs of each IPOE step.

4.2 GENERAL INTELLIGENCE PREPARATION OF THE OPERATIONAL ENVIRONMENT CONSIDERATIONS

This section provides an overview of IPOE illustrated by figure 4-1 and describes its relationship to the intelligence process, the operational and tactical levels or war, and its application to the range of naval operations.

4.2.1 Intelligence Preparation of the Operational Environment and the Intelligence Process

Like the intelligence process that was described in chapter 1, IPOE is a continuous and dynamic process. There are many intersections between IPOE and the intelligence process as each mutually supports the other. These intersections occur within each phase of the intelligence process.

4.2.1.1 Planning and Direction

IPOE provides basic data and assumptions regarding the adversary and other relevant aspects of the intelligence environment that help the naval commander and staff identify intelligence information and collection requirements. An analysis of the mission defines the information and intelligence requirements vital to mission accomplishment that drive IPOE efforts. In some cases, a review of existing IPOE products may provide very little relevant intelligence while in other cases there may be so many potentially relevant IPOE products available that organizing and evaluating this data in a timely manner may be an issue. Regardless of the maturity of the IPOE products available to the naval intelligence professional, the review and analysis of these products within the context of mission planning serves to identify gaps in understanding. This analysis provides the basis needed to assist the commander in developing prioritized, synchronized, specific, and detailed PIRs and information needs that is used to drive collection and analysis efforts. IPOE products are also used in wargaming and other planning activities to determine potential adversary COAs and COGs which aids in the discovery of additional gaps and intelligence requirements.

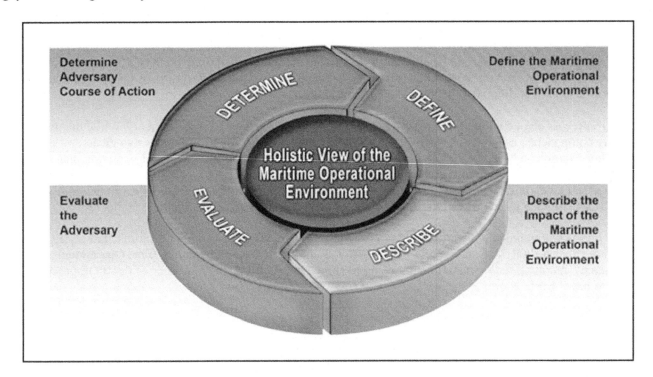

Figure 4-1. Intelligence Preparation of the Maritime Operational Environment—The Process

4.2.1.2 Intelligence Collection

IPOE provides the foundation for the development of an optimal intelligence collection strategy by enabling analysts to identify the time, location, and type of anticipated adversary activity corresponding to each potential adversary COA. It also employs analytic tools such as geospatial overlays, templates, and matrices that are used to analyze and identify ACOAs. From these ACOAs analysts derive NAIs that indicate the time and location when adversary activity may occur. By determining the indicators that are associated with these adversary activities, naval intelligence professionals can build collection strategies and plans that are used to confirm adversary actions and from this information provide a more refined understanding of adversary capabilities and intent. The IPOE analysis, by ordering potential adversary activity in time, space, and force dimensions, enables efficient collection planning by supporting sequencing and synchronization of the most effective collection capabilities available to address gaps and confirm intelligence estimates.

4.2.1.3 Processing and Exploitation

IPOE provides a disciplined yet dynamic time-phased methodology for optimizing the processing and exploiting of large amounts of data. By developing standardized templates, overlays, matrices, and graphics, IPOE enables analysts to remain focused on the critical aspects of the adversary and the maritime operational environment that require attention. Use of these standardized intelligence products promotes efficient processing of information as newly exploited and analyzed intelligence can be used to rapidly update these products and provide the commander, staff[1], supported organizations, and the intelligence team with a convenient medium for displaying the most up-to-date information, identifying critical information gaps, and supporting operational and campaign assessments.

4.2.1.4 Analysis and Production

IPOE products provide the foundation for the intelligence estimate. Requirements established in the planning and direction phase of the intelligence process drive the analytic effort and the production of basic, current, and estimative intelligence assessments in formats that meet the needs of the commander, staff, and supported organizations. The IPOE process is closely linked to the development of standardized intelligence estimates, which typically address mission, adversary situation, adversary capabilities, analysis of adversary of capabilities, and conclusions. Figure 4-2 provides an overview of the relationship between IPOE and the intelligence estimate.

4.2.1.5 Dissemination and Integration

The intelligence estimate, supported by the IPOE process, provides vital information used by the staff and subordinate commanders to complete their own estimates and to conduct concurrent planning activities. In cases where time does not permit the development of a formal intelligence estimate, verbal reports and the direct dissemination of IPOE templates, matrices, and graphics should be used to meet the planning and execution timeliness requirements.

4.2.1.6 Evaluation and Feedback

It is the responsibility of naval intelligence professionals to continuously evaluate IPOE products and solicit feedback from the commander, staff, and supported organizations to determine if these products are meeting the criteria for intelligence excellence described in chapters 1 and 2. IPOE products must anticipate the needs of the commander and must meet the standards for relevance, timeliness, accuracy, usability, completeness, and objectivity required to support the planning, execution, and assessment of operations. Shortfalls in meeting these standards must be addressed immediately as failure to do so could directly result in mission failure.

---

[1] The term "staff" is used in this publication to generically describe the warfighting or functional leads that are directly assigned to and support the commander in the planning and execution of operations at any level of war.

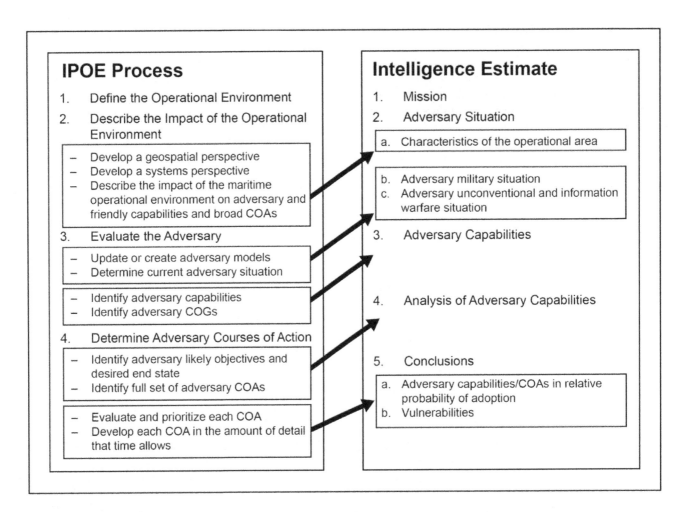

Figure 4-2. Intelligence Preparation of the Operational Environment and the Intelligence Estimate

4.2.2 Intelligence Preparation of the Operational Environment Relationship to the Levels of War

The IPOE process is fundamentally the same regardless of the level of war or type of operation. Planning considerations may vary depending on the level of war due to obvious differences in mission, available resources, and size of the operational areas and AOIs. Specific IPOE considerations for each level of war are as follows:

1. Strategic Level. Joint doctrine states that IPOE supports activities at the strategic level such as "establish national and multinational military objectives; develop global plans or theater war plans to achieve these objectives; sequence operations; define limits and assess risks for the use of military and other instruments of national security policy; and provide military forces and other capabilities in accordance with strategic plans." Strategic concerns may be global in nature given factors such as international law and the pervasiveness of information technologies and space-based capabilities that are not constrained by national boundaries. At the strategic level, the operational environment concerns geographic regions, nations, and climate vice the issues of oceanography and meteorology that are primary concerns at the operational and tactical levels. Nonmilitary factors of the operational environment may be of more concern to the strategic planner who looks at current and long-term trends in areas like technological development, economics, politics, sociology, and civil-military relations to determine the types of forces a nation may develop and field. The strategic planner also attempts to understand the adversary's will to employ its instruments of national power (e.g., DIME) for offensive or defensive purposes.

2. Operational Level. At the operational level, the analysis of the operational environment depends on such varied factors as the location of adversary political and economic support structures, military support units, force generation capabilities, potential third-nation or third-party involvement, logistic and economic infrastructure, political treaties, press coverage, adversary propaganda, and the potential for information warfare. Much of the IPOE analysis done at the operational level assesses PMESII systems and their interrelationships. The emphasis placed on examining individual PMESII components may vary depending on the nature of the operations being planned and supported. For example, IPOE efforts for major combat operations in the maritime operational environment will likely focus on developing an understanding of how the adversary uses sea, land, and air lines of communications (LOCs) to support operations in the littorals; how best to interdict and disrupt LOCs; determining the best approaches and standoff distances for the naval force in the operational area and the AOI; developing an understanding of adversary doctrine; and describing operational level ACOAs involving the large scale movement of forces to counter friendly naval forces; and the sequencing of adversary movements in space and time. For stability, FHA, or defense support of civil authorities (DSCA) operations, the focus of the IPOE effort may be substantially different and requires less emphasis on military threats and more emphasis on understanding cultural, civil, political, and informational factors and networks in the effected region.

3. Tactical Level. Analysis of the maritime operational environment at the tactical level is typically focused on adversary land, air, maritime, space, and other forces as well as other relevant aspects of the operational environment that could pose a direct threat to the security of the friendly force or the success of its mission. Given the tighter time and resource constraints inherent at the tactical level, IPOE efforts should, at a minimum, analyze the maritime operational environment in terms of: military objectives; air, land, and maritime avenues of approach; and the impact of METOC and geographic conditions on personnel, military operations, weapons systems, and force mobility. Analysis of military forces should focus on developing an understanding of the composition, disposition, tactics, and readiness status of units that could pose a threat to operations. Nonmilitary factors must also be considered as they impact how an adversary employs military forces. Just as at the operational level, nonmilitary factors also have increased importance in irregular warfare, FHA, DSCA, and other naval missions that do not involve traditional, force-on-force combat operations. COA analysis at the tactical level requires a higher level of detail than that done at the strategic and operational levels.

4.2.3 Intelligence Preparation of the Operational Environment Considerations across the Range of Naval Operations

Intelligence preparation of the maritime operational environment applies to the entire range of naval operations. While the focus of the IPOE effort is on developing an understanding of adversary capabilities and intentions and the impacts of the maritime operational environment on friendly and adversary forces, the term "adversary" is used in this publication to broadly refer to those organizations, groups, decision makers, or even physical factors that can delay, degrade, or prevent the . . . [naval] force from accomplishing its mission. In cases of FHA or DSCA missions, the adversary (or focus of the IPOE effort) may be a situation or condition such as a man-made or natural disaster, a pandemic, or the secondary effects of a disaster such as starvation conditions for a local populace. In the case of maritime security operations, the adversary could be smugglers, drug traffickers, pirates or other groups that can blend in with a civilian population. These criminal or irregular forces typically pose a far greater collection and analytic challenge than what is generally encountered for conventional military forces. IPOE, while a powerful tool, does not solve deficiencies in strategic guidance or planning. IPOE is useful for determining potential ACOAs and addresses the question of what the adversary may do or how an event in the operational environment may unfold. It provides planners a starting point for collection planning and wargaming. IPOE cannot predict the future and define what the adversary will do or how a situation in the maritime operational environment will develop. Estimates of ACOAs are based on assumptions, which may later prove invalid, and on "factors and conditions in the operational environment that may change," at times with great rapidity and in entirely unexpected directions. The main focus of the analytic effort, the adversary, possesses a thinking mind and a human will that is determined to gain, whenever and wherever possible, an operational advantage over the naval force. Naval intelligence professionals must work with their operations counterparts to explore the implications of and responses to a variety of probable and possible ACOAs. Focusing only on the most likely ACOA could result in surprise for friendly forces given the adversary is only constrained by what they consider possible vice what the analyst determines the adversary can do.

4.3 DEFINE THE MARITIME OPERATIONAL ENVIRONMENT

The first step of the IPOE process is to define the maritime operational environment. Given the complex, multidimensional nature of the maritime operational environment, the needs of the mission must serve to bound the IPOE process and will define the aspects and characteristics of the operational environment that are the focus of the IPOE effort. Successfully defining the relevant aspects of the maritime operational environment is critical to the rest of the IPOE process as it enables the efficient and effective use of intelligence resources to support the commander. Figure 4-3 provides an overview of the specific activities involved in this IPOE step.

Defining the relevant aspects of the maritime operational environment includes identification of physical and nonphysical aspects that are relevant to the commander's decision-making process and the development of COAs for naval forces. Failure to focus on only the relevant aspects of the maritime operational environment leads to inefficient use of intelligence resources while failure to focus on all relevant aspects may lead to a naval force that is unprepared to meet the challenges of combat or other operations. Determining the relevant aspects of the operational environment requires experienced naval intelligence professionals who are engaged in consistent and open dialog with their commander, other staff elements, and supported organizations concerning friendly force objectives, requirements, and perspectives.

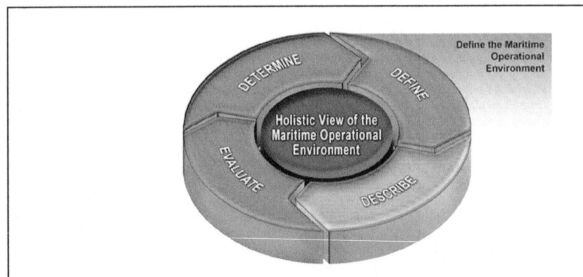

1. Identify the naval force's area of operations (AO)
2. Analyze the mission and joint force commander's intent
3. Determine the significant characteristics of the maritime operational environment
4. Establish the limits of the naval force's area of interest
5. Determine the level of detail required and feasible within the time available
6. Determine intelligence and information gaps, shortfalls, and priorities
7. Collect material and submit requests for information to support further analysis

Figure 4-3. Intelligence Preparation of the Maritime Operational Environment—Step One

4.3.1 Identify the Naval Force's Area of Operations

area of operations. An operational area defined by the joint force commander for land and maritime forces that should be large enough to accomplish their missions and protect their forces. (JP 1-02. Source: JP 3-0)

The geographic boundaries of these areas may be derived from operation orders or OPLAN from the higher headquarters (CCDR, joint force commander (JFC), fleet commander, etc.) that assigned the mission to the naval force. Areas of operations do not typically encompass the entire operational area of the joint force commander but are a subset of that area. The size of the AO depends on the scope and nature of the crisis and the projected duration of operations. These same factors may also influence the term used to designate the AO and can include specific types of AOs such as joint operations areas, joint security areas, or amphibious objective areas (AOAs). AOs may also be contiguous, which refers to AOs separated by boundaries, or noncontiguous, which are AOs that do not share common boundaries.

4.3.2 Analyze the Mission and Joint Force Commander's Intent

Once the mission is received from higher headquarters, the commander and staff must conduct a mission analysis. The commander's intent and characteristics of the mission that could affect COA selection are critical inputs to the IPOE process. While an AO is defined by higher headquarters, the characteristics of concern to the commander typically extend well beyond the designated AO. In addition to conceptualizing the AO and other areas of concern, there are other characteristics of the mission that inform the IPOE process. These other mission characteristics include:

1. The type of military operation being considered or planned; the purpose of the operation; the amount of time available for planning and execution

2. The expected duration of the operation

3. The risks to be managed

4. Whether allied or coalition forces will be involved.

Intelligence planners must also be cognizant of operational limitations identified in the mission order. These include constraints and restraints. Constraints are actions directed by higher headquarters (things the force must do) while restraints are prohibitions against certain types of action (things the force cannot do). For example, higher headquarters direction to use a specific tactic or capability or the imposition of restrictions, such as no-strike targets and zones, also have an impact on the IPOE effort.

4.3.3 Determine the Significant Characteristics of the Maritime Operational Environment

Naval doctrine states that this substep "is an initial review of the factors of space, time, and forces and their interaction with one another." Given it is an initial review, joint doctrine amplifies on this guidance stating this is "a cursory examination of each aspect of the operational environment in order to identify those characteristics of possible significance or relevance to the . . . force and its mission." This step aims to develop an overarching list of aspects of the maritime operational environment that may be relevant. In steps two and three of the IPOE process, these aspects are analyzed in-depth to determine their significance in the planning effort.

Since the maritime operational environment is complex and multidimensional, a more detailed discussion of the characteristics of that operational environment is warranted here so that the naval intelligence professional can fully appreciate the challenge involved with determining the significant characteristics of the maritime operational environment. Obtaining an understanding of the physical (geospatial and environmental), system (PMESII), and informational aspects of the maritime operational environment provides the foundation needed to determine the impact of the operational environment. During most operations commanders will require not only military intelligence, but also intelligence on nonmilitary aspects of the operational environment such as economic, informational, social, political, diplomatic, biographic, human factors, and other types of intelligence. Equally important is knowledge of how all these aspects interrelate to form a systems perspective of the adversary and other relevant aspects of the operational environment. The need to identify these interdependencies is especially relevant to naval operations as the maritime domain is recognized as one of the "global commons"[2]—further reinforcing the notion that maritime matters are inherently linked to many nonmilitary aspects of the operational environment.

JP 2-01.3 defines the operational environment "... the composite of the conditions, circumstances, and influences that affect the employment of capabilities and bear on the decisions of the commander." The environment is described as a framework of three interactive aspects—physical areas and factors (or geospatial perspective), a systems perspective, and the information environment. Some of the characteristics of the maritime operational environment to consider may include:

1. Geographic and hydrographic features as well as the impact of climate, weather, and other environmental factors such as pollution, disease vectors, and potential for earthquake or volcanic activity

2. Population characteristics such as ethnicity, religious and ideological groups, demographics, income distribution, cultural issues, and attitudes

3. Political landscape and socioeconomic factors

4. Infrastructure such as power distribution and transportation

5. Operational limitation imposed by international laws, treaties, or agreements

6. Friendly, neutral, and adversary military forces (conventional, unconventional, paramilitary), their capabilities, disposition, readiness, tactics, and objectives

7. Psychological factors that impact adversary decisionmaking

8. Locations of foreign embassies and international or nongovernmental groups.

Each operation is unique, and entails different combinations of information and intelligence requirements to address the CCIRs on the maritime operational environment. However, identifying those requirements through a deliberate intelligence planning process is the first, and arguably the most critical, stage of the IPOE process.

4.3.3.1 Geospatial Environment (Physical Areas and Factors)

The most basic question about any operational environment is "What is it like there?" As illustrated by figure 4-4, this information requirement encompasses geography, hydrography, spatial features, climate, and weather—and is especially relevant considering the environmentally complex nature of operations in the maritime environment. Defining the geospatial aspects of the maritime operational environment is an especially broad effort—since the nature of naval operations includes the maritime domain and may well include many, if not all, of the other domains. Figure 4-5 depicts the relationship between these domains in a fleet-specific context.

---

[2] The term "global commons" is further described in The National Military Strategy of the United States, Joint Chiefs of Staff, 08 Feb 2010, p. 3.

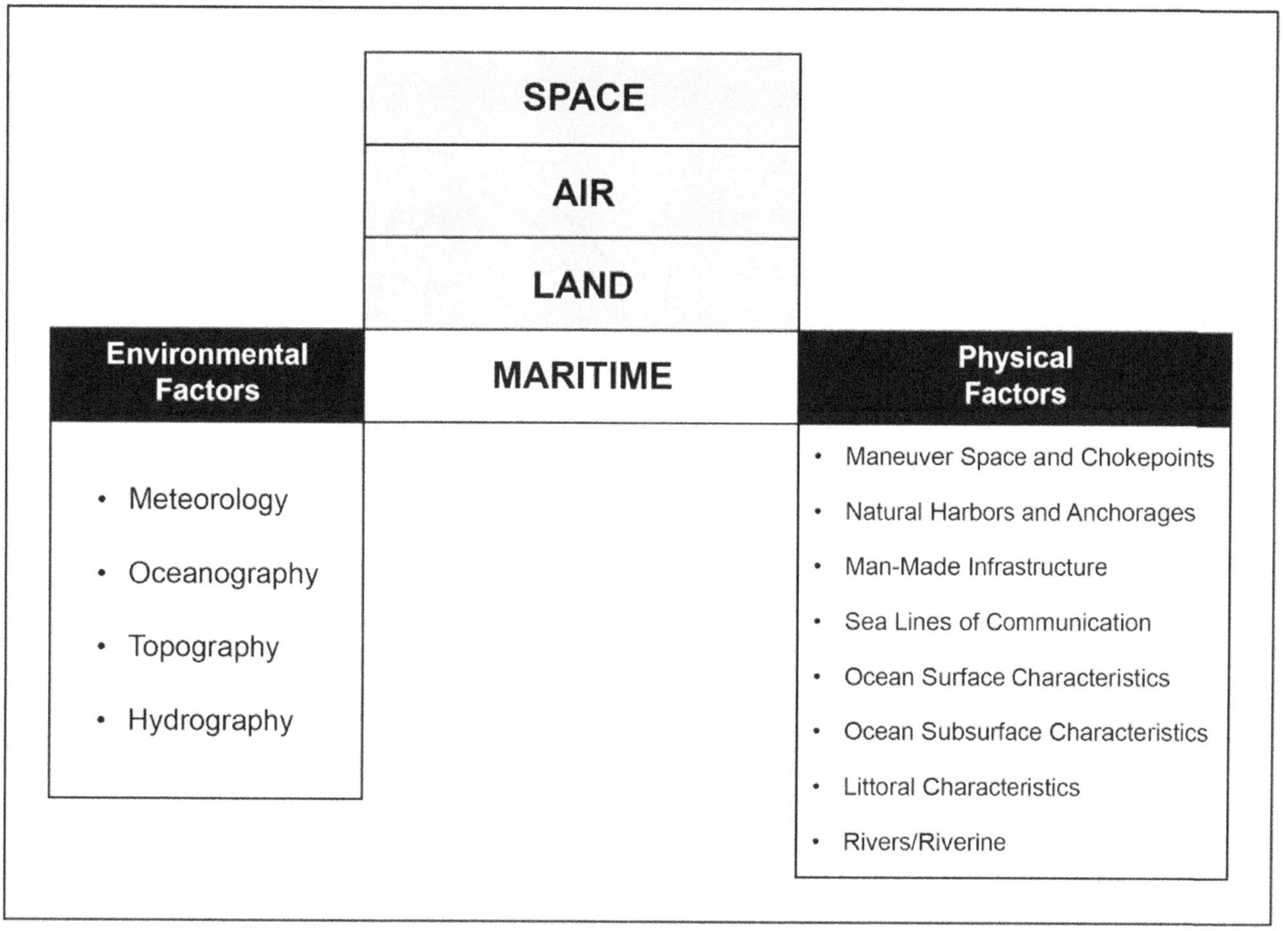

Environmental Factors		Physical Factors
	SPACE	
	AIR	
	LAND	
	MARITIME	
• Meteorology • Oceanography • Topography • Hydrography		• Maneuver Space and Chokepoints • Natural Harbors and Anchorages • Man-Made Infrastructure • Sea Lines of Communication • Ocean Surface Characteristics • Ocean Subsurface Characteristics • Littoral Characteristics • Rivers/Riverine

Figure 4-4. Geospatial Areas and Maritime Domain Factors

Notes

- Graphics in this chapter depicting aspects of the maritime operating environment (figures 4-5, 4-6, 4-7, and 4-10) are portrayed in a "top and side view" to show the progressive integration of domains, systems, and dimensions comprising the final three-dimensional depiction (figure 4-11).

- While cyberspace is not included in current joint doctrine as being coequal with the geospatial domains, JP 2-01.3 mentions it as being " . . . pervasive to all activities worldwide, and is a common backdrop for the air, land, maritime, and space physical domains of the . . . operational environment." Additionally, planners should look beyond the obvious geospatial limits of the physical areas by identifying any nonphysical factors that may impact the mission. Many of these factors transcend the traditional concept of physical boundaries and have worldwide implications and relevance. The Internet, for example, is a nonphysical aspect that should be considered. Therefore, while cyberspace is not explicitly included as a physical area in joint doctrine it is highlighted as a necessary consideration for nonphysical influences. That linkage is addressed in a following section detailing the information environment.

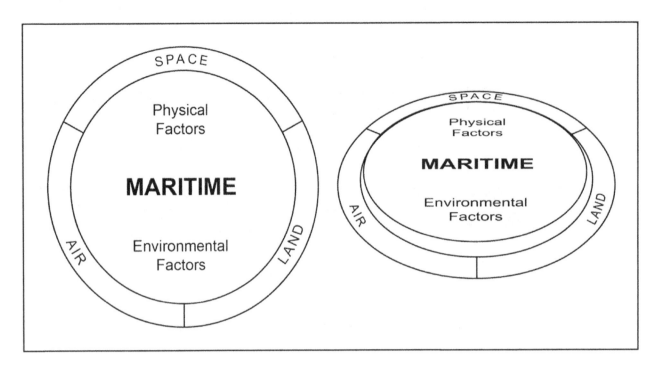

Figure 4-5. The Geospatial Domains in Fleet Intelligence Preparation of the Battlespace

4.3.3.2 The Systems Perspective

This aspect of the maritime operational environment strives to provide an understanding of significant relationships within interrelated friendly, adversary, or neutral PMESII systems that are potentially relevant to the success of a naval operation. This assessment seeks to identify strengths, weaknesses, and interdependencies amongst those systems so that COGs and critical vulnerabilities may be determined.

PMESII analysis helps intelligence analysts identify potential sources from which to gain indications and warning, and understand the continuous and complex interaction of friendly, adversary, and neutral systems, while also helping operational planners identify nodes in each system, the links (relationships) between the nodes, critical factors, and potential decisive points so that commanders may consider a broader set of options to focus limited resources, create desired effects, and achieve objectives.

This viewpoint of interacting systems is critical for fleet intelligence personnel attempting to develop information and intelligence requirements that may be unique to naval missions and the maritime operational environment. Figure 4-6 depicts how these six systems could be arranged to show their interdependencies.

Further details on the PMESII methodology is provided in JP 2-01.3, Joint Intelligence Preparation of the Environment.

4.3.3.3 A Composite of Domains and Systems

Joint doctrine (JP 2-01.3) states that "[a] holistic view of the operational environment encompasses physical areas and factors (of the air, land, maritime, and space domains) . . . , [and included] within these are the adversary, friendly, and neutral PMESII systems and subsystems that are relevant to a specific . . . operation." Therefore, a composite of these two aspects of IPB—the geospatial domains and the PMESII subsystems within the maritime domain—is depicted in figure 4-7. This figure underscores that the focus of PMESII analysis by naval intelligence professionals should be on the maritime domain (with any influence by other domains that pertain to naval operations). This construct is also relevant to assessing friendly, adversary, or neutral force influences—and for determining the information and intelligence requirements needed to inform commanders and staff planners of such influences.

NWP 2-0

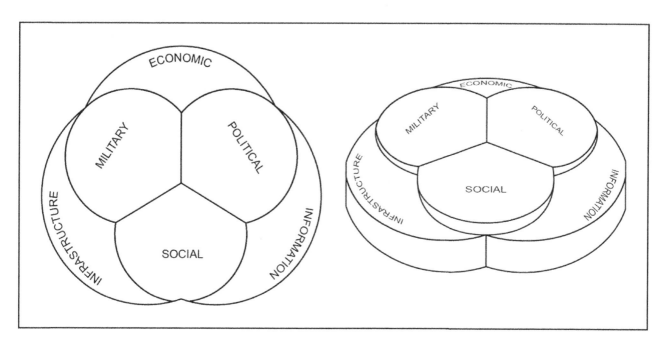

Figure 4-6. The Political, Military, Economic, Social, Information, and Infrastructure "Systems Perspective" with Interacting Subsystems

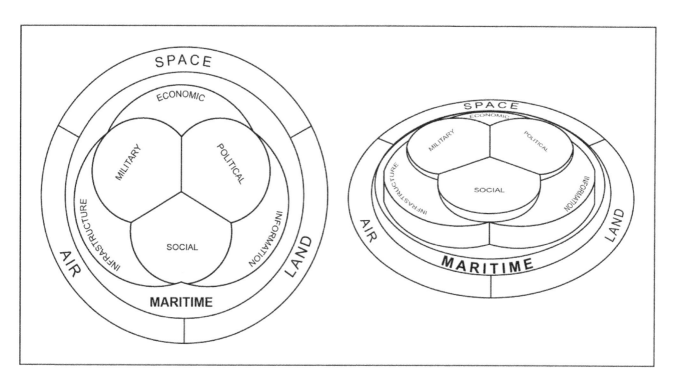

Figure 4-7. Relationship between Geospatial Domains and Political, Military, Economic, Social, Information, and Infrastructure Subsystems

4.3.3.4 Information Environment

Cyberspace is often referred to as a key consideration in operational environment. That factor has more recently been identified as an element of the "information environment." JP 2-01.3 describes the information environment as "... the aggregate of individuals, organizations, and systems that collect, process, disseminate, or act on information. It is made up of three interrelated dimensions: physical, informational, and cognitive." This definition clearly highlights that the environment includes more than cyberspace operations.

4.3.3.4.1 Dimensions of the Information Environment

Figure 4-8 provides a graphic depiction of the dimensions of the information environment.

4.3.3.4.2 Physical Dimension

The physical dimension is composed of C2 systems, key decision makers, and supporting infrastructure that enable individuals and organizations to create effects. It is the dimension where physical platforms and the communications networks that connect them reside. The physical dimension includes, but is not limited to, human beings, C2 facilities, newspapers, books, microwave towers, computer processing units, laptops, smart phones, tablet computers, or any other objects that are subject to empirical measurement. The physical dimension is not confined solely to military or even nation-based systems and processes; it is a defused network connected across national, economic, and geographical boundaries.

4.3.3.4.3 Informational Dimension

The informational dimension encompasses where and how information is collected, processed, stored, disseminated, and protected. It is the dimension where the C2 of military forces is exercised and where the commander's intent is conveyed. Actions in this dimension affect the content and flow of information.

4.3.3.4.4 Cognitive Dimension

The cognitive dimension encompasses the minds of those who transmit, receive, and respond to or act on information. It refers to individuals' or groups' information processing, perception, judgment, and decisionmaking. These elements are influenced by many factors, to include individual and cultural beliefs, norms, vulnerabilities, motivations, emotions, experiences, morals, education, mental health, identities, and ideologies. Defining these influencing factors in a given environment is critical for understanding how to best influence the mind of the decision maker and create the desired effects. As such, this dimension constitutes the most important component of the information environment.

4.3.3.4.5 Interactions across the Information Environment

These three dimensions can be arranged to depict how a friendly force would interact with an adversary across the information environment. This arrangement is depicted in figure 4-9. The figure shows the human and computer network activities across the various dimensions.

Figure 4-8. Dimensions of the Information Environment

NWP 2-0

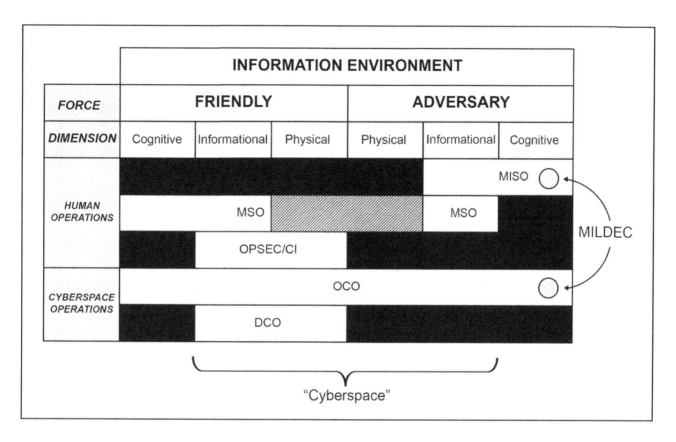

Figure 4-9. Activities Across the Information Environment

This figure shows how a friendly force's dimensions would be leveraged through human and computer networks to:

1. Influence the adversary's informational and cognitive dimensions through military information support operations (MISO)—"planned operations to convey selected information and indicators to foreign audiences to influence their emotions, motives, objective reasoning, and ultimately the behavior of foreign governments, organizations, groups, and individuals in a manner favorable to the originator's objectives." (JP 1-02)

2. Assist friendly force decisionmaking by military source operations—"the collection, from, by and/or via humans, of foreign and military and military-related intelligence." (JP 1-02)

3. Protect the friendly force's information dimension through operations security (OPSEC) and counterintelligence (CI). JP 1-02 defines OPSEC as "a process of identifying critical information and subsequently analyzing friendly actions attendant to military operations and other activities" and CI as "information gathered and activities conducted to identify, deceive, exploit, disrupt, or protect against espionage, other intelligence activities, sabotage, or assassinations conducted for or on behalf of foreign powers, organizations or persons or their agents, or international terrorist organizations or activities."

4. Infiltrate the adversary's physical dimension and influence the decisions of an adversary through offensive cyberspace operations (OCO)—"cyberspace operations intended to project power by the application of force in or through cyberspace." (JP 3-12)

5. Protect the friendly force's information and physical dimensions through defensive cyberspace operations (DCO)—"passive and active cyberspace operations intended to preserve the ability to utilize friendly

cyberspace capabilities and protect data, networks, net-centric capabilities, and other designated systems." (JP 3-12)

6. Assist friendly force decisionmaking by infiltrate the adversary's physical dimension and extracting information through cyberspace ISR—"an intelligence action conducted by the JFC authorized by an execution order or conducted by attached SIGINT units under temporary delegated signals intelligence operational tasking authority. Cyberspace ISR includes activities in cyberspace conducted to gather intelligence from target and adversary systems that may be required to support future operations, including OCO or DCO." (JP 3-12)

In figure 4-9, military deception (MILDEC) operations by friendly forces against an adversary could be accomplished through either MISO or OCO activities.

Notes

- JP 1-02 defines cyberspace as a "global domain within the information environment consisting of the interdependent network of information technology infrastructures and resident data, including the Internet, telecommunications networks, computer systems, and embedded processors and controllers." Within the information environment construct, cyberspace overlaps the physical and informational dimensions of the information environment. Figure 4-9 depicts that overlap.

- Intelligence support to cyberspace operations is discussed in JP 3-12, Cyberspace Operations. Like any other operation, intelligence support to the planning, execution, and assessment of cyberspace operations is a multidiscipline intelligence problem that depends heavily on SIGINT, GEOINT, HUMINT, OSINT, and MASINT capabilities. SIGINT units attached to JFCs may be authorized to conduct cyberspace ISR. The focus of cyberspace ISR is on tactical and operational intelligence and on mapping adversary cyberspace to support military planning, execution, and assessment of operations, to include OCO and DCO. Cyberspace ISR requires appropriate deconfliction, is conducted by personnel trained to IC standards, is conducted pursuant to military authorities, and must be coordinated and deconflicted with other United States Government departments and agencies in accordance with the Trilateral Memorandum of Agreement Among the Department of Defense, the Department of Justice, and the Intelligence Community Regarding Computer Network Attack and Computer Network Exploitation Activities, 9 May 2007, and Executive Order 12333, United States Intelligence Activities.

4.3.3.4.6 A Composite of Domains, Systems, and Dimensions

All three aspects of the maritime operational environment—the geospatial environment, PMESII systems, and the information environment—can be visualized as one integrated construct by the following linkages:

1. One factor of the maritime domain is the man-made infrastructure which can connect to the infrastructure system of PMESII.

2. One dimension of the information environment is the physical domain can also connect to the infrastructure system of PMESII.

3. Cyberspace is the overlap of the informational and physical domains of the information environment.

4. An implied element of the maritime domain, and all physical domains, is cyberspace which connects the maritime domain and information environment.

The linkages between the three operational environment aspects are depicted in figure 4-10 which ties together the connecting elements of the definition of cyberspace and figures 4-4, 4-6, and 4-8.

Note

> Figure 4-10 suggests that that the information environment is already embedded within the information system of PMESII, with a direct linkage to the infrastructure system of PMESII and with implied linkages to the maritime domain through cyberspace. This figure does not portray the information environment as an equally prominent part of the overall operational environment. This portrayal is not intended to minimize the role of the information environment (and potential information operations planning) in IPOE efforts. Rather, this figure demonstrates that these relationships culminate in the cognitive dimension of the information environment—which directly impacts the political, military and social systems of PMESII.

The merger of the three aspects of the operational environment is depicted in figure 4-11. The purpose of this figure is to visually represent the various domains, systems, and dimensions with significance to naval operations. This model highlights considerations in joint doctrine that may assist in developing the naval commander's information and intelligence requirements for the range of military operations.

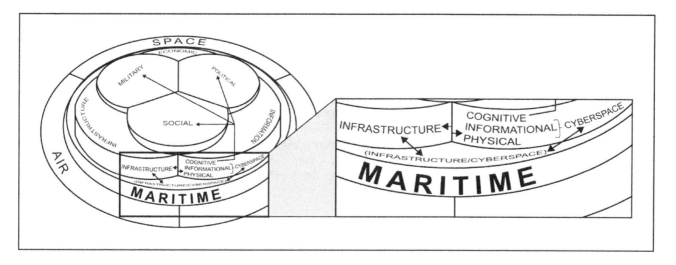

Figure 4-10. Intersection of Operational Environment Domains, Systems, and Dimensions

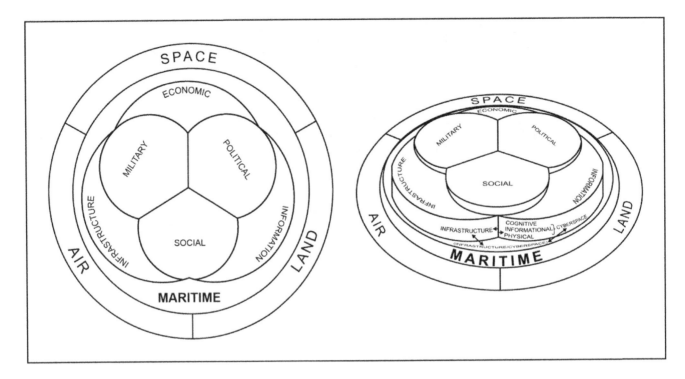

Figure 4-11. The Domains, Systems, and Dimensions of the Operational Environment

4.3.4 Establish the Limits of the Naval Force's Area of Interest

Once the potentially relevant characteristics of the maritime operational environment are understood, the naval intelligence professional must work with their commander, other staff elements, and supported organizations to define the physical and nonphysical aspects pertinent to the IPOE effort and mission accomplishment.

4.3.4.1 Physical Aspects

As previously discussed, planning directives typically assign an AO that defines geospatial boundaries which should be large enough for the commander to accomplish the mission and protect the naval force. While significant, because it is the focus of operational activity, the AO is just one area of concern for the IPOE process. In addition to the AO, the commander must also designate an associated area of influence (AI) and as many AOIs as are needed to ensure mission success. AI and AOI are defined as follows and illustrated by figure 4-12:

4.3.4.1.1 Area of Influence

area of influence. A geographical area wherein a commander is directly capable of influencing operations by maneuver or fire support systems normally under the commander's command or control. (JP 1-02. Source: JP 3-0)

The AI should encompass the AO and defines the area where the commander can use the naval force to influence the adversary or local populace. Defining this area is relevant to the IPOE process in two ways. First, IPOE efforts should focus on understanding aspects of the AI that may threaten operations in the AO. For example, adversary airfields outside of the AO could be used against naval forces in the AO and will need to be closely monitored so that the threat they pose can be deterred or neutralized if required. Second, since the AI is adjacent to the AO, there is the potential that follow on combat or other operations may occur in the AI once objectives in the AO are accomplished. Obtaining an understanding of relevant activities in the AI significantly benefits future planning steps by providing an established set of IPOE products to support those efforts.

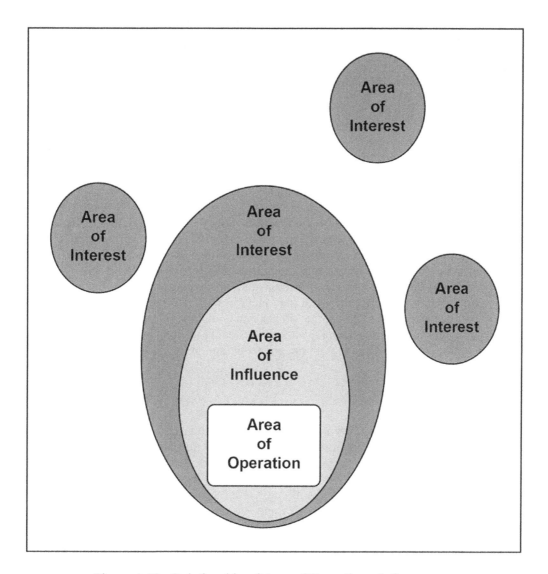

Figure 4-12. Relationship of Area of Operations, Influence, and Interest

4.3.4.1.2 Area of Interest

area of interest. That area of concern to the commander, including the area of influence, areas adjacent thereto, and extending into enemy territory. This area also includes areas occupied by enemy forces who could jeopardize the accomplishment of the mission. (JP 1-02. Source: JP 3-0)

The AOI determines the maximum scope of intelligence-gathering activities and also encompasses the pertinent parts of cyberspace. The AOI focuses intelligence activities on friendly, neutral, adversary, or other concerns that could potentially impact current or future operations.

A commander may designate multiple AOIs. In some cases these AOIs may be well outside the AO and AI. AOIs are not restricted by political boundaries. While typically represented graphically, as a geospatial overlay, the AOI does not represent a boundary or other control measure. The geospatial depiction is meant to show the physical region where IPOE efforts are focused, even if that focus is on a nonphysical factor such as how a populace in another country perceives U.S. operations.

In defining an AOI, the power projection capabilities of the adversary and potential allies, paramilitary, insurgent, or terrorist forces sympathetic to the adversary must be considered. In addition to examining ports, airfields, garrisons, and missile bases from which adversary forces may attack, IPOE efforts should also focus on logistics and support networks by which these forces could move capabilities into targeting range of the naval force or by which they could reinforce or resupply the adversary.

The element of time is another important factor to consider in defining AOIs. Considering the mobility of the adversary and sympathetic forces in relation to the projections for how long the mission is anticipated to take helps narrow the focus of the IPOE effort. For example, if an amphibious raid designed to destroy a pirate base is expected to take 24 hours, then adversary forces who are unable to mobilize and arrive within the AO during that period should be outside the AOI and not analyzed further within the IPOE process.

AOIs are always mission-dependent. A short term tactical operation may have very few AOIs focused on very small areas while a long-term counterdrug operation may have many AOIs focused on a number of physical and nonphysical factors. Figure 4-12 shows an example of the relationship between the AO, the AI, and AOIs. In this depiction a naval force could be given a mission to seize a specific objective within the AO. The AI would include those areas outside the AO where the naval force would be capable of interdicting and deterring adversary attempts to reinforce or oppose the seizure. The AOI may include areas deep within adversary territory from which the adversary may generate additional combat power, but could also include the border countries that might provide logistical bases for friendly forces, take advantage of friendly military action to conduct cross-border raids or seize territory, or who may be aligned with the adversary and generate threats to the naval force in the AO.

4.3.4.2 Nonphysical Aspects

Naval intelligence professionals must look beyond the geospatial boundaries of the areas of concern to identify any nonphysical aspects of the maritime operational environment that may impact the accomplishment of the naval force's mission. In some cases, these nonphysical aspects may be global in nature. For example, the Internet not only provides adversaries with advanced communications capabilities but also provides access to advanced satellite imagery and other information sources. Additionally the Internet can be used to rapidly disseminate propaganda, challenging U.S. control of "the narrative," and is often a vector for cyber attacks such as denial of service attacks, data theft, and data corruption. Due to the nature of Internet protocol-based communications, these propaganda efforts of cyber attacks can occur from anywhere in the world and countering them may require coordination with forces or agencies in other theaters of operations. Other examples of nonphysical factors include friendly and adversary employment of the electromagnetic spectrum; the attitudes, perceptions, and morale of leadership, military forces, and populations; and the links within and among adversary PMESII system nodes.

4.3.5 Determine the Level of Detail Required and Feasible within the Time Available

The time available for developing IPOE products is a significant factor in determining how the IPOE process is applied to naval force planning and execution efforts. To be relevant, IPOE must apply to the needs of the commander and supported activities. Naval intelligence professionals must determine what elements of the process are the highest priorities and, within each element, definitively state the expected output and desired level of detail. These priorities and level of detail requirements, along with clearly stated deadlines, must be communicated within the intelligence team and understood and accepted by the commander, staff, and supported organizations.

Scope of the effort has a significant impact on the IPOE process and its application. Long-term strategic and operational planning efforts may use every step in the IPOE process while crisis operations and tactical-level activities may require applying the steps to a greater or lesser degree based on the circumstance. Similarly, long-term planning efforts may require high-level assumptions that prevent IPOE products from being as detailed as those required at the tactical level.

The commander's PIRs must always serve to focus the IPOE effort. For example, some adversary forces may have no capability to interfere with the naval force so the commander may see no need in expending intelligence resources to further analyze any potential threat they may pose. Discussions with the commander, staff elements, and others supported by the IPOE process should also serve to define how the IPOE process is applied to support planning and execution activities. Naval intelligence professionals must be forthright in articulating their capabilities and limitations, which includes identifying unrealistic expectations about what intelligence can provide in the timeframes given by the commander, planners, and operators.

4.3.6 Determine Intelligence and Information Gaps, Shortfalls, and Priorities

Naval intelligence professionals must evaluate intelligence holdings, databases, and raw data, information, and finished intelligence available from other units, theater elements, Services, agencies, coalition partners, or open sources to determine what information is available to continue the IPOE process.[3] Regardless of the situation, there are gaps identified in this step and often addressing these gaps requires the support of external collection or analytic capabilities. The naval intelligence professional must use the commander's intent, CCIRs, and PIRs to determine priorities for intelligence collection, processing, exploitation, analysis, and dissemination activities associated with the IPOE process. In many cases gaps may be addressed through additional research activities, RFIs to organizations having greater familiarity with a given intelligence problem, or by tasking organic or external collection and analytic resources to address intelligence shortfalls. Gaps that cannot be addressed within the timeframe given for the IPOE effort must be identified to the commander, staff, and supported organizations as early in the process as possible since these gaps have impacts on the planning effort and the commander's decisionmaking.

4.3.7 Collect Material and Submit Requests for Information to Support Further Analysis

Collecting data and incorporating it into the IPOE process is a continuous effort. The naval intelligence professional must gather or produce relevant IPOE products and disseminate these products within the timeframes and formats required by the commander and supported activities. Intelligence gaps can be addressed through a variety of means to include:

4.3.7.1 Requests for Information

request for information (RFI). 1. Any specific time-sensitive ad hoc requirement for intelligence information or products to support an ongoing crisis or operation not necessarily related to standing requirements or scheduled intelligence production. A request for information can be initiated to respond to operational requirements and will be validated in accordance with the combatant command's procedures. 2. The National Security Agency/Central Security Service uses this term to state ad hoc signals intelligence requirements. (JP 1-02. Source: JP 2-0)

In Navy usage, RFI is also a general term for an information request that can be used to meet an information need associated with an operation. While each unit, theater, Service, or agency's policy may differ, RFIs are typically generated using informal e-mail, chat, or verbal means with the expectation that the supporting organization will provide existing, off-the-shelf products or can address the requirement with a short, informal response from an experienced subject matter expert. RFIs that require employment of a supporting organization's collection, processing, exploitation, analysis, production, or dissemination resources typically require a formal request in accordance with policies articulated in documents such as the theater-specific and Navy-wide operational task (OPTASK) intelligence message or the annex B of an operation plan. For crisis operations or emergent situations, RFIs are typically initiated via informal means and a followed by a formal request as time permits.

---

[3] A best practice is to define areas or topics of concern to the commander prior to deployment and determine the Navy, theater, other Service, agency, coalition, and open source sites, feeds, and databases that can provide relevant intelligence on these issues. Regular review of these online holdings is encouraged as it significantly speeds the IPOE effort as the naval intelligence professional will already have a clear sense of the relevant intelligence available, its quality and age, and from this determine their gaps.

4.3.7.2 Production Requests

While many IPOE products can be developed using organic resources, some production requirements may require more time or analytic resources than are available to the naval intelligence professional and are best produced by an external organization with more capacity or expertise in a given topic. For example, a CSG team, having completed its review of potentially relevant IPOE materials, may determine that there is little available information on new UAS capabilities of a potential adversary; intelligence that is needed to conduct counter-ISR and deception planning efforts for the naval force. The CSG N-2 would make a formal request to the NCC for their theater of operations stating their intelligence production requirement and the timeframe, format, and dissemination path for the product. The NCC would either address the requirement, if capable, or forward the requirement to the theater, Service, national, or coalition organization best suited to address the requirement.

4.3.7.3 Collection Requirements

collection requirement. 1. An intelligence need considered in the allocation of intelligence resources. Within the Department of Defense, these collection requirements fulfill the essential elements of information and other intelligence needs of a commander, or an agency. 2. An established intelligence need, validated against the appropriate allocation of intelligence resources (as a requirement) to fulfill the essential elements of information and other intelligence needs of an intelligence consumer. (JP 1-02. Source: JP 2-01.2)

In cases where intelligence is lacking to address the IPOE production requirement, collection of information may be required to address intelligence shortfalls. As new data, information, and intelligence is received, the naval intelligence professional must integrate that intelligence into existing IPOE products and update them to ensure they are accurate, complete, and thorough. In some cases, new intelligence may confirm or repudiate assumptions made in the planning process and in these cases the commander and supported activities must be notified. In cases where the new intelligence repudiates assumptions, the commander and supported activities must review any assessments, evaluations, or decisions based on those assumptions and determine whether changes are needed to address the change in assumptions.

4.4 DESCRIBE THE IMPACT OF THE MARITIME OPERATIONAL ENVIRONMENT

The second step in the IPOE process assesses the impact of the maritime operational environment on adversary, friendly and neutral capabilities, and broad COAs (figure 4-13). Analysts examine all relevant physical and nonphysical aspects of the maritime operational environment to develop a geospatial perspective that is used to evaluate and disseminate assessments concerning the interplay of time, space, force, and environmental factors. A systems perspective is developed by analyzing relevant sociocultural aspects of the maritime operational environment as well as through assessment of relationships and dependencies within and among the nodes and links of relevant systems and subsystems. IPOE products created during this phase may include graphical overlays showing the disposition of forces and the impact of geography on maneuver; matrices; demographic overviews; and other products tailored to providing insights and understanding needed to inform the next step of the IPOE process, which is to evaluate the adversary. The discussion below provides additional detail on the substeps involved with step two of the IPOE process.

4.4.1 Develop a Geospatial Perspective of the Maritime Operational Environment

A geospatial perspective supports all views of the operational environment by helping to analyze relevant physical, nonphysical, and locational aspects of the operational environment. This analysis is a two-step process. First, each relevant characteristic of the maritime operational environment is analyzed and then evaluated for its potential impact on naval operations. Typically, the analysis of the AO requires the highest level of detail to support operational planning and execution but aspects of the AI and AOIs must also be addressed to support force protection, future operations, and future planning efforts. Since the physical aspects of the maritime operational environment vary widely, some maritime domain characteristics may require more or less analysis depending on the complexity of the geospatial and hydrological features of a region.

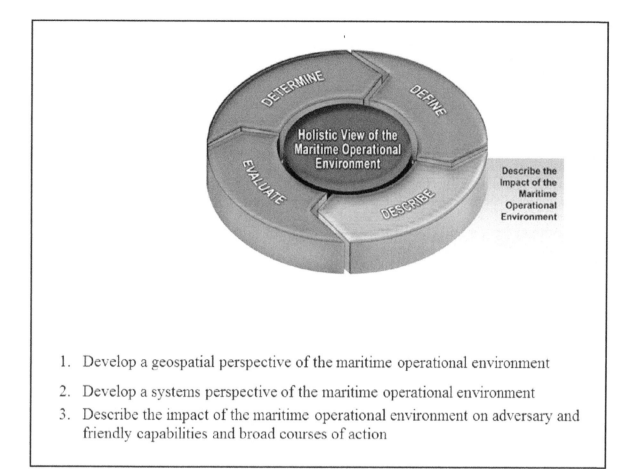

1. Develop a geospatial perspective of the maritime operational environment
2. Develop a systems perspective of the maritime operational environment
3. Describe the impact of the maritime operational environment on adversary and friendly capabilities and broad courses of action

Figure 4-13. Intelligence Preparation of the Maritime Operational Environment—Step Two

Age of information may also play a role in the time and resources needed to develop the geospatial perspective and assess its impacts. For example, an amphibious force may have a highly detailed, but dated, beach study to support an amphibious landing in a particular AOA. Examination of the study may show that it does not account for new roads, breakwaters, and other infrastructure improvements made in the years since the beach study was last updated and which may, upon further analysis, require the selection of a new landing area and the development of a new beach study.

Transient or cyclical events, such as meteorological or climatic conditions may also have impacts on the maritime operational environment that need to be assessed. For example, monsoon rains and winds in certain theaters may significantly degrade mobility, could severely impede the use of various ISR collection assets, and reduce the effectiveness of communication and weapons systems.

Joint doctrine also emphasizes the need to understand the effects of the maritime operational environment on chemical, biological, radiological, and nuclear (CBRN) weapons. Use of these weapons by adversaries or destruction of these capabilities by friendly forces presents special problems for analysis since assessing potential target damage and collateral effects, such as contamination, are highly dependent on terrain, environmental, and weather factors.

4.4.1.1 The Maritime Domain

maritime domain. The oceans, seas, bays, estuaries, islands, coastal areas, and the airspace above these, including the littorals. (JP 1-02. Source: JP 3-32)

littoral. In naval operations, that portion of the world's land masses adjacent to the oceans within direct control of and vulnerable to the striking power of sea-based forces. (JP 1-02)

The unique factors of the maritime domain require analysis as part of the IPOE process. The maritime domain is multidomain and multidimensional, so analysis of the maritime operational environment and development of the geospatial perspective often includes analysis of land, sea, air, and space domains. Given the capabilities of modern sensor and weapons systems, adversary land, air, surface, subsurface, and space capabilities can detect or engage naval forces that operate in open ocean areas which may necessitate development of IPOE products on distant landmasses to support naval force operations. Closer to shore, littoral areas, which may include straits or other chokepoints, offer significant challenges in terms of defining the relevant aspects of the maritime operational environment and in collecting and analyzing the data, information, and intelligence needed to understand the impact of the littorals on adversary and friendly forces.

4.4.1.1.1 Maneuver Space and Chokepoints

Straits and other chokepoints have a significant impact on naval operations as they significantly reduce the advantage that maneuver provides in frustrating adversary detection and targeting efforts. Close proximity to land not only increases the likelihood of detection and engagement but in the confined waters associated with chokepoints there may be substantial restrictions in the ability to tactically maneuver a unit or strike group to best counter the threat. Chokepoints are also typically high traffic areas. Analogous to the problems associated with operations in an urban environment, many of the world's straits experience high volumes of sea and airborne activity which provides adversaries with advantages for deception and concealment and significantly complicates friendly force detection and targeting considerations. Given the proximity to various land, air, sea, and subsurface forces in these confined areas, naval forces experience significantly reduced reaction times to adversary hostile actions which could include use of shore antisurface missile batteries, strike aircraft, small boat swarm attacks, or use of naval mines, the effectiveness of which are greatly enhanced in the shallower depths typically associated with operations in chokepoints and littoral areas. Narrowing down the relevant aspects of the maritime operational environment in these complex, restricted areas is critical to successful application of the IPOE process in meeting naval force mission requirements.

4.4.1.1.2 Natural Harbors and Anchorages

Naval intelligence professionals should examine natural harbors and anchorages in areas of concern to determine what operational advantages they can provide to the adversary and friendly forces. For example, certain types of natural harbors, such as fjords, may provide adversary or friendly forces with limited concealment opportunities that can be used to evade detection and launch attacks at the time most advantageous to the attacking force. IPOE products should clearly delineate the natural features in and surrounding these harbors and anchorages so that the commander and supported activities can visualize the operational advantages they may hold for adversary and friendly forces.

4.4.1.1.3 Man-Made Infrastructure

All relevant man-made infrastructure capable of influencing naval operations in the AOI should be identified and analyzed. This includes civilian port facilities, naval bases, airfields, and occupied and unoccupied antiship missile sites. IPOE products focused on civilian ports, for example, may detail how the logistic capabilities of these ports support adversary operations or how these same capabilities could be used to support a naval FHA mission focused on unloading and transshipping large amounts of relief supplies to affected regions of a country. Similarly, analysis of naval bases could detail how these facilities contribute to adversary force generation and power projection capabilities.

4.4.1.1.4 Sea Line of Communications

A sea line of communications (SLOC) refers to a water route that facilitates regional or international transit of maritime commercial and military forces. These routes should be analyzed to assess their relative importance to adversary, neutral, and friendly forces and IPOE products should highlight the impacts of disrupting or seizing control of these SLOCs on those forces. Potential interdiction areas, such as chokepoints, should be identified along the relevant areas of the SLOCs of concern as well as the naval and air bases and coastal defense facilities from which interdiction may occur. Additionally, characterizing the types and density of sea and airborne traffic along these SLOCs must be understood so that the commander and other supported activities have a clear understanding of the baseline, normal levels of activity that is needed to identify potentially anomalous activity and to develop offensive and defensive tactics required to identify, target, and engage adversary forces while minimizing collateral damage effects.

4.4.1.1.5 Ocean Surface Characteristics

Although the world's oceans and seas provide a vast maneuver space for naval operations, meteorological effects and temperature can have a significant impact on naval forces. Large storms may produce high waves and sea states that extend well beyond the storm center. These storms could impede or curtail naval operations and dramatically reduce the effectiveness of sensors and weapons systems. Some regions of the world are also prone to intense weather conditions that form with little warning and may have localized impacts on naval operations. Temperature may also be a factor, as it determines the extent of ice formation and the strength and direction of ocean currents. Seasonal variations in ocean ice may impact maneuverability and directly impacts navigation and port operations and availability. Similarly, excessive heat may have adverse impacts on equipment and may create challenging conditions for Sailors and Marines as hydration and heat stroke concerns limit activity. Given the impacts that weather and temperature can have on naval operations, naval intelligence professionals must work closely with their counterparts in naval oceanography to understand seasonal climatic conditions in potential AOs, AIs, and AOIs and factor these conditions into the development of IPOE products.

4.4.1.1.6 Ocean Subsurface Characteristics

There are many subsurface characteristics that affect the ability of the naval force to conduct submarine, antisubmarine, and mine warfare operations (collectively called undersea warfare). The sensitivity and range of sonar, one of the principle sensors used in the undersea environment, is impacted by various characteristics such as temperature and salinity (which changes with depth), currents and eddies, the composition of the sea floor, and ambient noise caused by natural and man-made activity. The sea floor and ocean bottom contours may provide submarines an opportunity to use the maritime equivalent of concealment or may constrain movement into narrow avenues of approach. Ocean depth has a significant impact on surface and subsurface operations. Shallow water may severely restrict the use of certain capabilities, such as submarines and large draft vessels (i.e., carriers, large deck amphibious ships, oilers, etc.), but may enable the use of other capabilities, such as moored or bottom mines. Since shallow waters are often associated with littoral regions, they provide an especially challenging environment for undersea warfare, especially in areas with significant maritime traffic, and may affect the use of sensors and weapons systems. Deep water provides submarines with greater three-dimensional maneuver space and has less of an impact on undersea sensors and weapons. Ocean ice is also a consideration for undersea forces as the depth and extent of surface ice constrains the horizontal and vertical maneuvering room of submarines. Understanding ocean depth and seasonal/current environmental conditions in the AO, AI, and AOIs is critical to creating IPOE products that detail how adversary and friendly forces may use this aspect of the maritime operational environment to their best advantage.

4.4.1.1.7 Littoral Characteristics

The littorals provide a particularly challenging and complex area of operations for naval forces. IPOE products should account for relevant aspects of the littorals such as the littoral gradient and its composition, coastal terrain features, and transportation infrastructure, tides, and currents as these factors are critical in the planning and execution of naval operations. Unlike the open ocean, adversary land-based forces can make use of hilly or mountainous terrain or forested areas to mask the composition, size, and disposition of their forces and may use this terrain to gain defensive or offensive advantages. Given the population density within the littorals, identification and engagement of adversary forces is a complex task and collateral damage considerations are significant. Littorals may also offer substantial operational advantages. For example, seizure of a modern port or airfield facility can provide seaports of debarkation (SPODs) and aerial ports of debarkation (APODs) that are typically linked to transportation networks that can facilitate rapid force generation and movement deep into adversary territory. In addition to understanding natural and man-made features of the littoral, naval intelligence professionals must also evaluate seasonal averages and climate conditions in these areas and their potential impact on operations as part of the IPOE process.

4.4.1.2 Evaluating Other Physical Domains and Maritime Operational Environment Characteristics

Some characteristics of the operational environment are an inherent part of all domains. Given the multidomain and multidimensional nature of the maritime operational environment, naval intelligence professionals must appreciate the complexities inherent in narrowing the scope of the IPOE effort to support the planning and execution of naval operations.

4.4.1.2.1 Land Domain

Analysis of the operational environment's land domain concentrates on terrain features such as transportation systems (road and bridge information), surface materials, ground water, natural obstacles such as large bodies of water and mountains, the types and distribution of vegetation, and the configuration of surface drainage. Terrain analysis must always consider the effects of weather as well as changes that may result from military action. The focus of the IPOE effort for the land domain is on how terrain, weather effects, and military action favor or impede the conduct of military operations. Characteristics considered in analyzing the land domain include: observation and fields of fire; concealment and cover; obstacles; key terrain; identification and categorization of mobility corridors; and identification, evaluation, and prioritization of avenues of approach.

4.4.1.2.2 Air Domain

IPOE characteristics that may be evaluated for the air domain include: target characteristics and configuration; airfields and support infrastructure; missile launch sites; potential carrier-based aviation and sea-launched cruise missile locations and operating areas; natural and man-made surface features and service ceilings; and air avenues of approach. The air domain is partially influenced by surface characteristics. For example, some military air operations may take advantage of terrain masking. Additionally, the effects of METOC conditions on the air domain are particularly crucial.

4.4.1.2.3 Space Domain

Forces that have access to the space domain are afforded a wide array of options that can be used to leverage and enhance military capabilities. Every country has access to either its own satellites or to those of another country or commercial entity through the purchase of services. Thus the monitoring and tracking of friendly, hostile, and even neutral space assets is necessary for a complete understanding of the operational environment. Characteristics of the space domain that may be relevant to the IPOE process include: orbital mechanics; propagation; orbit density and debris; and solar and geomagnetic activity.

4.4.1.2.4 Information Environment

Significant characteristics of the information environment that may be evaluated by the IPOE process include: physical dimensions, such as computer hardware and networking gear; informational dimensions, such as computer software, data, and procedures; cognitive dimensions, such as the minds of the populace and the leadership.

The information environment was discussed in section 4.3.3.4. Joint doctrine describes the information environment as an environment where humans observe, orient, decide, and act upon information, and is therefore the principal environment of decisionmaking. This environment is pervasive to all activities worldwide, and is a common backdrop for the air, land, maritime, and space physical domains of the JFC's operational environment. The actors in the information environment include military and civilian leaders, decision makers, individuals, and organizations. Resources include the information itself and the materials and systems employed to collect, analyze, apply, disseminate, and display information and produce information-related products such as reports, orders, and leaflets.

4.4.1.2.5 The Electromagnetic Spectrum

Characteristics of the electromagnetic spectrum that analysts should consider for the IPOE may include: military use of the infrared band; military use of multispectral or hyperspectral imagery; adversary passive and active electronic warfare (EW) capabilities; radio wave directionality; and radio wave attenuation (the effects of atmospheric conditions and natural and man-made features on radio wave propagation). The electromagnetic aspect of the operational environment includes all militarily significant portions of the electromagnetic spectrum, to include those frequencies associated with radio, radar, laser, electro-optic, and infrared equipment. It is a combination of the civil electromagnetic infrastructure; natural phenomena; and adversary, friendly, and neutral electromagnetic order of battle. The electromagnetic spectrum provides the operating medium for communications; electro-optic, radar, and infrared imagery; signals intelligence; measurement and signature intelligence; and EW operations. Use of the electromagnetic spectrum for military or civilian purposes is constrained by a variety of factors, ranging from international agreements on frequency usage to the physical characteristics of electromagnetic waves.

4.4.1.2.6 Weather and Climate

Weather and climate may have significant impacts on naval operations and must be accounted for in the IPOE effort. Characteristics of weather and climate impacts that the naval intelligence professional may evaluate as part of the IPOE process include: visibility; winds; precipitation; cloud cover; temperature and humidity; and sea state. Close coordination with oceanographic counterparts is essential to evaluating these factors.

4.4.1.2.7 Time

Understanding how the duration of operations may impede or advance friendly and adversary objectives should be accounted for in the IPOE effort. Adversary decision and reaction time is a major factor in the planning and conduct of operations. For example, a garrison force may not have time to mobilize and respond to a relatively quick strike or raid into adversary territory. For adversaries with limited capabilities, particularly forces engaged in irregular warfare, an extended conflict may be of benefit as the adversary may perceive that popular support for friendly military operations will erode over time and force the withdrawal of a the superior military force. The duration and timing of operations may also have a significant impact on how those operations are perceived. For example, a rapid naval force response to a humanitarian crisis that is followed by an expeditious and orderly withdrawal of forces once the situation is stabilized may engender good will from a populace that once harbored suspicions or a hostile attitude toward U.S. forces. Both adversary and friendly forces will seek to control the tempo of operations in all domains to include the information environment. Despite success on the battlefield, naval forces may find themselves on the defensive in the "war of the narrative" by which adversaries will rapidly propagandize and distort the true nature of activities within the maritime operational environment through the use of modern communications capabilities and news outlets. Finally, analysts must consider how cultures perceive and experience time as this is a key aspect of how groups and individuals will function and interact.

4.4.1.2.8 Sociocultural Factors and Country/Group Characteristics

There are many factors to consider when analyzing the sociocultural aspects of adversary, neutral, and friendly countries or groups. Relevance of various sociocultural factors is highly dependent on the operation but given the complex interrelationship of these factors in societies it is often difficult to analyze one specific factor to the exclusion of others. These factors have a significant impact on decisionmaking. Understanding how historical, social, cultural, religious, and political forces influence a society and individuals is typically a major component in analyzing intent and determining adversary objectives and "red lines." The mission also drives how sociocultural factors are used. Sociocultural factors that may be relevant to the IPOE process include: political and military limitations, such as territorial waters, excessive maritime claims, air defense identification zones; environmental and health hazards; history; politics and diplomacy; infrastructure; agriculture; economics; and religion. Analysts must also take care that they do not paint countries and regions with too broad a brush. For example, the American experience in Iraq and other places show that what may appear to be superficial or nuanced distinctions among different ethnic, cultural, or religious groups may actually represent very deep fissures in society that may be considered challenges or opportunities by planners.

4.4.2 Leveraging Intelligence Disciplines

IPOE efforts to determine the nature of the maritime operational environment will most likely generate some information requirements that are best addressed by nonintelligence organizations. For example, leveraging METOC sources, databases, and expertise reduces the naval intelligence "burden" to satisfy information requirements in these areas. In addition, insight on some of the physical factors (e.g., ocean characteristics) is also available through METOC channels.

Chapter 3 and appendix D contain comprehensive overviews of the intelligence disciplines and how each is best suited to address specific intelligence requirements. Figure 4-14 provides an assessment of the range of intelligence disciplines able to address geospatial environment requirements.

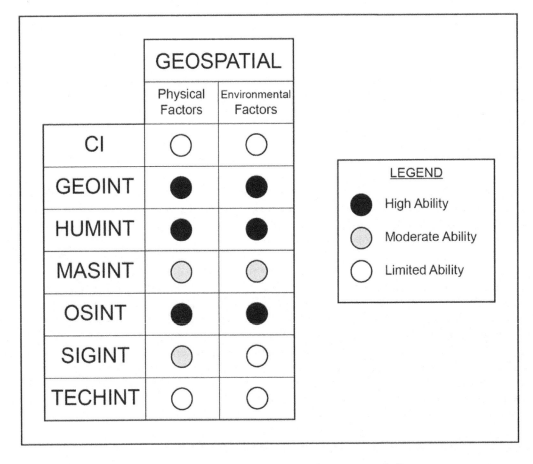

Figure 4-14. Assessed Ability of Intelligence Disciplines to Collect Against the Geospatial Environment

Note

Figures depicting the ability of certain intelligence disciplines to collect against aspects of the operational environment are for illustrative purposes only. These graphics portray the general capability of each intelligence discipline to collect against the various categories of requirements—understanding that specific subintelligence disciplines or classified programs may have a greater ability than summarized in these figures. The overall intent of these graphics is to demonstrate that: (1) there are numerous intelligence disciplines that may be marshaled against high-priority requirements; (2) certain intelligence disciplines have more relevance than others vis-à-vis select requirements; (3) intelligence disciplines not normally a part of the fleet ISR inventory may be available to address fleet requirements. These figures attempt to summarize these points, and should not be considered authoritative or all-inclusive presentations of specific intelligence capabilities.

Determining the nature of the information environment will likely generate very specific information and intelligence requirements. Naval intelligence professionals should leverage the expertise of other Navy or IC communities to either further refine requirements or determine best approaches to tasking collection assets to address gaps. For requirements involving human networks, that expertise may reside in the Department of Defense HUMINT community. For requirements involving computer networks, that expertise resides in the information warfare community. In addition, many equities in the information environment (e.g., MISO and OCO) reside with national agencies—such as the CIA and NSA—involving authorities not delegated to fleet commanders. Ensuring early liaison with such organizations not only leverages the appropriate expertise but ensures deconfliction with other IC activities.

The information environment is generally viewed in very technical terms; however, many facets of that environment involve human factors and other nontechnical effects as discussed in sections 4.3.3.4 and 4.4.1.2.4. Figure 4-15 provides an assessment of the range of intelligence disciplines able to address information environment requirements.

4.4.3 Evaluate the Impact of the Maritime Domain

The naval intelligence professional must evaluate military and other characteristics of the maritime domain to determine how they may affect adversary and friendly COAs. This analysis should include developing an understanding of how controlling or denying the maritime domain affects friendly and adversary plans and operations. Factors such as locations of naval, air, and land bases in relation to the AO and the identification of key geographic features such as chokepoints, islands, rivers, and harbors should be evaluated with regard to their potential impact on operations. For example, the Straits of Gibraltar and the Suez Canal are two vital chokepoints which control the ability to generate, reinforce, and resupply naval forces in the Mediterranean Sea. In addition to adversary capabilities and geographic features, analysts must also evaluate how environmental conditions can impede or facilitate operations as winds, seas, currents, and sea bottom characteristics may have substantial impacts on maneuverability and operational employment of specific platforms, sensors, and weapons systems.

Early in the IPOE process the naval intelligence professional should discuss collection, analysis, production, dissemination, and integration requirements with the commander, staff, supported organizations, the intelligence team, and potential supporting intelligence organizations. Prior to deployment, the intelligence team should become practiced at developing graphics, matrices, and textual products that meet the needs of those they support. These products should then be used as templates that can be easily adapted to meet a variety of mission requirements. While lengthy textual reports and studies may be necessary for large, long-term planning efforts, all efforts should be made to convey intelligence in a manner that is concise and is readily understood by the commander and other users of intelligence. Graphics, geospatial overlays, and matrices provide a way to convey intelligence assessments that provide the depth and breadth of the analytic effort in a way that planners and those executing operations can swiftly integrate to inform their understanding of the maritime operational environment.

	INFORMATION		
	Physical	Informational	Cognitive
CI	Limited	High	Limited
GEOINT	High	Limited	Limited
HUMINT	High	High	High
MASINT	High	Limited	Limited
OSINT	High	Moderate	Moderate
SIGINT	Moderate	High	High
TECHINT	High	Limited	Limited

LEGEND
- ● High Ability
- ◐ Moderate Ability
- ○ Limited Ability

Figure 4-15. Assessed Ability of Intelligence Disciplines to Collect Against the Information Environment

Figure 4-16 is an example of a maritime modified combined obstacle overlay (MCOO) which displays a wealth of intelligence concerning adversary disposition, infrastructure, terrain, lines of communication, restricted areas, and operational objectives in a single product. While products must be tailored to the needs of the intelligence user, the end result should be an evaluation of how the maritime domain aids or hinders forward presence, deterrence, sea control operations, maritime power projection, maritime security, FHA, DSCA, and other operations in and around the key areas of the maritime domain that are identified as crucial to adversary and friendly COAs.

4.4.4 Develop a Systems Perspective of the Maritime Operational Environment

Understanding the operational environment's systems and their interactions can help the commander and staff visualize and describe how military actions can affect other agency and coalition partners as well as how those partners' actions can affect operations. Consequently, naval planners and intelligence professionals must take a systems perspective to the maritime operational environment as this perspective is critical for understanding relationships and dependencies within and among systems. This understanding is critical to efficient and effective operations. Given the military is just one component of national power, the systems perspective also provides a useful means for coordinating and integrating actions with those agencies, organizations, or coalition partners that are attempting to influence events in the maritime operational environment through use of their diplomatic, economic, and informational capabilities. Figure 4-17 provides a conceptual overview of this systems perspective using the PMESII framework discussed earlier in this chapter. While conceptual, this figure highlights the fact that a graphical product that represents complex interrelationships within and among systems can aid commanders, staffs, and supported organizations with visualizing potential or actual strengths, weaknesses, key nodes, COGs, and other factors that affect the development and analysis of COAs.

NWP 2-0

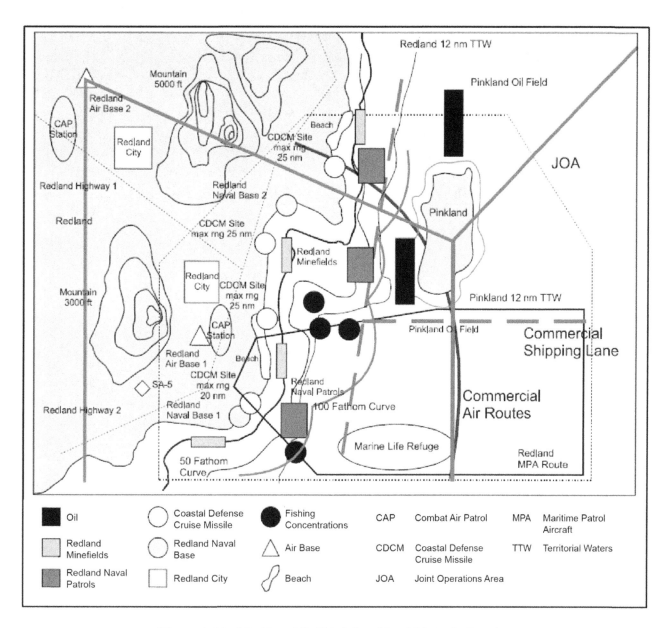

Figure 4-16. Maritime Modified Combined Obstacle Overlay

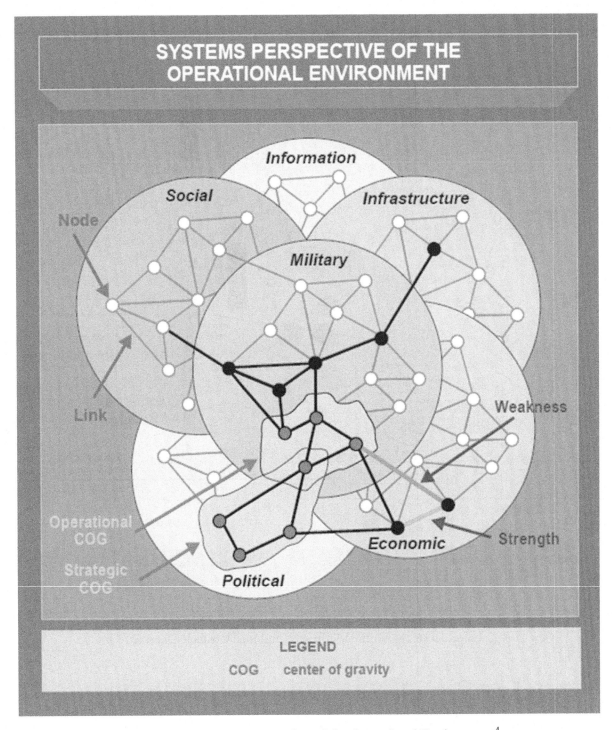

Figure 4-17. Systems Perspective of the Operational Environment[4]

---
[4] JP 2-03.1.

4.4.4.1 Systems Perspective Key Terms

Describing the systems perspective in detail requires establishing definitions of the key terms used:

1. System—"a functionally, physically, and/or behaviorally related group of regularly interacting or interdependent elements; that group of elements forming a unified whole." (JP 1-02) Given the potential complexity of this type of analysis, naval intelligence professionals must coordinate closely with their commander, other staff elements, and supported organizations to identify which PMESII systems (and components within those systems) are relevant to the mission. This discussion should also define the level of detail required in the analysis. Understanding the interactions within and among systems enables the commander, staff, and supported organizations to visualize how friendly and adversary actions on a single system component affect that system and others.

2. Node—"an element of a system that represents a person, place, or physical thing." (JP 1-02) These are the components of the system that can be targeted and influenced by the naval force.

3. Link—"a behavioral, physical, or functional relationship between nodes." (JP 1-02) Links can include such things as the relationship of a naval vessel to its homeport or a maintenance facility; the administrative and operational control arrangements that connect that naval vessel to command and control elements; the communication network path that allows for the transmission of order from the C2 node; or the ideological and religious alignment of the adversary naval strike group's commander to a larger societal group. Links enable the visualization of how internal system components interconnect internally and externally. They are also useful in facilitating the understanding of how nodes behave to accomplish specific tasks and functions within a system and how the system being studied influences other systems.

4.4.4.2 Network Analysis

Since there are insufficient resources to analyze every node and link in the maritime operational environment, a process must be used to determine which nodes and links are relevant and then analyze them to a sufficient level of detail to determine adversary vulnerabilities and opportunities for the naval force. Both the mission objectives and the analysis of the maritime domain aid in narrowing the scope of this IPOE step. Ideally, analysis should result in the identification of critical nodes, which if degraded or disabled, would have impacts within and across multiple adversary systems. Network analysis is an IPOE process step that provides a structured approach to determining and evaluating systems in the maritime operational environment. The following paragraphs list and describe the substeps of the network analysis process.

4.4.4.2.1 Identification of Relevant Nodes

Limiting the scope of the systems analysis effort begins with determining the relevant nodes that the analyst will examine. While at their basic level nodes consist of discreet people, places, or things, the definition of what constitutes a node is largely dependent on the level of war within which the commander and staff operate. Figure 4-18 provides a listing of sample nodes from a PMESII perspective. Related functional groupings of nodes and links have both horizontal and vertical aspects. For example, a radar assigned to an air defense unit provides that unit with the capability to detect and track potential threats but that same radar may be relied upon by sector and national C2 centers to develop situational awareness of events within a larger context. Analysts must also be conscious of the fact that many individual systems are complex as are the interactions among systems. The value in identifying relevant nodes is it provides a focus for predictive analysis and is the beginning step in answering the question "What will happen if . . . ?"

> Analysis must also consider the element of time. For example, a short-term FHA mission may engender good will with local leaders and groups yet a protracted naval force presence in an area may engender long-term resentments among those same leaders and groups.
>
> ### SAMPLE PMESII NODES
>
> **Political:** advisors, governors, mayors, political interest groups, cabinet officials, courts, policy documents.
>
> **Military:** individual leaders at all levels, plans and orders, defense ministry, C2 headquarters, air defense system, artillery maintenance facility, ammunition storage point, key terrain.
>
> **Economic:** banks, corporations, trade unions, market places, shipping facilities, smugglers, commercial depots.
>
> **Social:** ethnic groups, clans, tribes, religious groups, unions, associations, schools, cultural centers, health, and welfare facilities.
>
> **Infrastructure:** nuclear power plants, hydroelectric dams, gas pipelines, aqueducts, pumping stations, rail yards, airports, port facilities, relevant factories, hospitals, schools, civil defense shelters.
>
> **Informational:** plans and orders, newspapers, information ministry, television networks, computer networks, information technology centers, intelligence agencies, postal facilities, radio stations, national or influential specialty magazines or periodicals, and other existing information infrastructure and information-dissemination capabilities.

Figure 4-18. Sample Political, Military, Economic, Social, Information, and Infrastructure Nodes

4.4.4.2.2 Determining and Analyzing Node-Link Relationships

Links represent the relationship between and among nodes. Graphically portraying node-link relationships helps the commander, staff, and supported organizations visualize the impact of actions against specific nodes. For example, a single node with links to many other nodes (a one-to-many relationship) may indicate that single node is critical to the functioning of a subsystem, system, or network. The strength of a single link may also be relevant in determining the importance of a given node. Since the number and strength of links between and among nodes is critical to identifying key nodes and possible COGs, the naval intelligence professional must determine the best way to convey this analysis to the user of intelligence. Two of the most common techniques used to disseminate network analysis results are the association matrix and the network analysis diagram.

4.4.4.2.3 Association Matrix

Figure 4-19 provides an example of an association matrix. The association matrix shows the link (or suspected relationship) among nodes. Relevant nodes are always plotted along the diagonal axis and the association between any two nodes is indicated at the intersection of the horizontal and vertical axes.

4.4.4.2.4 Network Analysis Diagram

Figure 4-20 provides a conceptual example of a network analysis diagram that shows nodes and links between and among various PMESII systems. These links, or relationships, are properties of the group and can represent several different forms of interaction, including kinship (parent of, sibling of), role-based (boss of, rival of), interactive (travels with, meets with), and affective (trusts, likes) relationships. Links typically represent directions of influence among nodes rather than a linear progression.

Depending on the complexity of the systems being analyzed, naval intelligence professionals may begin by examining high-level relationships among nodes and then drill down to specific system components and subsystems to discover key nodes and vulnerabilities (a top-down approach). Analysts may also take a bottom-up approach by conducting network analysis at a system level and then build a consolidated overview of interactions among systems based on that analysis.

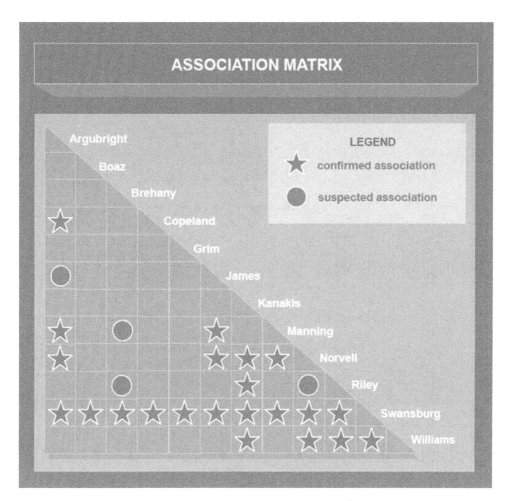

Figure 4-19. Association Matrix[5]

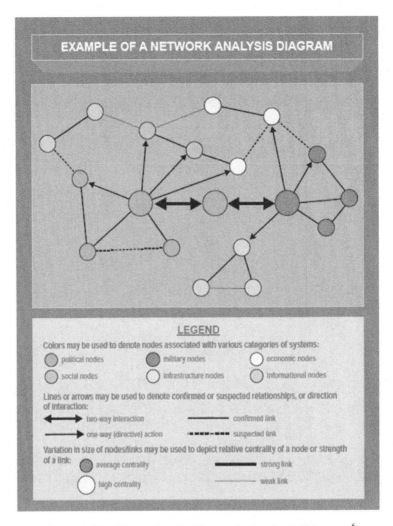

Figure 4-20. Example of a Network Analysis Diagram[6]

When creating a network analysis product, three general rules are typically observed:

1. Nodes are represented by circles and color-coded, shaded, or patterned based on their association with a specific system. Nodes can be of different sizes to represent relative quantitative or criticality differences among nodes within or among systems.

2. Links are represented by lines with solid lines depicting confirmed relationships and dotted or dashed lines indicating suspected relationships. These lines may also terminate in arrowheads to indicate the direction or flow of information, influence, or other characterizations of the relationship.

3. Whenever possible, the arrangement of nodes and links should be done in a way to prevent or minimize the crossing of lines.

4.4.4.2.5 Identify Key Nodes

Every major system and subsystem contains key nodes, (i.e. nodes that are critical to the functioning of an associated system). Key nodes may present naval commanders with decisive points for their operations as seizing, degrading, disrupting, or neutralizing a key node can produce a significant advantage for the naval force. Often a

---

[6] JP 2-01.3.

key node is linked to multiple systems and so affecting the key node could have impacts beyond the system within which the node operates. For example, given the dependency of many sectors on modern, networked communications capabilities, the degradation or elimination of a major telecommunications hub may have regional or national impacts that affect other infrastructures (such as power or transportation), military forces, political, economic, and social activities.

A method to analyze the criticality of a particular node is to determine that node's relative centrality by analyzing the measurable characteristics of degree, closeness, and betweenness. Figure 4-21 provides a conceptual model to illustrate these points.

1. "Degree" refers to the number of direct links that one node has to others. So, in figure 4-21, node D has the highest degree centrality and, because it appears central to the operation of the system, it may also be referred to as hub. In this example, node D would be designated a key node since elimination of this node would most likely cause significant degradation to this system or neutralize it entirely. Nodes that have one or few direct connections have a low degree of centrality and are sometimes referred to as "peripheral." The amount of degree centrality does not always correlate to importance. Peripheral nodes, like an early warning radar station or ISR asset, may provide a critical function for the overall system that is not readily apparent from an analysis of degree.

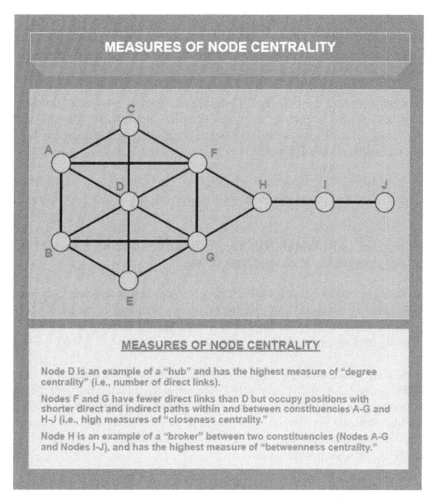

Figure 4-21. Measures of Node Centrality[7]

---

[7] JP 2-01.3.

2. "Closeness" evaluates the node's overall position in the network. It differs from degree in that, while a particular node may have many links to other nodes, those nodes may not be as influential as other nodes in the network. An analyst can determine closeness by counting the number of hops between a single node and all other nodes in a network. The lower the score, the closer that node is to others in the network. Such a node is in a strong position to monitor overall network activity. In Figure 4-21 the nodes F and G have fewer links than D but they have shorter paths to other nodes in the system.

3. "Betweenness" is a measure of the number of times a node lies within the shortest path between two other nodes. In figure 4-21, node H would be an example of a node with high betweenness. This type of node is called a "broker" node as it may facilitate exchanges (data, money, power, etc.) between different components within or among systems. Degradation or elimination of a broker node could significantly reduce the effectiveness of a system or network of systems by breaking the linkages among them.

4.4.4.3 Leveraging Intelligence Disciplines

Determining the nature of PMESII systems generates some information and intelligence requirements that are best addressed by IC organizations with cognizance over the nonmilitary systems depicted in figure 4-17. For example, expertise on foreign political or diplomatic factors ("P") may be available through the Department of State, and insight on economic ("E") or social ("S") factors might be available through the CIA.

The utility of the PMESII model is that it reminds planners that information and intelligence requirements directly affecting the planning and execution of naval operations may fall outside the capabilities of organic ISR systems that are generally focused on maritime domain awareness (MDA) at the tactical level.

In addition, PMESII analysis also reveals to the naval intelligence professional what interagency, international and nongovernmental influences may be present to impact maritime matters. By identifying these influences early in the planning process, the naval intelligence professional may expand their tasking to other IC members with greater roles and insight regarding such influences.

PMESII systems span a broad array of subject matter; therefore, information and intelligence requirements will also be wide in scope. An assessment of the range of intelligence disciplines able to address such requirements is provided in figure 4-22.

4.4.4.4 Evaluate the Impact of Political, Military, Economic, Social, Information, and Infrastructure Systems on Military Operations

Development of the systems perspective in the PMESII process provides a structured approach to identifying nodes and links within and among systems. The systems analysis reveals key adversary vulnerabilities and potential opportunities for the naval force. The impact of these PMESII systems on military operations can be measured in terms of density and distance.

1. "Density" refers to how well connected a network is by comparing the number of ties in a network to the number of total ties possible. A highly connected network may indicate high resiliency, if there are multiple and redundant links among various key nodes. Since each network is unique, a high number of connections may also indicate inefficiencies or single points of failure that can be exploited by the naval force. A network with low interconnectivity may indicate that there is high network segregation, such as along military Service, clan, or political lines.

2. "Distance" measures the number of hops between any two network nodes. Analyzing distance can provide understanding regarding how such things as information, influence, or materials flow through a system and may be an important factor in assessing the cohesiveness or brittleness of a system or network of systems. For example, large distances may indicate poor communication flows between and among nodes.

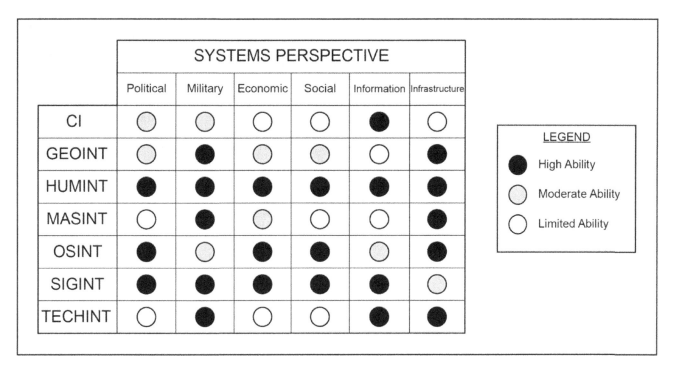

Figure 4-22. Assessed Ability of Intelligence Disciplines to Collect Against the Political, Military, Economic, Social, Information, and Infrastructure Systems

4.4.5 Describe the Impact of the Maritime Operational Environment on Adversary and Friendly Capabilities and Broad Courses of Action

Specific products developed during the analysis of the maritime operational environment and the relevant systems of concern may be provided to the commander, staff, and other supported organizations at any stage in the IPOE process. The ultimate goal of this phase of the IPOE effort is to develop a single integrated assessment that focuses on the overall impact of the maritime operational environment on all COAs available to both friendly and adversary forces. This assessment consolidates the intelligence produced during this phase of the IPOE process and may take the form of a briefing, set of overlays, written analysis of the operational environment, intelligence estimate, or any other format the commander deems appropriate.

Regardless of the format in which it is delivered, the assessment has two main purposes. First, it enables the commander, staff, and other supported elements understand the operational environment while planning and conducting operations. This includes identifying land, sea, and air lines of communication and avenues of approach; key nodes and links; potential threats and opportunities; significant terrain, oceanographic, and hydrographic features; and details the potential impacts of weather on the friendly force. Secondly, this assessment provides the foundation for an evaluation of the maritime operational environment from the perspective of the adversary. When combined with intelligence on adversary capabilities this evaluation enables the analyst to determine potential ways the adversary may employ its military forces and other instruments of national power to accomplish its objectives. For example, an adversary may have an objective to influence events in a neighboring island country but lacks the seaborne lift or amphibious capability to seize territory in that country. Given these circumstances, the naval intelligence professional would turn their attention to other available ACOAs such as use of air assets or information operations to coerce the neighboring island nation.

While no specific format for the overall assessment of the maritime operational environment and its impact on friendly and adversary COAs is proscribed, naval intelligence professionals should deliver their assessment with consideration of how the operational factors of space, time, and force within the maritime operational environment affect friendly and adversary COAs. The key elements of these operational factors are provided in figure 4-23 and this list is provided to show that the IPOE analysis conducted in this step supports the assessment of time, space, and force considerations required by Navy planning doctrine. While the factor of force is the focus of IPOE steps three and four, it is useful to discuss the concept of operational factors here as the three factors must be considered together in assessing the adversary and the impact of the maritime operational environment.

Per NWP 5-01, Navy Planning, a commander's freedom of action is described as being " . . . achieved primarily by properly balancing the factors of space, time, and forces." The portrayal of these three factors is provided in figure 4-24.

Note that the three operational factors (space, time and force, shown in gray) have three interrelationships—indicated by the dashed lines. These relationships culminate in the overall balancing of factors that lead to military planning and direction, particularly at the operational level of war (OLW). These factors and, increasingly, information are pivotal for making sound decisions at all levels of warfare, which is why understanding these factors is so critical to the broad adversary and friendly COA analysis done within this phase of the IPOE process. This statement also underscores that the proper assessment of space, time, and force is highly dependent on relevant information (shown in black) to determine the right balance between these three operational factors.

Accordingly, the three components of threat—intent, capability, and will—can be aligned to the relationships between the three operational factors, as portrayed by the dashed lines in figure 4-24.

1. Time-Space—This relationship pertains to force mobility. Although primarily a maneuver and logistics calculation, the threat factor of enemy "intent" weighs heavily in the consideration of when an engagement may begin.

2. Space-Force—This relationship pertains to force composition. This calculation equally considers the strategic or operational objectives against the threat factor of enemy "capability" to determine friendly force capabilities.

3. Force-Time—This relationship involves force duration. As an extension of the strategic or operational objectives, the threat factor of enemy "will" to oppose those objectives is a key consideration.

SUMMARIZING THE KEY ELEMENTS OF OPERATIONAL FACTORS (SPACE, TIME, AND FORCE)

Space: military geography (area, position, distances, land use, environment, topography, vegetation, hydrography, oceanography, climate, and weather), politics, diplomacy, national resources, maritime infrastructure and positioning, economy, agriculture, transportation, telecommunications, culture, ideology, nationalism, sociology, science and technology.

Time: preparation, duration, warning, decision cycle, planning, mobilization, reaction, deployment, transit, concentration, maneuver, accomplish mission, rate of advance, reinforcements, commit reserves, regenerate combat power, redeployment, reconstruction.

Force: combat potential, combat power, friction, fog of war, numerical strength, number of weapons, quality of weapons, firepower, mobility, maneuverability, speed, command organization, logistics, communications, operational leadership, morale, discipline, loyalty, small-unit cohesion, social or group cohesion, task cohesion, combat motivation, doctrine, unity of effort, training.

Figure 4-23. Summarizing the Key Elements of Operational Factors[8]

---

[8] NWP 5-01; Milan Vego, Joint Operational Warfare: Theory and Practice, pp. III-33–III-46.

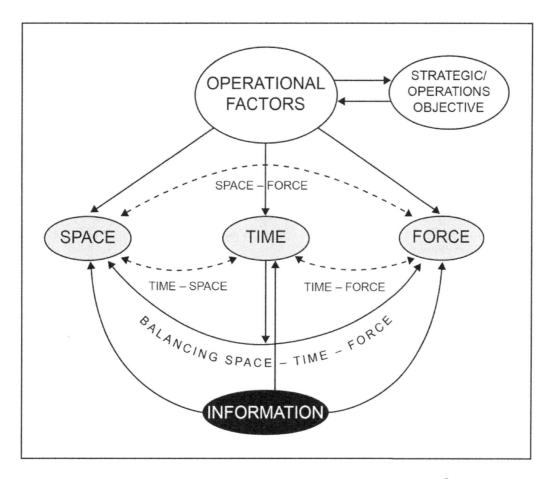

Figure 4-24. Operational Factors—Space, Time, and Force[9]

Figure 4-25 provides an example of an intelligence assessment that incorporates the IPOE analysis of the maritime domain and systems and applies it to assessing the impact of the maritime operational environment on adversary and friendly COAs in terms of time and space factors. While simple in presentation, the matrix shown in figure 4-25 should be seen as the output of a detailed and systematic application of the intelligence process and the analytic methodologies discussed above.

The individual intelligence products and overall assessment of adversary and friendly COAs must be tailored to the needs of the commander, the staff, and supported organizations. It flows from the objective and requirements of the mission. The goal for the intelligence team is to provide a prioritized set of preliminary ACOAs within the constraints imposed by the adversary's capabilities and the operational advantages and limitations of the maritime operational environment. The understanding of adversary capabilities is developed further in step three of the IPOE process while the prioritized list of potential ACOAs is further refined in step four of the IPOE process.

4.5 EVALUATE THE ADVERSARY

The third step of the IPOE process evaluates the adversary's capabilities and limitations, current situation, and potential COGs. It consists of four main substeps listed in figure 4-26 through which the adversary's capabilities and limitations, current situation, and potential COGs are identified and assessed. Naval intelligence professionals must first develop an adversary model that provides insights into how an adversary may employ their capabilities based on their doctrine, TTP, and observed operational behavior. Analysts can then assess the adversary's current situation to develop an understanding of adversary readiness and intent as informed by how adversary actions

---

[9] Vego, p. III-3.

conform to the model combined with current intelligence reporting and evaluation. During this phase, a more detailed examination of the factor of force occurs to determine the tangible and intangible aspects that enable or inhibit adversary execution of their potential COAs. Finally, this analysis of adversary capabilities and limitations is also used to identify adversary COGs, which are "the source of power that provides moral or physical strength, freedom of action, or will to act" for adversary forces. (JP 1-02)

This capability analysis must be conducted within the context of the broad COAs defined in the previous step of the IPOE process. Any new COAs that present themselves during this stage of the analysis should be provided to the commander and staff at the earliest possible opportunity and included for further evaluation. In analyzing adversary capabilities, naval intelligence professionals are cautioned against "mirror-imaging" and assuming that an adversary will employ capabilities or react to friendly operations in the same way U.S. forces would. Red teaming during this phase of the IPOE process provides an effective means of detecting defects in analysis and identifying COAs that were not considered in the early phases of the IPOE effort. Failure to properly evaluate the adversary could lead to surprise or the expenditure of resources against nonexistent or inconsequential adversary capabilities and can result in mission failure.

4.5.1 Update or Create Adversary Models

Adversary models are analytic products that provide the analyst and users of intelligence with an understanding of adversary doctrine and their observed operational employment of capabilities. Typically, they are graphical assessments that depict adversary dispositions for specific types of forces (air, naval, land, special warfare, etc.) and their employment within the maritime operational environment. Examples of these models include graphical templates that describe how an adversary conducts antisubmarine warfare or employs its air defense capabilities during periods of heightened alert. Prior to deployment, naval intelligence professionals should determine likely operations they will expect to conduct and create adversary models that can be quickly updated as new intelligence is gathered and evaluated. Adversary models typically consist of three major analytic products—the adversary templates, which show how the adversary employs forces within specific COAs; descriptions of an adversary's preferred TTP and options; and a list of adversary high-value targets (HVTs) and high-value individuals.

TIME/SPACE FACTOR IMPACTS ON ADVERSARY AND POTENTIAL FRIENDLY COAS		
Situation	Effects on Adversary COAs	Effects on Friendly COAs
Item: Redland holds central position.	Short lines of communication; should be relatively easy to supply or resupply.	Must commit significant assets if adversary lines of communication are to be interdicted.
Item: Oil and gas platforms in NW sea.	Potentially attractive, high-value, and currently undefended targets within striking range.	Joint force maritime component commander (JFMCC) must be prepared to provide for the defense of the North Sea oil and gas assets.
Item: Redland is bounded on three sides by neutral nations.	Redland can minimize force defenses on those neutral borders.	The lines of operation into Redland will be predictable; some form of deception will be required.

Figure 4-25. Example Intelligence Preparation of the Operational Environment Step Two Assessment Product[10]

---

[10] NWP 5-01.

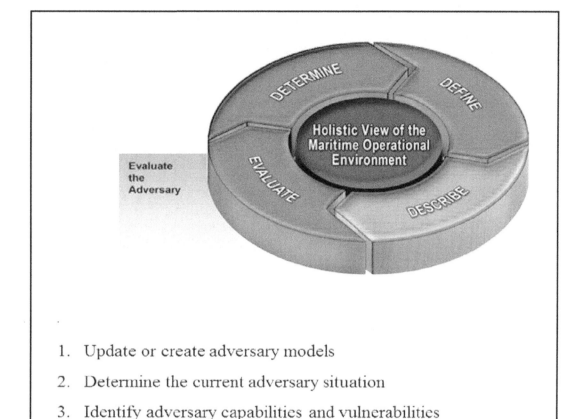

Figure 4-26. Intelligence Preparation of the Maritime Operational Environment—Step Three

4.5.1.1 Adversary Templates

Adversary templates depict the dispositions and employment patterns of adversary forces within a specific or similar maritime operational environment. They are usually scaled graphic depictions of adversary dispositions for specific types of military (conventional or unconventional) operations such as movements to contact, antisurface warfare operations, insurgent attacks in urban areas, combat air patrols, and aerial ambushes.[11] These templates may represent a single service to reduce complexity, or can depict an adversary's employment of joint capabilities. For example, a joint adversary template for a defensive naval force mining operation may also include the air bases, types of aircraft, and the employment pattern these air assets would use to provide air cover for the mining effort. Adversary templates should be constructed for all adversary broad COAs (such as attack, defend, reinforce, or retrograde) and must be informed by thorough analysis of adversary doctrine, observed operations, military exercise, and TTP. Figure 4-27 provides the specific factors that should be addressed in an adversary template. This list is not all-inclusive as requirements for the adversary template are driven by mission requirements and the needs of the commander, staff, and supported organizations.

Operational needs and mission objectives also drive the format of the adversary template. Figure 4-28 provides an example of adversary template constructed from a geospatial perspective that shows how an adversary's naval force may be used to defend land force operations.

---

[11] Doctrinal templates may also be used, which show the disposition and employment of forces without regard to the constraints imposed by a specific operational environment. NWP 5-01

Adversary Template Components

- Organization for combat
- Distances (such as frontages, depths, boundaries, spacing between ships, and intervals between march units or waves of attacking aircraft)
- Functions (such as disruption, assault, exploitation, fixing, contact, shielding, or counterattack) that various parts of the adversary force are intended to perform in order to accomplish objectives in a certain type of operation
- Engagement areas
- Patterns for the use of terrain and weather
- Timing and phasing of operations
- Relative locations and groupings of forces and support units

Figure 4-27. Adversary Template Components[12]

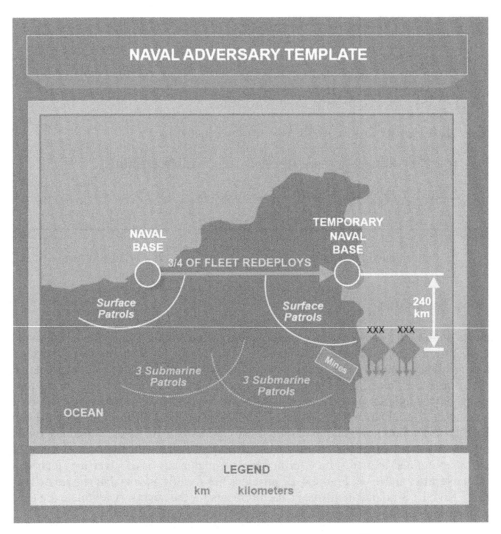

Figure 4-28. Naval Adversary Template[13]

---

[12] NWP 5-01; JP 2-01.3.
[13] JP 2-01.3.

4.5.1.2 Threat Template Depicting Systems Network Perspective

While the geospatial template is an effective tool for depicting military force disposition and employment, other techniques may be needed to represent anticipated changes that may occur in node-link relationships such as the establishment of potential new links. Such a view is useful as it may provide indications of future ACOAs. Changes to the current (or routine) node-link structure may be indicators of adversary intentions and should be grounded in an understanding of adversary past practices or patterns of operation. For example, the standup of special communications circuits and increased logistic activity around a naval base may be precursors to adversary intentions to sortie a number of their fleet units. A modified version of an association matrix is one way to depict potential node-link changes associated with different ACOAs and provides a systems perspective that may either complement a geospatially oriented adversary template or it may stand alone as its own product. The value of the systems perspective is that it also forms the basis for a set of indicators that may be used to drive collection and analysis efforts that will confirm or refute whether the adversary has selected a particular COA. Figure 4-29 provides an example of a threat template depicting a systems network perspective.

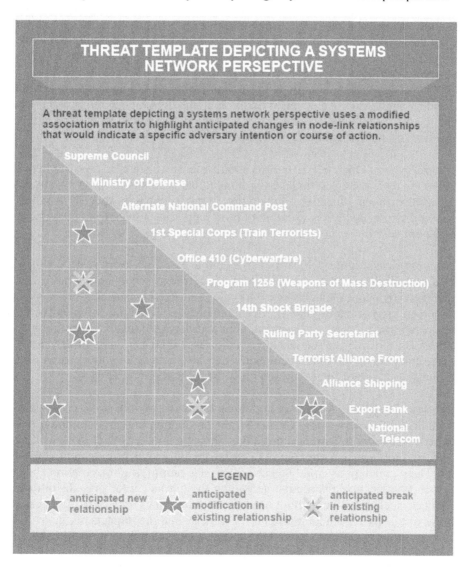

Figure 4-29. Threat Template Depicting a Systems Network Perspective[14]

---

[14] JP 2-01.3.

4.5.1.3 Describe Adversary Tactics and Options

While an effective tool for visualizing adversary disposition and potential employment of forces, both adversary and threat templates should be supplemented with a written description of the adversary's preferred tactics. This description should address the types of activities and supporting operations the adversary units depicted in the templates are expected to conduct. The description should also contain an analysis of the options (branches and sequels) available to the adversary should their efforts succeed or fail.

Branches are "contingency options built into the base plan used for changing the mission, orientation, or direction of movement of a force to aid success of the operation based on anticipated events, opportunities, or disruptions caused by enemy actions and reactions." (JP 1-02) For example, should an adversary succeed in an attack, their preferred tactic may be to exploit success and pursue. Potential branches available to the adversary when an attack begins to fail may include committing reserves, calling for reinforcements, or a covered withdrawal.

While branches detail the how the adversary will deal with contingencies that arise in a particular phase of an operation, sequels are the "subsequent major operation or phase based on the possible outcomes (success, stalemate, or defeat) of the current major operation or phase." (JP 1-02) For example, if an adversary has completed a phase objective of delaying an opposing force, they may now move on to the next phase of operations that involves committing a large force to the attack. In addition to detailing branches and sequels, the assessment of how the adversary will potentially employ their forces should also include an understanding of how the adversary uses weather and terrain. An adversary with no night-fighting capabilities and whose only observed activity occurs during the day is unlikely to employ forces at night. Suggested techniques for building IPOE products to describe adversary tactics and options include:

1. Identify a Specific Type of Operation. Analysts should identify a specific type of operation and then create a scenario within the maritime operational environment that accounts for how the adversary would employ their forces to meet particular objectives. In addition to describing the expected activities of the main force, this analysis should also include an evaluation of how supporting forces from all Service components will participate in the operation.

2. Use Time-Event Matrices. Figure 4-30 shows an example of a time-event matrix, which is used to describe how an adversary typically conducts specific types of operations sequenced in time. These may be operations conducted by major elements of a particular Service (i.e., surface forces, submarines, expeditionary forces, etc.) or major joint Service components. Given the complexity of multielement or multi-Service operations, it is very difficult and time consuming to display operations over time using a graphical method. By showing the sequencing of major and supporting operations by the responsible joint or single Service component, the time-event matrix can also be used to show changes in organization, composition, and disposition of forces at each phase of their overall operation.

3. Annotate Adversary Templates. Adversary and threat templates are not static products. While examples are given in this publication, there is no set format for the production of these templates. In some cases, annotating these templates with marginal notes that are tagged to key events, dispositions, or locations may provide the user of intelligence greater depth of understanding concerning potential adversary employment of their forces. These annotations could be used to provide amplifying information about a specific item on the template, such as indicating a particular defending force is an "elite" brigade, or details about adversary doctrine and tactics, such as "all remaining operational submarines will sortie if opposing naval forces are within 150 miles of this base." Annotation must be concise and the impacts on the overall template must be considered. For example, excessive annotations could be overwhelming or distracting for users of intelligence and may cause important details to be overlooked.

4. Identify Decision Criteria. If there are decision criteria that the adversary uses to determine which options are preferred over others, then these should be identified and listed. Understanding how the adversary responds to certain types of events is useful for wargaming adversary and friendly COAs, targeting, and deception planning.

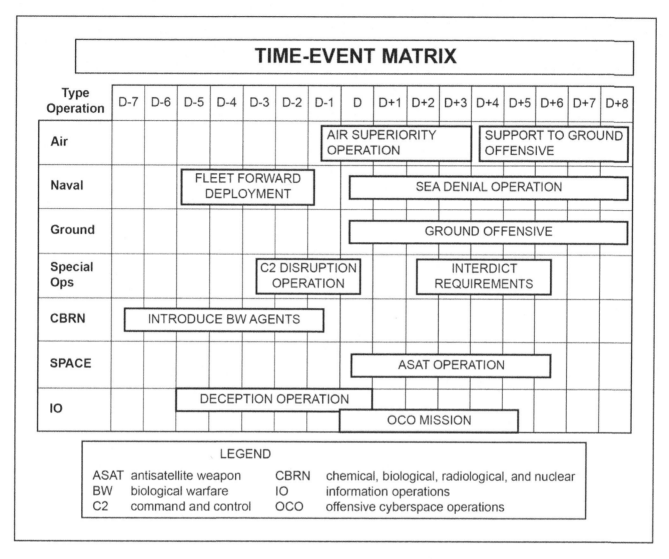

Figure 4-30. Time-Event Matrix[15]

5. **Breakdown Component or Element Actions.** To the extent possible, the actions of each major Service element or joint force component in each phase should be described in sufficient detail to facilitate the analysis of HVTs that occurs next in the IPOE process. Each phase should be analyzed separately since a target's value usually varies depending on its role within a phase. For example, mine-laying forces may be very valuable to the adversary in a prehostility phase but during hostilities an adversary may place greater value on antiship coastal missile sites.

4.5.2 Identification of High-Value Targets

high-value target (HVT). A target the enemy commander requires for the successful completion of the mission. (JP 1-02. Source: JP 3-60)

Adversary models must also include a list of high-value targets. The loss of high-value targets would be expected to seriously degrade important enemy functions throughout the friendly commander's area of interest. HVTs are identified by combining operational judgment with an analysis of the information contained in the adversary template and the supporting documentation discussed above. Assets that are critical to the success of the main

---

[15] JP 2-01.3.

mission and each component or element supporting the main mission should be assessed to determine their value to the adversary in accomplishing their objectives.

Analysis should also include how relative value may change with the adoption of branches and sequels by the adversary and further analyzed by the naval intelligence team. For example, an adversary ground force defending a front across a peninsula may be vulnerable to amphibious flanking attacks in its rear area. In this situation, the adversary's ability to deny access to its rear area coastal waters may be crucial, and therefore its coastal defense assets (artillery, antiship cruise missiles, local surface and subsurface combatants) may constitute HVTs. Given the joint nature of operations, HVTs are typically identified through collaboration within a larger joint targeting community and the responsible producers (i.e., agencies, Service intelligence centers, etc.) for various intelligence product category codes. This collaboration should be conducted by any available secure communications means (e.g., JWICS, video teleconferencing (VTC), secure voice, SIPRNET). The following techniques may be useful in identifying and evaluating HVTs.

4.5.2.1 Conduct Mental Wargaming

Naval intelligence professionals should mentally wargame the operation under consideration and think through how the adversary will use the assets of each component or major service element. The output of this analysis is a list of potential HVTs.

4.5.2.2 Determine Adversary Response

Once a list of potential HVTs is obtained, continue the wargaming process to determine how the adversary may respond to the loss of each identified HVT. An important consideration in this analysis is determining whether the adversary can substitute another capability if the HVT is lost or if they are able to adopt different options.

4.5.2.3 Rank High-Value Targets

Analysts should then assess all HVTs and place them in a priority order in accordance with their relative worth to the adversary. Given the relative worth of a HVT may change depending on the phase of the overall operation or campaign, this analysis should be conducted for each phase.

4.5.2.4 Construct a Target Value Matrix

HVTs should be arrayed in a target value matrix that is produced by grouping HVTs according to their function and then indicating the relative worth of each HVT category. As shown in figure 4-31, the matrix should also describe how an attack on each specific category (to include the timing of the attack) would affect the adversary's operation.

4.5.3 Determine the Current Adversary Situation

Naval intelligence professionals should use all available intelligence resources to analyze and determine the current adversary situation. Since the adversary is dynamic, analysts must provide a current snapshot of adversary forces and then update these IPOE products as the situation changes. The analytic efforts is primarily focused on OB factors for adversary forces (ground, air, naval, SOF, etc.) operating within the AO, the AI, and the AOIs and which may otherwise be capable of interfering with the friendly mission.

Current information on adversary composition and disposition is particularly important and may come from a variety of sources such as the local COP and current intelligence products, theater and agency daily intelligence products and COP overlays, and MIDB. The current adversary situation is usually based on an analysis of the OB factors listed in figure 4-32.

NWP 2-0

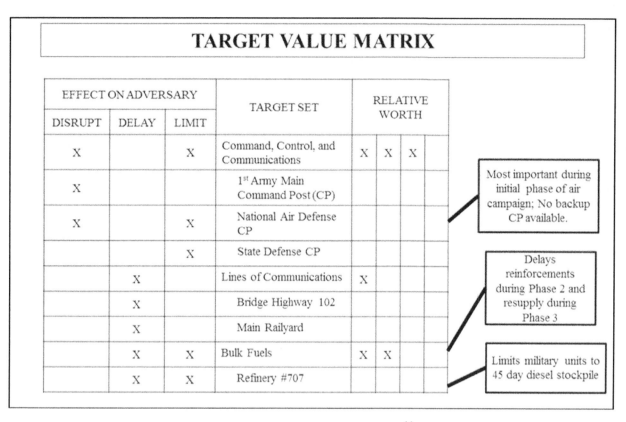

Figure 4-31. Target Value Matrix[16]

ORDER OF BATTLE FACTORS

- Composition
- Disposition
- Strength
- TTP
- Training status
- Logistics
- Effectiveness
- Electronic technical data
- Personalities
- Miscellaneous data (information that contributes to knowledge; historical studies, cultural idiosyncrasies, civil-military relations)

Figure 4-32. Order of Battle Factors

---

[16] JP 2-01.3.

4.5.4 Identify Adversary Capabilities and Vulnerabilities

Adversary capabilities are expressed in terms of the broad COAs and supporting operations that the adversary can take to interfere with the accomplishment of the friendly mission. For conventional operations, these are typically defined as offense, defense, reinforcement, or retrograde.[17] Each broad COA can be broken down into more specific COAs. For example, an offense may take the form of an envelopment or a penetration. In addition to conventional capabilities, adversaries may employ a variety of capabilities to accomplish their mission to include chemical, biological, radiological, nuclear, and high-yield explosives (CBRNE), nonlethal fires, information operations, or deception.

The process for determining adversary capabilities consists of comparing the current adversary situation with the adversary models and then assessing the adversary's ability to meet the requirements of their model. In general, there will be differences in the actual or observed capability of an adversary from the ideal found in the adversary model. Adversary deficiencies in meeting the requirements of their model should be labeled as vulnerabilities while the adversary's ability to meet or exceed the requirements of their model should be labeled as strengths. If time or another factor are assessed to be a critical element in an adversary capability, that circumstance should be specifically stated in the capability statement. Some example of capability statements are shown in figure 4-33.

Intelligence concerning adversary capabilities, strengths, and weaknesses should be disseminated to the commander, staff, and supported organizations as soon as possible. Typically, the intelligence estimate is used to provide this evaluation. Based on the situation and the requirements of the commander and the mission, this analysis may be disseminated in whatever form is required to facilitate planning and execution of operations.

4.5.5 Identify Adversary Centers of Gravity and Decisive Points

Centers of gravity are what Clausewitz called "the hub of all power and movement, on which everything depends . . . the point at which all our energies should be directed." For this reason, an objective is always linked to a COG. Given the criticality of COGs to both adversaries and friendly forces, one of the IPOE analyst's most important responsibilities is to identify potential adversary COGs and to develop an understanding of the adversary's perception of friendly force COGs. Typically, COGs are associated with the strategic and operational levels of war. Strategic level COGs may be a military force, an alliance, key leaders, critical capabilities, or national will. Operational level COGs are often associated with military capabilities but could include other aspects of the maritime operational environment. While COGs may exist at the tactical level, these are often nested within higher level COGs and usually constitute objectives and decisive points at that level of war.

In conducting COG evaluations, analysts must continuously assess and reassess a number of factors, such as leaders, military forces, resources, infrastructure, population, and internal/external relationships, to determine which of these factors or combination of factors is truly critical to the adversary's strategy. COGs may also be driven by the type of conflict. For example, in an irregular warfare scenario, the focus of operations is on gaining legitimacy and influence with regard to a particular population. Given this fact, it is possible that for irregular warfare, the same population is the COG for both adversary and friendly forces. Figure 4-34 provides a further overview of characteristics associated with COGs that analysts may consider in making their assessments.

---

[17] Navy planning doctrine also uses the term DRAW-D (defend, reinforce, attack, withdraw, and delay), NWP 5-01.

Examples Of Adversary Capability Statements

- The adversary has the capability to attack with up to 6 destroyers and 4 frigates supported by 25 daily sorties of land-based fixed-wing aircraft, but is capable of penetrating no further than line BRAVO due to supply limitations and extent of air cover.
- The adversary has the capability to interdict friendly SLOCs at chokepoints GREY and BLUE after repositioning units of the 4th Fleet. Current naval deployments preclude an attack before 4 August.
- Adversary insurgents will have the capability to resume offensive action after the fall harvest is completed in October.

Figure 4-33. Examples of Adversary Capability Statements

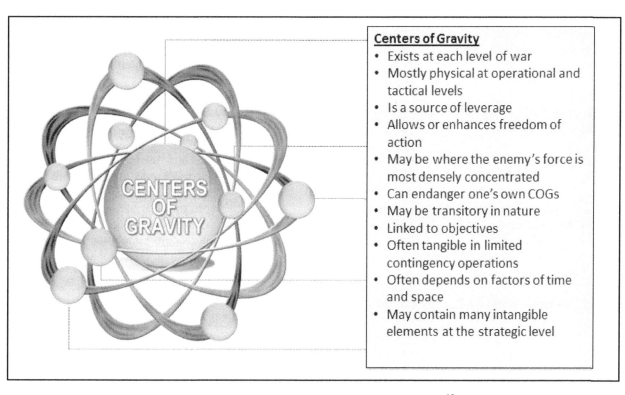

Figure 4-34. Centers of Gravity Characteristics[18]

Note

NWP 5-01 and JP 5-0 provide detailed information on COGs and the analytic process for determining both adversary and friendly COGs. The discussion below is an overview of the COG analysis process as it pertains to IPOE. Readers desiring a more in-depth understanding of this phase of the IPOE process are encouraged to review NWP 5-01 and JP 5-0.

4.5.5.1 Center of Gravity Identification and Analysis Process

center of gravity (COG). The source of power that provides moral or physical strength, freedom of action, or will to act. (JP 1-02. Source: JP 5-0)

---

[18] JP 2-01.3.

Figure 4-35 provides an overview of the COG identification and analysis process.

4.5.5.1.1 Identify Adversary Objectives

The first step of that process is to identify the adversary's strategic, operational, and major tactical objectives. This is a critical step. Errors made in this step perpetuate through the rest of the analysis and cause it to be flawed. Objectives should be nested—there should be clearly definable linkages from strategic to operational to major tactical objectives. At the strategic level, an adversary's objective may be the focus of a nonmilitary element of national power but these objectives must still be considered as part of the analysis.

4.5.1.1.2 Identify Critical Factors

The second step in the process is to identify critical factors. Critical factors are defined as the attributes which are considered crucial for the accomplishment of the objective. These factors must be classified as either sufficient (critical strength) or insufficient (critical weakness) and can be characterized as quantifiable (tangible) or unquantifiable (intangible). Critical factors are present at every level of war and require constant attention as they may change in response to adversary or friendly actions. To determine critical factors, the following two questions should be asked for each level of war:

1. What are the attributes, both tangible and intangible, that the enemy has and must use in order to attain his objective?

2. What are the attributes, both tangible and intangible, that the enemy has and must use in order to achieve his objective, but which are weak and may impede the enemy while attempting to attain his objective?

The answer to the first question identifies critical strengths while the answer to the second question identifies critical weaknesses. Given the fact that objectives are linked, there are often linkages among critical strengths and weaknesses as well. For example, a strategic critical weakness may be a tenuous communication link between a strategic leader and his deployed forces. This same critical weakness may result in an operational weakness for the deployed forces.

4.5.1.1.3 Identification of Centers of Gravity

Identification of COGs is the next step of the process. Since COGs are a source of power for the adversary, an initial list of potential COGs is drawn from the analysis of the adversary's critical strengths. Identifying adversary COGs is best done through comparative analysis of these critical strengths in which the analyst should draw on the IPOE analysis done to this point and visualize each potential COG's role/function relative to each of the various systems and subsystems.

Analysts should also adopt a systems approach to this analysis, evaluating the nodes, links, and functions of these critical strengths to determine the specific elements that protect, sustain, integrate, or enable these potential COGs. Some COGs may exist within a single system but more likely, a COG consists of a set of nodes and their respective links within and external to a system. The analyst must determine which critical strength or strengths is most critical to the accomplishment of the adversary's objectives at each level and these define the adversary COGs. Designation of some critical strengths as COGs does not diminish the importance of the other critical strengths but is necessary to focus the planning effort and carry out the final steps of the COG analysis process.

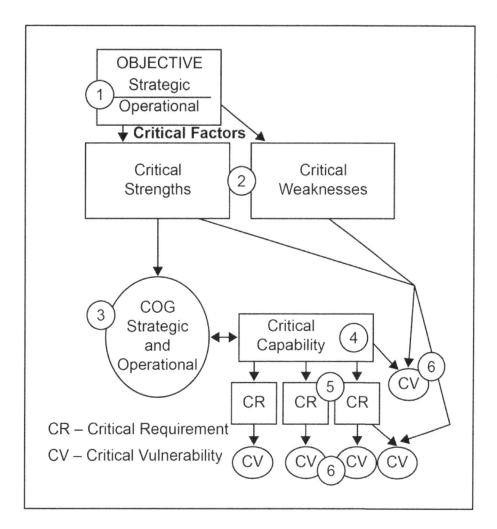

Figure 4-35. Center of Gravity Identification and Analysis Process

4.5.1.1.4 Identify Critical Capabilities

After COGs are identified, the analyst must identify critical capabilities, which JP 1-02 defines as a "means that is considered a crucial enabler for a center of gravity to function as such and is essential to the accomplishment of the specified or assumed objective(s)." Once again, the analysis of a capability as a system of nodes, links, and functions is useful here to determine the means that are enablers of a particular COG. For example, if the COG is a military force, a typical objective at the operational level, then a potential analytic framework is to evaluate the adversary's combat power in terms of standard military functions—C2, intelligence, sustainment, force protection, fires, movement, and maneuver—to determine that force's critical capabilities. Often these capabilities have been previously identified during the IPOE analysis done to this point and were most likely identified as critical strength and weaknesses in step two of this process.

There are two rules with regard to designation of an adversary capability as a critical capability. First, even if the capability is a critical strength, it must be related to a COG to be considered a critical capability. Second, even if a capability is perceived as a critical weakness, if it is an essential enabler for an adversary COG then it must be considered a critical capability, although it is weak in nature. Given the C2 example above, the communications link between the commander and deployed forces, while tenuous, is still considered a critical capability since it is an essential enabler of that commander's need to exert C2 over his forces. Insights regarding critical weaknesses are valuable and are used in a later step of this process.

4.5.1.1.5 Identify Critical Requirements

Once critical capabilities are identified, analysts must then identify critical requirements, which JP 1-02 defines as an "essential condition, resource, and means for a critical capability to be fully operational." Critical requirements may be tangible and intangible and they may also be identified by employing a systems analysis framework at a discreet level. For example, using the C2 scenario above, the critical requirements of the communications system discussed can be broken down to numerous smaller components such as the communication nodes, command posts and centers, antennas, spare parts, bandwidth allocations, frequencies, and satellite transport assets. Intangible factors may include perceptions or morale issues related to adversary C2. While this step provides insights into specific components of the adversary's capabilities that may be targets of friendly forces, it is crucial to continue to the next step of the COG analysis process to ensure the most efficient and effective employment of friendly capabilities.

4.5.1.1.6 Identification of Critical Vulnerabilities

Identification of critical vulnerabilities is the next step in the COG analysis process. JP 1-02 defines a critical vulnerability as an "aspect of a critical requirement which is deficient or vulnerable to direct or indirect attack that will create decisive or significant effects." Identification of critical vulnerabilities is crucial to the efficient application of combat power and other resources as the potential effects achieved through successful engagement of a critical vulnerability are significant relative to the effort expended. Critical vulnerabilities must have a direct relationship to the critical capabilities that enable a particular COG.

As part of the IPOE effort, analysts must also compare the criticality of these vulnerabilities with other factors such as accessibility, redundancy, ability to recuperate, and impact on the civilian population. The analyst first looks for critical vulnerabilities within the critical capabilities and supporting critical requirements but this step must also consider if there are opportunities found in critical strengths that may also provide decisive results disproportionate to the amount of military resources applied. For example, an integrated air defense system (IADS) protecting a COG may be considered a strength but the neutralization of this capability through direct attack may be the most efficient and effective means of opening other COG critical capabilities open to attack.

Figure 4-36 provides an example that depicts the critical capabilities, critical requirements, and critical vulnerabilities associated with the adversary's strategic and operational COGs.

JP 2-01.3 states that a "proper analysis of adversary critical factors must be based on the best available knowledge of how adversaries organize, fight, think, make decisions, and their physical and psychological strengths and weaknesses." The IPOE analysis done to this point is critical to informing the identification and evaluation of COGs and requires analysts to penetrate the adversary's mindset to determine the factors that may influence an adversary to abandon or alter their objectives and to understand how the adversary will interpret and respond to friendly force actions. Analysts must always use caution to avoid falling into the trap of mirror imaging in their analysis.

4.5.5.2 Identification and Analysis of Decisive Points

decisive point. A geographic place, specific key event, critical factor, or function that, when acted upon, allows commanders to gain a marked advantage over an adversary or contribute materially to achieving success. (JP 1-02. Source: JP 5-0)

The identification and analysis of decisive points stems from the COG analysis and every decisive point must have a direct link to a COG.

NWP 2-0

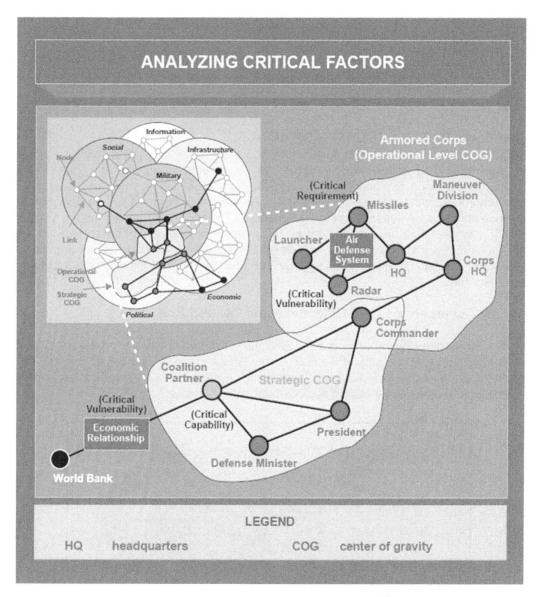

Figure 4-36. Analyzing Critical Factors[19]

A decisive point is by definition important to both the adversary and friendly commander. For example, an APOD/SPOD complex may be critical to the adversary's freedom of action as it provides the infrastructure needed to resupply military forces and provide the civilian populace with essential goods. Given its criticality, the APOD/SPOD is a decisive point for the adversary commander that must be defended and so it presents a focus of effort for the friendly commander, who may have denial or seizure of this capability as an objective.

Decisive points must be constantly reassessed as the dynamic nature of operations is such that these points may change in response to changing conditions in the maritime operational environment. Identification and comparative analysis of potential decisive points allows the IPOE analyst to determine which of these potential decisive points provides the best opportunity for indirect attack on the adversary's COGs, extends friendly operational reach, or enables the application of friendly forces and capabilities.

---

[19] JP 2-01.3.

NWP 2-0

Note

Appendix H provides a worksheet that may be used to present the results of the COG and decisive point identification and analysis processes along with a completed example of this type of analysis applied to the Iraqi IADS during Operation DESERT STORM.

4.6 DETERMINE ADVERSARY COURSES OF ACTION

The first three steps of the IPOE process were used to scope the elements of the maritime operational environment that were relevant to the mission; evaluate the impacts of the operational environment; assess adversary capabilities and limitations; and determine adversary COGs and decisive points. The fourth step of the IPOE process builds upon the previous steps and is focused on determining adversary intent and the COAs the adversary may adopt to meet their objectives. Figure 4-37 provides an overview of the substeps involved with this step. The goal of the process is to determine the full set of COAs available to the adversary and then analyze those COAs to determine, at a minimum, the most likely COA the adversary will adopt and the COA that is most dangerous to friendly forces.

4.6.1 Identify the Adversary's Likely Objectives and Desired End State

Although already addressed during the COG analysis, the analyst should review the adversary's objectives for each level of war and confirm the validity of the assessment. An adversary's likely objectives and desired end state may be determined by analyzing a number of factors such as the adversary's political and military situation, national and operational level capabilities, and sociocultural characteristics.

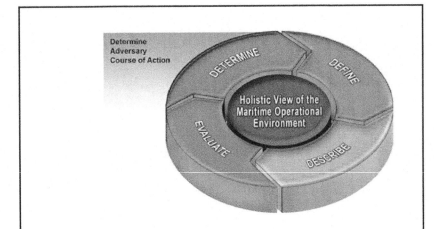

Figure 4-37. Intelligence Preparation of the Maritime Operational Environment—Step Four

MAR 2014
4-54

The analyst should begin determining the adversary's strategic objectives as all operational and tactical objectives flow from the strategic level down. Naval intelligence professionals should identify the likely objectives of all major adversary forces in the AO, the AI, and AOIs. While it is unlikely these objectives will be known with certainty, analysts must make assumptions based on their understanding of the adversary and the maritime operational environment and must identify and coordinate these assumptions with the commander and the staff. Objectives may be defined as a military force that the adversary will decisively engage or defend against, such as a carrier strike group or riverine forces, or as a key geographic feature to be seized or retained, such as a port or bridge. Objectives may have more than one purpose. For example, during the Battle of Midway, Japanese forces were interested in seizing Midway to deny its use to U.S. forces and employ it as a staging base for other operations but the Japanese were equally interested in drawing the bulk of the remaining U.S. Pacific Fleet into a decisive engagement.

4.6.2 Identify the Full Set of Adversary Courses of Action

ACOAs flow from their likely objectives and the conditions of the maritime operational environment. At a minimum this list will include all COAs that the adversary's doctrine or pattern of operations indicates are appropriate to the current situation and accomplishment of likely objectives; all ACOAs that could significantly influence the friendly mission, even if the adversary's doctrine or pattern of operations indicates they are suboptimal under current conditions; and all ACOAs indicated by recent activities or events. While time and resource constraints limit the number of ACOAs that can be effectively evaluated, analysts should consider as many ACOAs as possible given the adversary will likely consider all options available to them. To narrow down the list of ACOAs for evaluation and prioritization, the next stage in the process, analysts should ensure each ACOA meets the criteria from figure 4-38.

Once the consolidated list of ACOAs is developed the analyst should use the evaluation of adversary capabilities developed in step three of the IPOE process to winnow down the list. Any ACOA that the adversary is incapable of executing is removed from the list. Since the adversary will do what they think is possible, which may result in battlefield innovation, analysts should have a high degree of confidence in their assessment before eliminating an ACOA from the list due to perceived lack of adversary capability.

Next, the analyst uses the adversary templates created in IPOE step three and analyzes the remaining ACOAs with regard to the impact of the maritime operational environment developed in IPOE step two. The goal of this step is to further winnow the list of ACOAs by determining how the operational environment impedes or enables the actual implementation of the adversary model. If the operational environment provides significant impediments to particular ACOAs, then these ACOAs should be further analyzed to determine if they should be removed from the list due to their low feasibility.

The remaining broad ACOAs are then further evaluated and developed into specific ACOAs that include details such as timing or phasing of operations and the delineation of adversary main and supporting efforts. Naval intelligence professionals should also consider the factors that may cause an adversary to adopt a "wildcard" ACOA. These factors may include: adversary perception of friendly force capabilities, limitations, vulnerabilities, disposition, and intentions; a lack of understanding of military art and science; a lack of understanding of international law and customs; immature decision-making processes; the relative importance the adversary places on other aspects of the operational environment, such as politics or social issues; and desperation.

> **COURSE OF ACTION EVALUATION CRITERIA**
>
> 1. **Suitability:** Does this COA have the potential to accomplish the adversary's likely objective or end state?
> 2. **Feasibility:** Does the adversary have sufficient time, space, and resources to execute the COA? In answering this question, analysts should explore all the options available to the adversary to address potential gaps. For example, while an adversary fleet may not have the capability to establish sea control in a given area due to lack of assets, there may be sufficient time for the adversary to marshal forces from other fleets to address this shortfall and field a credible force. The analyst must always look for innovative or even radical measures that the adversary may use to accomplish their objectives.
> 3. **Acceptability:** Does the amount of risk associated with the COA exceed the adversary's risk tolerance level? If so, then this would indicate that the COA is not valid. While determining risk tolerance may be as difficult as determining intent, an analyst can derive the adversary's acceptance of risk based on doctrine, observed patterns of behavior, current activity, and psychological or sociocultural factors influencing key leaders. Once again, analysts must avoid mirror imaging in assessing risk tolerance since some adversaries may accept high levels of risk if it is perceived as the only means to accomplishing the objective.
> 4. **Uniqueness:** Does this COA significantly differ from the others being evaluated? If the answer to this question is "no," then the COA could be a variation on an existing COA but should not be considered a distinct COA. Factors that may be considered in evaluating uniqueness include the COA's effect on friendly forces, task organization, use of reserves, location of the main effort, and scheme of maneuver.
> 5. **Consistency with Adversary Doctrine or Patterns of Operation:** Is this COA consistent with the adversary's doctrine, TTP, and observed patterns of operation? While this is an important consideration, analysts must be conscious of the fact that an adversary may deliberately employ forces in new ways to achieve surprise. New capabilities or desperation may also cause the adversary to act outside of their doctrine or operational patterns so analysts must consider the potential for the use of new tactics in light of their understanding of the adversary's capabilities and limitations, their risk tolerance, and the conditions of the maritime operational environment.

Figure 4-38. Course of Action Evaluation Criteria

4.6.3 Evaluate and Prioritize Each Course of Action

Once a full set of potential ACOAs is developed, these ACOAs must be assessed and put in a priority order according to the likelihood of their adoption by the adversary. The purpose of this step is to identify the most likely ACOA that the adversary will adopt and also to identify the ACOA that is most dangerous to friendly forces. Naval intelligence professionals must always be mindful that these are assessments of potential adversary action, not facts. The adversary may swiftly change ACOAs based on their perception of friendly force dispositions and the operational environment. Additionally, adversaries may use deception, providing indicators they have adopted a particular ACOA while actually planning to execute a different ACOA to maximize surprise. ACOA evaluation is a dynamic process so collection and analytic resources should be focused on indicators that will confirm or deny adversary preparations or execution of a particular ACOA and to ascertain adversary deception efforts. Figure 4-39 provides a list of steps analysts should use to prioritize ACOAs.

4.6.4 Develop Each Course of Action in the Amount of Detail Time Allows

Within the amount of time available for this analysis step, each adversary COA is developed in sufficient detail to describe: the type of military operation; the earliest time military action could commence; the location of the action, and the objectives that make up the COA; the OPLAN, to include scheme of maneuver and force dispositions; and the objective or desired end state. Analysts should develop the most likely and most dangerous ACOAs first and then address each of the remaining ACOAs in order of probable adoption by the adversary. A developed ACOA should contain a situation template, an ACOA description, and a list of HVTs.

Adversary Course Of Action Prioritization Process

1. Analyze each ACOA to identify its strengths and weaknesses, COGs, and decisive points.
2. Evaluate how well each ACOA meets the criteria of suitability, feasibility, acceptability, uniqueness, and consistency with doctrine. Avoid cultural bias by considering these criteria in the context of the adversary's culture.
3. Evaluate how well each ACOA takes advantage of the operational environment.
4. Compare each ACOA and determine which one offers the greatest advantages while minimizing risk.
5. Consider the possibility that the adversary may choose the second or third most likely ACOA while attempting a deception operation portraying adoption of the best ACOA.
6. Analyze the adversary's current dispositions and recent activity to determine if there are indications that one ACOA has already been adopted.
7. Guard against being "psychologically conditioned" to accept abnormal levels and types of adversary activity as normal. Identify and focus in greater detail on those adversary preparations not yet completed that are, nevertheless, mission-essential to accomplish a specific ACOA.

Figure 4-39. Adversary Course of Action Prioritization Process

4.6.4.1 Situation Template

Situation templates depict assumed adversary dispositions, based on that adversary's preferred method of operations and the impact of the operational environment if the adversary should adopt a particular course of action. Like adversary models, these templates can be from a geospatial or a systems perspective and the use of either or both types depends on the commander's needs and the mission. In addition to situation templates, situation matrices may also be developed to show phasing of operations over time.

These templates depict a point in time. For relatively simple operations, a single situation template that shows the adversary's disposition and scheme of maneuver to reach the most critical point in their operations may be sufficient. For more complex operations, a series of situation templates depicting where and how an adversary may adopt branches and sequels to the main COA may be required. A systems perspective situation template is constructed by comparing a consolidated systems overlay with a modified association matrix that shows anticipated changes in system networks for different COAs. Situation templates are built to facilitate wargaming by the commander and staff. Figure 4-40 provides a graphical depiction of the technique used to build a situation template and these techniques are discussed as follows:

4.6.4.1.1 Geospatial Perspective

Construction of a situation template from a geospatial perspective begins with selecting the adversary template or model that most closely matches the COA being analyzed. The adversary template should be overlaid on or compared with the MCOO or other products that have been developed to depict the impact of the operational environment on adversary forces. Using the adversary's doctrine, observed operations, and TTP as guides, adjust the forces found on the adversary template to account for the impact of the maritime operational environment on the adversary's operations. Analysts should ensure that no major adversary assets are lost or duplicated when conducting this step and that the situation template depicts the locations and anticipated activities of all adversary HVTs.

Using this view of adversary dispositions and the impact of the operational environment, analysts should then wargame the scheme of maneuver the adversary will employ to attain their objectives (tactical, operational, or strategic). The analysis should demonstrate how each of the adversary's force components supports the scheme of maneuver. Once the scheme of maneuver is determined, the situation template should be updated with time phase lines that show the likely course of the adversary advance in terms of space and time factors. This analysis should be based on the understanding of adversary capabilities, doctrine, and patterns of operation developed in IPOE step three. Time phase lines should be modified, as needed, based on the anticipated effects of friendly force actions and the current situation on adversary force maneuver capabilities. Figure 4-41 provides an example of a situation template.

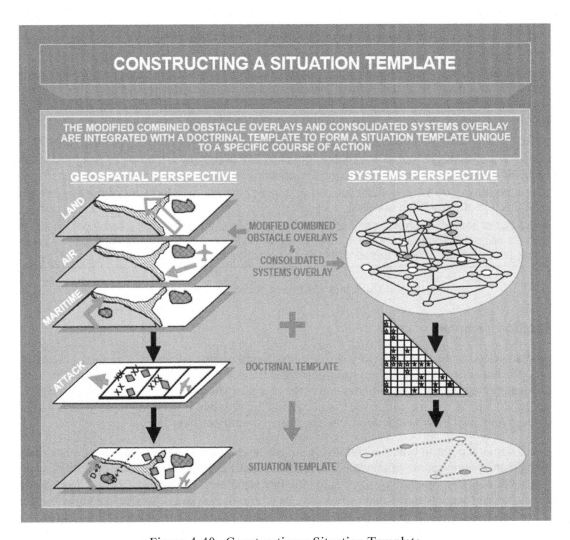

Figure 4-40. Constructing a Situation Template

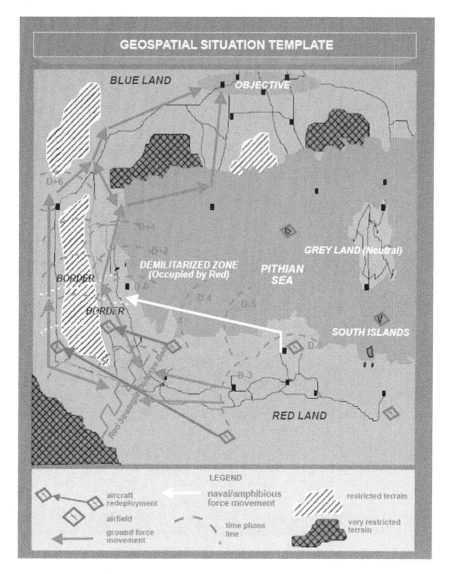

Figure 4-41. Geospatial Situation Template

4.6.4.1.2 Systems Perspective

To construct a situation template from a systems perspective, the analyst should select the adversary template (modified association matrix) that best suits the COA being evaluated. The analyst should then focus on a relevant subset of the consolidated system network diagram and plot any anticipated node-link changes that are expected to occur as the COA unfolds. These anticipated node-link changes should be shown using dotted lines (or other locally established convention) and color-coded for each specific COA. Figure 4-42 provides an example of a situation template from a systems perspective, which essentially shows an expected future state and the node-link changes that may occur for each specific COA. JP 2-01.3 contains an excellent historical case study on the 1992–1993 U.S. operations in Somalia that goes into greater detail concerning how to construct this type of template.

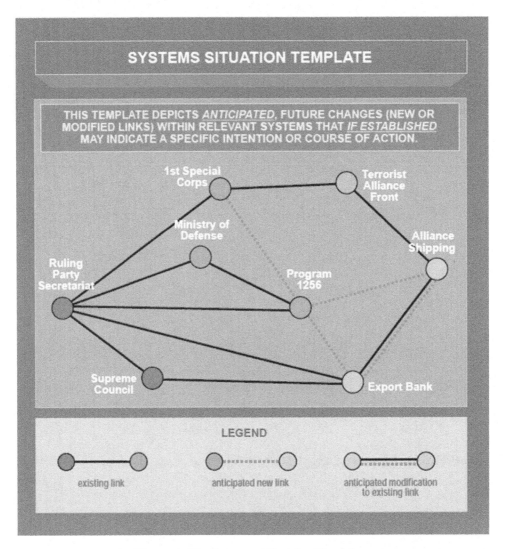

Figure 4-42. Systems Situation Template

4.6.4.1.3 Situation Matrix

Depending on the relative complexity of the situation, some ACOAs may be best represented in a matrix format vice a geospatial or systems template. A situation matrix, such as shown in figure 4-43, may be very useful in depicting how each of the adversary's supporting forces may phase their operations to support the main effort.

4.6.4.2 Course of Action Description

The COA description is typically a narrative report that accompanies each COA analyzed. It details the activities of the adversary forces shown in the situation template or matrix. At a minimum it should include information on the earliest time the COA could be executed; the location of the main effort and supporting operations; and the time and phasing of operations within the maritime operational environment. Critical decisions that must be made by the adversary commander as the COA unfolds should be described in terms of their location in space and time (decisive points) and should discuss the relevant factors that will be considered in the adversary's decision-making process.

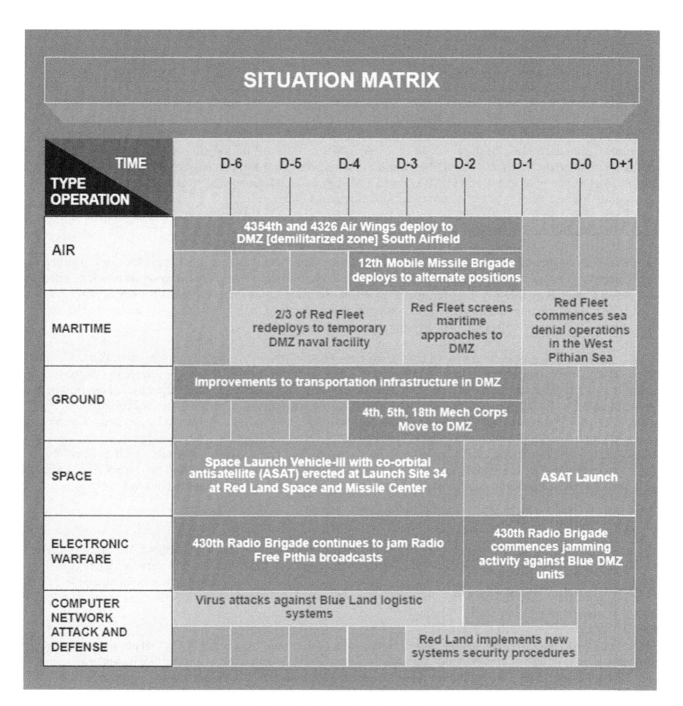

Figure 4-43. Situation Matrix

4.6.4.3 High-Value Targets

Analysts should refine and reevaluate the decisive points identified during the COG analysis step of the IPOE process and the HVTs recorded on the adversary or doctrinal templates associated with each COA. Depending on the specific situation and the duration of the COA's execution, HVTs may change in relative worth. Wargaming each COA aids in determining the potential disposition and employment of HVTs. Analysts can use this technique to derive when and where HVTs are likely to be of most value to the successful execution of the adversary's operations within each phase of the adversary's COA. This list of times and locations when particular HVTs will be of most value to the adversary should be passed to the targeting cell and the associated areas should be designated as target areas of interest (TAIs). These TAIs can be annotated on the situation template or maintained in a separate list and overlay as the commander's requirements dictate.

4.6.5 Identify Initial Collection Requirements

The COA analysis process results in an estimative intelligence product that attempts to predict how the adversary will balance the factors of time, space, and force to achieve their objectives. Given the analysis predicts where and when adversary activity is likely to occur, the analyst can use the assessment as the basis for developing an initial set of collection requirements that are focused on confirming which COA the adversary has adopted. The areas in which these adversary activities are expected to occur are designated as NAIs.

JP 2-01.3 defines NAIs as the "geospatial area or systems node or link against which information that will satisfy a specific information requirement can be collected. Named areas of interest are usually selected to capture indications of adversary courses of action, but also may be related to conditions of the operational environment." For example, if an adversary naval attack requires the adversary to sail a force through a chokepoint to engage friendly forces, the chokepoint would be designated an NAI and collection assets would be focused on observing activity there to discern if, when, and in what strength the adversary is moving their naval force. For an FHA or DSCA scenario, NAIs may be used to assess conditions within the operational environment such as flood levels or to monitor whether distributed aid is being provided to intended recipients. NAIs and their associated indicators are typically represented using event templates and event matrices, which are discussed below.

4.6.5.1 Event Template

Figures 4-44 and 4-45 provide an overview of the process for constructing an event template and an example of a completed event template. To produce an event template, the analyst must first take the various situation templates developed for each ACOA still under consideration. Each of these ACOAs must be analyzed to determine potential NAIs that could produce indicators of adversary intentions if collection resources were applied to observing them.

Each COA is then compared against the others with a goal of determining a set of NAIs that are unique to each COA. Adversary activity that is the same across all COAs should be eliminated from further consideration as an NAI since observing this activity tells the analyst nothing about which COA the adversary has chosen to adopt.

Once a unique set of NAIs is compiled, this consolidated list of NAIs is then depicted on the event template and each NAI is annotated with its COA association. An NAI can be a specific point, route, area, or network node or link and can match obvious geographic features or arbitrary features such as timed phase lines or engagement areas. It should be large enough to encompass the geospatial activity or network link that serves as the indicator of the adversary's COA. Event templates may also be created from a systems perspective and JP 2-01.3, Appendix B contains a case study from U.S. operations in Somalia that provides an example of a systems event template.

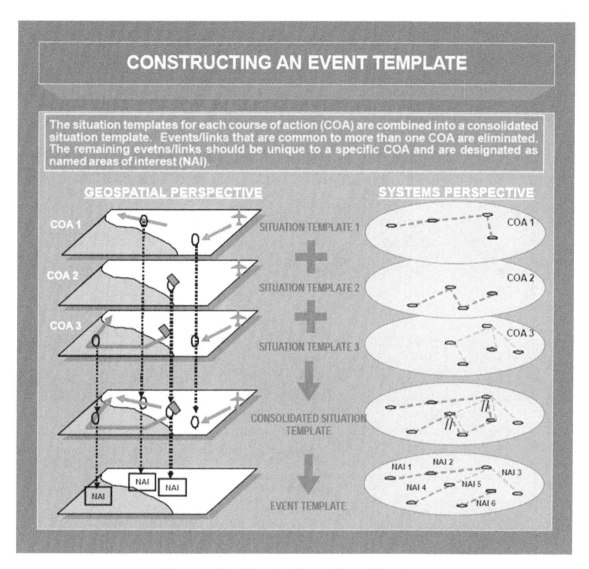

Figure 4-44. Constructing an Event Template

NWP 2-0

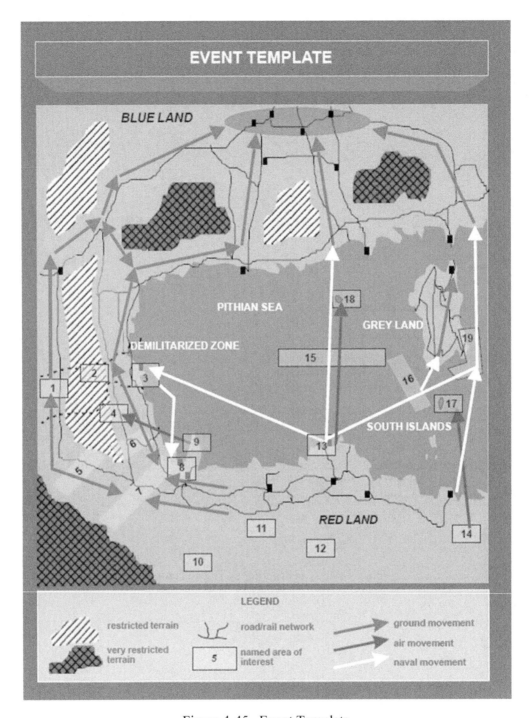

Figure 4-45. Event Template

4.6.5.2 Event Matrix

The event matrix augments the event template by providing details concerning the type of adversary activities and the expected changes in the conditions of the maritime operational environment expected at each NAI. It also provides an estimated timeframe during which this activity is expected to occur and links the NAI to the specific COA with which it is associated. Figures 4-46 and 4-47 provide information on how to construct an event matrix and give an example of a completed event matrix. While the event matrix is primarily used to facilitate intelligence collection planning, it may also have applicability in situation development, wargaming, and determination of intelligence gain and loss considerations.

4.7 APPLICATION

To summarize, the IPOE process provides a structured approach to applying the intelligence process and the command's intelligence resources to support the planning and execution of naval operations in the maritime operational environment. Appendix E provides sample worksheets that help the naval intelligence professional focus on the outputs of various phases of the IPOE effort. Understanding the IPOE process and the outputs of that process provides the underlying context for intelligence support to planning, execution, and assessment of naval operations.

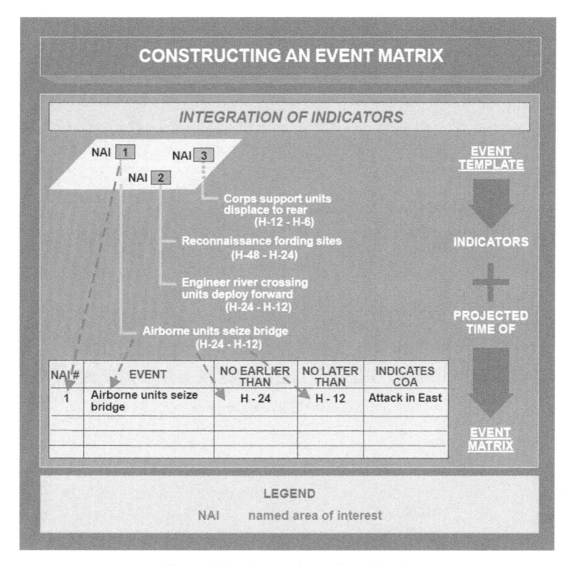

Figure 4-46. Constructing an Event Matrix

EVENT MATRIX

NAI	EVENT	TIME Earliest	TIME Latest	INDICATE COA
1	Laying of minefields and construction of obstacles in depth in the demilitarized zone (DMZ).	D-10	D-0	Defend
2	Improvements to transportation infrastructure in northern DMZ.	D-10	D-1	Attack (in West)
3	Presence of additional surface combatants and minelayers at DMZ port.	D-7	D-3	Reinforce
4	Deployment of additional combat aircraft at forward airfield.	D-7	D-1	Reinforce
5	Northward movement of red corps size force.	D-3	D-1	Attack (in West)
6	Northward movement of two more corps.	D-3	D-1	Attack (in West)
7	Occupation of red strategic defense belt by second echelon units.	D-2	D-1	Retrograde
8	Presence of red surface combatants and minelayers.	D-3	D-1	Retrograde
9	Deployment of additional combat aircraft near red strategic defense belt.	D-7	D-1	Retrograde
10	Departure of mobile missile units from garrison and local dispersal areas.	D-3	D-1	Attack
11	Concentration of additional aircraft at rear airfield.	D-2	D-1	Defend
12	Presence of intermediate-range ballistic missiles on or near launch pad.	D-3	D-1	Attack
13	Departure of surface combatants and amphibious support ships from port.	D-6	D-4	Attack
14	Concentration of additional combat aircraft.	D-2	D-1	Defend
15	Northward transit of amphibious task force.	D-6	D-3	Attack (in Center)
16	Eastward transit of amphibious task force.	D-6	D-4	Attack (in East)
17	Deployment of additional combat aircraft.	D-3	D-1	Attack (in East)
18	Deployment of additional combat aircraft.	D-3	D-1	Attack (in Center)
19	Northward transit of amphibious task force.	D-5	D-3	Attack (in East)

Figure 4-47. Event Matrix

NWP 2-0

CHAPTER 5

Naval Intelligence Support to Planning, Executing, and Assessing Operations

> In any problem where an opposing force exists and cannot be regulated, one must foresee and provide for alternative courses. Adaptability is the law which governs survival in war as in life. . . . To be practical, any plan must take account of the adversary's power to frustrate it; the best chance of overcoming such obstruction is to have a plan that can be easily varied to fit the circumstances met . . .
>
> *Sir Basil H. Liddel-Hart*

5.1 INTRODUCTION

This chapter describes naval intelligence support to the planning, conduct, and assessment of operations at the operational and tactical levels of war. It provides a general overview of where intelligence fits in planning, conducting, and assessing operations; introduces naval intelligence professionals to the Navy planning process; discusses intelligence support for the various phases of war and some of the unique considerations for intelligence support to naval missions; and examines the role intelligence plays in assessing the results of operations. For more specific information on how naval intelligence supports operational and tactical level planning, execution, and assessment, readers should consult NWP 2-01, Intelligence Support to Naval Operations; JP 2-0, Joint Intelligence; NWP 5-01, Navy Planning; and NTTP 3-32.1, Maritime Operations Center.

5.2 JOINT CONTEXT

All naval operations take place within a joint context. At the operational level of war, numbered fleet commanders are dual-designated as the NCC responsible to their respective combatant commanders for integrating Navy capabilities into joint operations. This may involve planning and directing joint operations as head of a JTF or as the JFMCC subordinate to the JTF commander.

5.3 NAVAL INTELLIGENCE SUPPORT TO PLANNING

5.3.1 The Navy Planning Process Overview

Figure 5-1 provides an overview of the Navy planning process (NPP) described by NWP 5-01. The NPP is a six-step process that enables commanders to effectively plan for and execute operations, ensure that the employment of forces is linked to objectives, and integrate naval operations seamlessly with the actions of a joint force. The NPP is fully aligned to the joint planning process and compatible with the planning processes used by other Services. It provides a framework for assessing the operational environment and distilling a mass of planning information in a way that enables commander decisionmaking. While the NPP is a deliberate planning process, experience, training, and use of the process increases proficiency in applying the process and the employment of templates and anticipatory actions can significantly enhance efficiency of its execution. The NPP is a flexible and adaptable process. In a time-compressed environment, such as a crisis situation, the NPP can be significantly streamlined to support operational requirements. For example, amphibious units often employ the Marine Corps rapid response planning process (R2P2) in planning their operations. The goal of the R2P2 process is to enable execution of certain operational tasks within 6 hours of orders receipt. In any case, while the time available to plan may be variable, the process itself does not change.

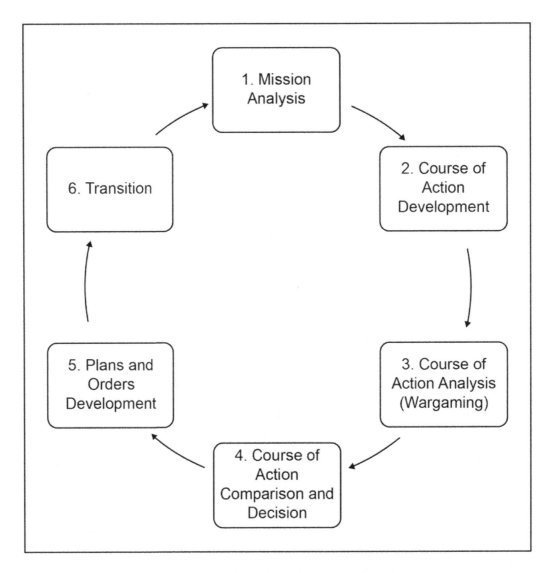

Figure 5-1. The Navy Planning Process

The NPP provides the procedures needed to analyze a mission, develop friendly COAs and wargame these COAs against ACOAs, compare friendly COA's against the commander's criteria and one another, select a COA, prepare an operation order (OPORD) for execution, and transition the plan to the subordinate units that will execute it. The output of the process is a military decision that is translated into a directive, such as an OPLAN or an OPORD. For intelligence, the results of planning are captured in the annex B of the OPLAN, contingency plan (CONPLAN), or OPORD. The format for an annex B is provided as appendix G.

5.3.2 Mission Analysis

Mission analysis drives the entire planning process and its purpose is to develop an overall assessment of the situation for use by the commander and staff. The process begins with a review of orders, relevant plans, intelligence, and higher headquarters guidance. The goal of this review is to develop an understanding of the possible AO, probable mission, forces available, and relevant aspects of the operational environment such as terrain, political, military, cultural, and other factors. Sources of input for this review include information, particularly IPOE products, from higher headquarters, national-level and other Service intelligence organizations, other government agencies, and nongovernmental organizations. This review is needed to develop an operations mission statement and to identify the specific tasks needed to accomplish the mission. The step ends when the commander issues the planning guidance and a warning order (WARNORD), which are needed to initiate COA development.

A thorough mission analysis focuses the activities of the commander, staff, and planning team and ensures the efficiency and effectiveness of other steps in the NPP. Planners should have access to any relevant on-the-shelf OPLANs or CONPLANs, standing rules of engagement, SOPs, and IPOE products. All staff elements develop preliminary staff estimates as early as possible during the mission analysis step and refine these estimates throughout the process.

Mission analysis should provide answers to the following questions:

1. What tasks must the command do for the mission to be accomplished?

2. What is the purpose of the mission received?

3. What limitations have been placed on our own forces' actions?

4. What forces/assets are available to support the operation?

5. What additional assets are needed?

6. What are the gaps of knowledge that inhibit planning?

7. What risks exist?

Higher headquarters intelligence products form the basis of the naval commander's own intelligence support. Higher headquarters intelligence products include JIPOE materials, which are derived from the intelligence estimate, intelligence summaries, and annex B of an OPLAN or OPORD. At a minimum, higher headquarters JIPOE should provide identification and analysis of the adversary's objectives, critical strengths and critical weaknesses, COGs, critical capabilities, critical requirements, and critical vulnerabilities. It also should estimate the ACOAs that are most likely to be encountered based on the current situation. JIPOE products from higher headquarters and those from the commander's intelligence staff may include the MCOO and threat situation templates. A format for producing an intelligence estimate is found in appendix I.

The planning team presents a mission analysis briefing to the commander and staff to obtain approval of the mission statement, intent, and follow-on planning guidance. The mission analysis briefing reviews the specific products that were developed and refined during mission analysis before proceeding to COA development. For the naval intelligence team, the focus is on presenting an initial intelligence estimate brief that includes IPOE products such as terrain analysis, METOC analysis, threat integration with situation templates, adversary's COGs, and ACOAs. The intelligence brief should also include any RFIs, collection and production requirements requiring the commander's approval and advocacy, and recommended PIRs.

The outputs of this step of the NPP are an approved mission statement, the commander's intent, the commander's planning guidance, and WARNORDs.

5.3.2.2 Guidance and the Commander's Intent

A commander's intent is broader than the mission statement; it is a concise, free-form expression of the purpose of the force's activities, the desired results, and how actions will progress toward that end. It is a clear and succinct vision of how to conduct the action. In short, the commander's intent links the mission and the concept of operations (CONOPS). The intent expresses the broader purpose of the action that looks beyond the why of the immediate operation to the broader context of that mission, and it may include how the posture of the force at the end state of the action will transition to or facilitate further operations (sequels). Understanding the commander's intent is critical to the intelligence process as it provides the focus needed to synchronize intelligence operations in support of the commander's requirements.

Commander's intent is not a summary of the CONOPS. It does not tell specifically how the operation is being conducted but is crafted to allow subordinate commanders sufficient flexibility and freedom to act in accomplishing their assigned mission(s) even in the "fog of war." While there is no specified joint format for commander's intent, a generally accepted construct includes the purpose, method, and end state.

1. Purpose. The reason for the military action with respect to the mission of the next higher echelon. The purpose explains why the military action is being conducted. This helps the force pursue the mission without further orders, even when actions do not unfold as planned. Thus, if an unanticipated situation arises, participating commanders understand the purpose of the forthcoming action well enough to act decisively and within the bounds of the higher commander's intent.

2. Method. The "how," in doctrinally concise terminology, explains the offensive form of maneuver, the alternative defense, or other action to be used by the force as a whole. Details as to specific subordinate missions are not discussed.

3. End State. Describes what the commander wants to see in military terms after the completion of the mission by the friendly forces.

The commander is responsible for formulating the single unifying concept for a mission. Having developed that concept, the commander then prepares an intent statement from the mission analysis, the intent of the higher commander, and his own vision to ensure that subordinate commanders are focused on a common goal. When possible, the commander delivers it, along with the order (or plan), personally (or via VTC). Face-to-face delivery ensures mutual understanding of what the issuing commander wants by allowing immediate clarification of specific points. While intent is more enduring than the CONOPS, the commander can, and will, revise intent when circumstances dictate. The following offers an example of a NCC's intent.

Example: Navy Component Commander's Intent

GENTEXT/EXECUTION//

(U) PURPOSE: NEUTRALIZATION OF THE REDLAND MARITIME CAPABILITY IN ORDER TO SUPPORT OPERATIONS AGAINST THE 23RD GUARDS DIVISION AND THE ELIMINATION OF THE TERRORIST FORCES AND INFRASTRUCTURE IN REDLAND.

(U) METHOD: OUR OPERATION MUST REMAIN FOCUSED ON FOUR KEY REQUIREMENTS. FIRST, WE MUST ASSIST IN SETTING THE CONDITIONS FOR THE JTF'S INTRODUCTION OF FORCES INTO REDLAND—THEY CANNOT BE HAMPERED BY ANY CHALLENGES FROM THE SEA. SECOND, THE EXPEDITIONARY STRIKE GROUP (ESG) MUST BE READY TO IMMEDIATELY EMPLOY THE AMPHIBIOUS READY GROUP (ARG)/MARINE EXPEDITIONARY UNIT (MEU) INTO EITHER OF THE BLOCKING POSITIONS AS SOON AS THE JFC DIRECTS ITS EXECUTION—WE CANNOT LOSE TIME FOR REPOSITIONING. THIRD, OUR DECEPTION MUST REMAIN CREDIBLE UNTIL THE AIRBORNE BRIGADE IS SECURE IN ITS LODGMENT IF WE ARE TO DRAW PRESSURE OFF OF THE FORCIBLE ENTRY UNITS. FOURTH, AND ABOVE ALL OTHERS, REMEMBER THAT THE TERRORIST ELEMENTS AND THEIR INFRASTRUCTURE IN REDLAND ARE THE PRIMARY OBJECTIVES—REMAIN FLEXIBLE TO EXPLOIT OPPORTUNITIES THAT MIGHT PRESENT THEMSELVES TO ALLOW US TO RENDER A DECISIVE BLOW.

(U) TASK FORCE OPERATIONS MUST RECOGNIZE THE TERRITORIAL WATERS AND AIRSPACE OF NEIGHBORING NEUTRAL COUNTRIES, PREVENT DAMAGE TO NEUTRAL COMMERCIAL SHIPPING, AND TAKE ALL NECESSARY STEPS TO MINIMIZE DAMAGE TO INFRASTRUCTURE WITHIN REDLAND.

(U) THE END STATE FOR OUR OPERATIONS WILL BE THE ESTABLISHMENT OF MARITIME SUPERIORITY AND A NEUTRALIZED REDLAND NAVAL FORCE THAT CAN RECONSTITUTE AND PROVIDE MARITIME SECURITY ONCE A NEW REDLAND REGIME, FREE OF TERRORISTS, IS IN PLACE.

The commander's intent focuses on further planning, and enables the commander to indirectly control events during the execution of the operation. It is crafted so that commanders two levels down have the flexibility to accomplish their mission in lieu of further guidance. For an NCC or JFMCC commander, this means that a level down to a task force (e.g., combined task force (CTF)) is considered when writing the commander's intent. For a CTF, the commander considers down to the unit level.

5.3.3 Course of Action Development

A COA is any concept of operation open to a commander that, if adopted, would result in the accomplishment of the mission. For each COA, the commander visualizes the employment of his forces and assets as a whole normally two levels down—taking into account externally imposed limitations, the factual situation in the AO, and the conclusions previously reached during mission analysis.

After receiving guidance, the planning team develops COAs for analysis and comparison. The commander usually involves the entire planning team in COA development. The commander's guidance and intent focus the planning team's creativity to produce a comprehensive, flexible plan within the time constraints. When possible, the commander's direct participation helps the staff gain quick, accurate answers to questions that arise during the process. Course of action development is a deliberate attempt to design unpredictable (difficult for the adversary to deduce) alternatives. A good COA positions the force for future operations and provides flexibility to meet unforeseen events during execution; it also provides the maximum latitude for initiative by subordinates.

Intelligence plays a key role in the COA development step by providing IPOE products that enable the commander and planning staff to analyze relative combat power by making a rough estimate of force ratios between friendly and adversary forces. Relative combat power analysis takes into account tangible (maneuver, firepower, etc.) and intangible factors (readiness, morale, leadership, etc.) and uses historical minimum-planning ratios for various combat missions that include careful consideration of relevant time, space, and force considerations. By developing ACOAs, the naval intelligence team also provides planners with a listing of ACOAs in probable order of adoption by the adversary, which aids in prioritizing planning actions, particularly during crisis action planning.

A significant output of the COA development step is the COA sketch and statement. Current joint military planning consists of six steps for a campaign or operation; the basic phases are: phase 0 (shape), phase I (deter), phase II (seize initiative), phase III (dominate), phase IV (stabilize), phase V (enable civil authority). The intelligence support considerations for each of these phases are discussed in section 5.4.1. At a combatant command echelon or a JTF, the COA sketch and statement incorporate the actions by each service or functional component in each phase of the campaign or operation. The JFMCC or NCC may be involved in each of the phases as well. Together, the statement and sketch cover the "who" (generic task organization), "what" (tasks and purposes), "when," "where," "how," and "why" (purpose of the operation) for each subordinate unit/component command, any significant risks, and where they occur for the force as a whole.

Figure 5-2 provides an example of a COA sketch and statement while figure 5-3 provides an example of a COA narrative for a JFMCC. The naval intelligence team must coordinate closely with other planners to ensure that collection operations and other intelligence operations are tightly synchronized with other operations across all phases of the friendly COA. Naval intelligence professionals must understand friendly force COAs so that they can develop corresponding collection, processing, exploitation, analysis, production, dissemination, integration, and evaluation and feedback plans that support the operational scheme of maneuver and the successful completion of the mission.

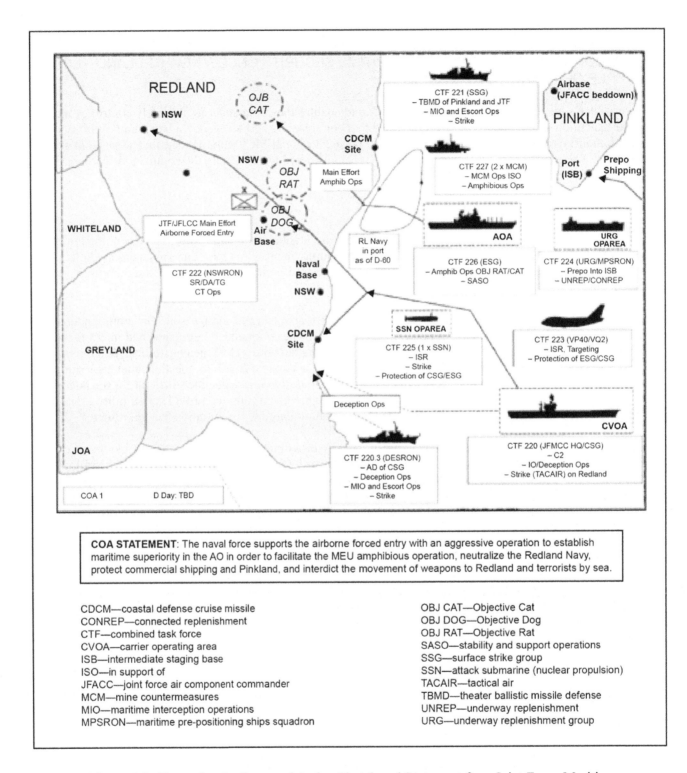

Figure 5-2. Example of a Course of Action Sketch and Statement for a Joint Force Maritime Component Commander or Navy Component Commander[1]

[1] NWP 5-01.

> **Phase 1 (deter):** Supported Commander: JFMCC. Main effort: Movement of pre-positioned shipping to intermediate staging base (ISB). Phase 1 begins with the flow of pre-positioned shipping to Pinkland ISB. During Phase 1, JFMCC continues to project power by conducting surface, air, subsurface, and amphibious operations off the coast of Redland and establishes MDA and local sea control in the AO. JFMCC HQ integrates, accepts, and establishes C2 with coalition maritime forces. IO and ISR conducted in conjunction with (ICW) JTF IO themes and collection plan. JFMCC be prepared to move to Phase 2 and conduct amphibious operations and strikes into Redland. Phase 1 ends with the completion of JFMCC achieving local sea control and MDA in the AO, pre-positioned shipping received at ISB, and coalition C2 structure completed.
>
> **Phase 2 (seize the initiative):** Supported Commander: JFMCC. Main effort: Maritime superiority. Phase 2 begins with the insertion of naval special warfare (NSW) into Redland to provide special reconnaissance and task group for strike operations. JFMCC maintains MDA and local sea control in the AO and continues to flow pre-positioned shipping into ISB. JFMCC beach party team (BPT) conduct amphibious operations and strikes into Redland in order to support joint force land component commander (JFLCC) airborne forced entry. Mine countermeasures (MCM) forces conduct mine clearance and establish Q-routes for amphibious operation. JFMCC BPT destroy Redland naval forces including ships, submarines, aircraft, and coastal defenses and BPT provide theater ballistic missile defense (TBMD) against Redland missiles launch against Pinkland or JTF. JFMCC conducts deception operation in south Redland to support MEU and JFLCC ops. On order, JFMCC surface forces conduct maritime intercept of designated suspicious vessels and provide escort of Pinkland shipping. IO and ISR conducted ICW with JTF IO themes and collection plan. Phase 2 ends with pre-positioned shipping at ISB, NSW inserted into Redland, Q-routes established, and JFMCC forces prepared to transition to Phase 3.
>
> **Phase 3 (dominance):** Supported Commander: JFLCC. Main effort: Airborne assault into Redland airfield. Phase 3 begins with JFMCC offensive operations to destroy Redland Navy and support joint force air component commander in strikes on Redland land targets to enable JFLCC airborne forced entry. MEU and NSW operational control (OPCON) to JFLCC and joint forces special operations component command(er), respectively. JFMCC air defense be prepared to provide TBMD, and surface forces be prepared to conduct maritime intercept and provide escort of Pinkland shipping. Naval mobile construction battalion and maritime special operations elements deployed to ISB. IO and ISR conducted ICW with JTF IO themes and collection plan. Phase 3 ends with Redland Navy defeated at sea and prevented from offensive operations and MEU and NSW OPCON shift complete.

Figure 5-3. Example of a Course of Action Narrative for a Joint Force Maritime Component Commander or Navy Component Commander[2]

Another key output of this step is the COA briefing. For this briefing, the naval intelligence team provides an updated intelligence estimate which should include refined IPOE products. Additionally the METOC impacts on adversary forces must be addressed and situation templates that provide refined and updated ACOAs are presented.

5.3.4 Course of Action Analysis (Wargaming)

The heart of the NPP is the analysis of opposing courses of action. In the previous steps of the planning process, ACOAs and COAs were examined relative to their basic concepts—ACOAs were developed based on adversary capabilities, objectives, and the estimate of the adversary's intent, and COAs were developed based on friendly mission and capabilities. In this step, the planning team conducts an analysis of the probable affect that each ACOA has on the chances of success of each COA. The aim is to develop a sound basis for determining the feasibility and acceptability of the COA. Analysis also provides the planning team with a greatly improved understanding of its COAs and the relationship between them.

---

[2] NWP 5-01.

Course of action analysis identifies which COA best accomplishes the mission while positioning the force for future operations. It helps the commander and staff to:

1. Determine how to maximize the effects of combat power while protecting the friendly forces and minimizing collateral damage.

2. Further develop a visualization of the operation, including posturing forces for follow-on operations.

3. Anticipate operational events.

4. Determine conditions and resources required for success.

5. Determine when and where to apply the force capability.

6. Focus IPOE on the adversary's strengths and weaknesses, important civil considerations, the adversary's desired end state, and other adversary information requirements.

7. Identify coordination needed to produce synchronized results.

8. Determine the COA that has the greatest chance of success against each adversary COA.

Course of action analysis is conducted using wargaming. The war game is a disciplined process, with rules and steps that attempt to visualize the flow of the operation. Specifically, wargaming:

1. Considers the results from the friendly COG determination, friendly COAs, adversary COG determination, ACOAs, as well as the characteristics of the physical environment.

2. Relies heavily on joint doctrinal foundation, tactical judgment, and operational experience.

3. Focuses the planning team's attention on each phase of the operation in a logical sequence.

4. Highlights critical tasks and provides familiarity with operational possibilities otherwise difficult to achieve.

5. Is an iterative process of action, reaction, and counteraction. Wargaming stimulates ideas and provides insights that might not otherwise be discovered.

At a minimum, each retained COA should be wargamed against both the most likely and the most dangerous ACOA. While the focus is on the analysis portion of the wargaming process, this stage also allows the various staff components to refine their estimates. Refined estimates help the planning team in determining feasibility and acceptability.

5.3.4.1 Naval Intelligence Team Role in Wargaming

The naval intelligence team has a dual role during the war game. First, naval intelligence professionals role-play as the adversary commander and develop critical adversary decision points. They project adversary reactions to friendly actions, and determine adversary losses. Naval intelligence professionals are also responsible for capturing the results of the adversary actions and counteractions in a war game worksheet. If the naval command is large enough, there may be a Red Cell that can assume the task of role-playing the adversary. Second, the naval intelligence team provides input such as RFIs, NAIs, TAIs, HVTs and high-payoff targets (HPTs) to the planning team and refines situation templates during this step.

Normally the senior naval intelligence professional or a selected Red Cell speaks for the adversary and responds to friendly actions. The senior naval intelligence professional or selected Red Cell uses an adversary synchronization matrix and event template to describe the adversary's activities. The event template is updated as new intelligence is received and as a result of the war game. These products depict the locations of NAIs and when to collect information that confirms or denies the adoption of a particular COA by the adversary; they also serve as a guide for collection planning. The adversary's actions are described by warfighting function. The adversary's CONOPS and ISR are presented, including the intelligence collection assets of the adversary and how and when the adversary might employ them. Also, the senior naval intelligence professional or selected Red Cell describes how the adversary would organize in the operational environment; identifies the location, composition, and expected strength of the adversary reserve; and determines the anticipated decision points and criteria that the adversary commander might use in committing reserves. Other adversary decision points that might be identified include likely times, conditions, and areas for the adversary use of WMD and friendly CBRNE defense requirements; when the adversary could begin a withdrawal; and where and when the adversary might use unconventional forces, etc. Based on the experience level of the senior naval intelligence professional or selected Red Team, insights on the likely effectiveness of friendly actions may be offered. The commander wants to know what decisions the adversary commander will have to make and when those decisions will be made, for example, are they event driven? When a deception plan is being wargamed, the senior naval intelligence professional or selected Red Team outlines target biases and predispositions, discusses how and when the adversary would receive the desired misleading indicators, and reviews adversary actions that indicate that the deception has been successful.

IPOE is a continuous, parallel process that takes place throughout all stages of planning. The planning team updates, refines, and prepares IPOE products to reflect the results of COA wargaming, facilitating the next step in the planning process, COA comparison and decision. By this point in the planning process, the following IPOE products and analysis must be well developed and detailed:

1. Operational environment

2. Analysis of the adversary

    a. Objectives

    b. Critical factors, decision points (DPs), and COGs

    c. High-value targets and HPTs, to include time-sensitive targets (TSTs). (Targets may and will change during the course of operations. Examples provided are for a notional adversary attempt to conduct aerial denial in an international strait. See figure 5-4.)

3. Adversary COAs: most likely and most dangerous.

Additionally, PIRs have been identified, and a collection plan that focuses on NAIs has been built and refined. Development of an indications and warning matrix, showing key indicators of adversary actions, may be helpful in providing scope and direction to the collection plan. MCOOs, doctrinal templates, and current adversary situation templates are updated.

5.3.4.2 Role and Responsibility of the Red Cell

Ideally, the Red Cell consists of individuals of varied operational backgrounds and specialties. Combining their own operational experience with adversary tactics, weapons, and doctrine, the Red Cell provides adversary reactions to the friendly COAs during the COA war game. To be successful, the Red Cell must function as an extension of the naval intelligence team. The primary purpose of the Red Cell is to provide additional operational analysis of the adversary, tailored to the needs of the planning team. During the war game, the Red Cell employs ACOAs against the friendly COAs. Although the Red Cell is used principally at the JFMCC and NCC level and above, it can also be scaled for use by smaller units such as CSG, destroyer squadron, or air wing.

High-Payoff Targets	High-Value Targets	Time-Sensitive Targets
• amphibious ships (for adversary amphibious assault) • adversary petroleum, oils, and lubricants storage • adversary commercial shipping bringing in logistics and military hardware • adversary IADS C2	• adversary mine-laying assets • adversary submarines • adversary coastal defense cruise missiles along a SLOC	• mobile surface-to-air missile • location of key adversary leader

Figure 5-4. Examples of High-Payoff Targets, High-Value Targets, and Time-Sensitive Targets

The objective of the Red Cell is not to defeat friendly COAs during the war game, but to assist the development and testing of friendly COAs. The Red Cell makes friendly COAs stronger and more viable for execution in battle. The naval intelligence team, in coordination with operations personnel, determines the composition of the Red Cell and often provides a number of its analysts. The senior naval intelligence professional oversees the functioning of the Red Cell, as its analysis of the adversary must be coordinated with the naval intelligence team. The naval intelligence team provides the Red Cell with the initial detailed information on adversary location, weapons, tactics, doctrine, order of battle, and assessed COAs. Differences in analysis between the Red Cell and

the senior naval intelligence professional must be identified and resolved. To be effective, the planning team and the Red Cell must exchange information and analysis continuously throughout the planning process. As the Red Cell conducts its own analysis, it should inform the senior naval intelligence professional of its findings regarding the adversary. For example, the Red Cell, through its own research and analysis, may determine that the adversary will employ electronic countermeasures and maritime patrol aircraft in a unique way. The Red Cell may also determine that a completely new COA is feasible for the adversary and more likely to be executed than the COA initially provided by the senior naval intelligence professional.

5.3.5 Course of Action Comparison and Decision

During the comparison step of the NPP, the planning team considers each retained COA for advantages and disadvantages. Each COA is evaluated in terms of the naval commander's previously established governing factors and, if modified, is tested a final time for feasibility and acceptability. This step is repeated until the naval commander selects the COA that offers the greatest prospect of accomplishing the mission.

The actual comparison of the COAs is critical. A number of varying techniques exists for conducting the comparison, but each of them must assist the commander in reaching a sound decision. While many of the comparison techniques offer a numerical value, staffs must remember that these are simply decision aids. The greatest utility of these comparison techniques is not which COA has the highest score; rather, it is the insight into the strengths and weaknesses of each COA relative to a given governing factor.

The naval intelligence team participates in the COA comparison and contributes to the recommendation to the commander concerning COA adoption. As part of the decision brief for COA adoption, the naval intelligence team must update IPOE products and refine their estimates of ACOAs, provide results of terrain and weather analysis, and discuss current adversary situation and intentions.

Once the commander has decided on a COA, the naval intelligence team takes the synchronization matrix developed in the wargaming step and refines their input to that product and the COA sketch. Using the synchronization matrix, the planning team expands and integrates the available information and provides the CONOPS—an elaboration of the selected COA. It should include the commander's vision of how major events are expected to occur in the forthcoming combat action and the commander's intent. The CONOPS must be developed quickly so that subordinate commanders have the time necessary to prepare their own plans and units for the impending action. Having already identified the risks associated with the selected COA, the naval commander refines what level of risk is acceptable to accomplish the mission and approves measures to reduce the risks. If time permits, there is a discussion of acceptable risks with lateral and senior commanders. However, the higher commander's approval must be obtained prior to accepting any risk that might imperil the higher commander's intent.

The CONOPS describes how arrayed forces will accomplish the commander's intent. It is the central expression of the commander's operational design and governs the development of supporting plans or annexes. Planners develop a scheme of maneuver by refining the initial array of forces and using graphic control measures to coordinate the operation and to show the relationship of friendly forces to one another, the adversary, and geography. During this step, units are converted from generic to specific units, such as the specific CSGs, ESGs, and ISR assets that will support and conduct the operation. The CONOPS includes:

1. The purpose of the operation

2. A statement of where the commander will accept risk

3. Identification of critical friendly events and phases of the operation (if phased)

4. Designation of the decisive operation, along with its task and purpose

5. Designation of shaping operations, along with their tasks and purposes, linked to how they support decisive operations

6. Naval warfare functions (i.e., undersea warfare, surface warfare, air defense, strike, etc.)

7. Intelligence, surveillance, and reconnaissance and protection operations

8. An outline of the movements of the force

9. Identification of options that may develop during an operation

10. Location of engagement areas (surface, air, and subsurface) and objectives

11. Responsibilities for AO and operating areas

12. Concept of fires (i.e., employment of Tomahawk land-attack missile, carrier-based tactical aviation).

Naval intelligence professionals must ensure they understand the CONOPS and that intelligence operations are synchronized and integrated with the plan.

5.3.6 Plans and Orders Development

The plans and orders development step in the NPP communicates the commander's intent, guidance, and decisions in a clear, useful form that is easily understood by those executing the order. Operation plans are normally produced at the combatant command or JTF level with subordinate Service or functional component commands (such as NCCs) producing supporting plans. In the case of a JFMCC or NCC, this would be the maritime supporting plan. Before proceeding, it is necessary to distinguish between plans and orders.

1. Plan. A plan is prepared in anticipation of operations and normally serves as the basis for an order. The procedures for producing a plan should therefore closely mirror the preparation of an order.

2. Order. An order is a written or oral communication that directs actions and focuses a subordinate's tasks and activities toward accomplishing the mission.

3. Various portions of the plan or order, such as the mission statement and staff estimates, have been prepared during previous steps of the NPP. The chief of staff or executive officer, as appropriate, directs plans or orders development. Plans or orders contain only critical or new information, not routine matters normally found in standing operating procedures. A good plan or order is judged on its usefulness, not its weight.

In the previous phase of the NPP, the planning team integrated the commander's selected COA with the staff estimates and planning support tools (developed in parallel) into a fully developed CONOPS. The planning team now translates the CONOPS into a clear, concise, and authoritative directive. This directive, whether it is a maritime supporting plan or an OPORD, is then backbriefed to the higher commander and crosswalked to other Service and functional components to ensure that it is synchronized, understood, and meets the higher commander's intent. A well-written directive possesses important characteristics that help assure understanding of the directive and the accomplishment of the mission:

1. Clarity. Each executing commander should be able to understand the directive thoroughly. Write in simple, understandable English and use proper military (doctrinal) terminology.

2. Brevity. A good directive is concise. Avoid superfluous words and unnecessary details, but do not sacrifice clarity and completeness in the interest of brevity alone. State all major tasks of subordinates precisely but in a manner that allows each subordinate sufficient latitude to exercise initiative. Short sentences are more easily and quickly understood than longer ones.

3. Authoritativeness. In the interest of simplicity and clarity, the affirmative form of expression should be used throughout all combat orders and plans.

4. Simplicity. This requires that all elements are reduced to their simplest forms. All possibilities for misunderstanding must be eliminated.

5. Flexibility. A good plan leaves room for adjustments that unexpected operating conditions might cause. Normally, the best plan provides the commander with the most flexibility.

6. Timeliness. Plans and orders must be disseminated in enough time to allow adequate planning and preparation on the part of subordinate commands. Through the use of WARNORDs, subordinate units can commence their preparation before the receipt of the final plan or order. Concurrent planning saves time.

7. Completeness. The plan or order must contain all the information necessary to coordinate and execute the forthcoming action. It also must provide control measures that are complete, understandable, and that maximize the subordinate commander's initiative. Only those details or methods of execution necessary to ensure that actions of the subordinate units concerned are synchronized with the CONOPS for the force as a whole should be prescribed.

5.3.6.1 Develop Base Paragraphs for Operation Plans and Orders

Plans and orders are produced in a variety of forms such as WARNORDs, OPORDs, and fragmentary orders (FRAGORDs). Plans and orders can be detailed written documents with many supporting annexes, or orders may be simple verbal commands. Their form depends on the time available, complexity of the operation, and levels of command involved. Supporting portions of the plan or order, such as annexes and appendixes, are based on staff estimates, subordinate commander's estimates of supportability, and other planning documents.

Directives most frequently use the standard five-paragraph format, briefly described below. In complex operations, much of the information required in the order is contained or amplified in the appropriate appendixes and annexes, such as synchronization and decision support matrices and logistics and sustainability analyses. However, the essential form of the commander's CONOPS, including the commander's intent, command and control, task organization, and essential tasks and objectives, should be contained in the body of the order. The format for OPLANs and OPORDs is contained in CJCSM 3122.03A JOPES Vol. II. The five basic paragraphs for all plans and orders are:

1. Paragraph 1. Situation. This paragraph, the commander's summary of the general situation, ensures that subordinates understand the background for planned operations. It often contains subparagraphs describing adversary forces, friendly forces, and task organization, as well as higher headquarters guidance. The naval intelligence team provides the input for the section of this paragraph that deals with adversary forces and the operational environment and uses a refined intelligence estimate and updated IPOE products to inform this section.

2. Paragraph 2. Mission. The commander inserts his own restated mission developed during mission analysis. This is derived from the mission analysis step and contains those tasks deemed essential to accomplish the mission.

3. Paragraph 3. Execution. This paragraph expresses the commander's intent for the operation, enabling subordinate commanders to better exercise initiative while keeping their actions aligned with the operation's overall purpose. It also specifies the objectives, tasks, and assignments for subordinate commanders. It should articulate not only the objective or task to be accomplished but also its purpose, so that subordinate commanders understand how their tasks and objectives contribute to the overall CONOPS.

4. Paragraph 4. Administration and Logistics. This paragraph describes the concepts of support, logistics, personnel, public affairs, civil affairs, and medical services. The paragraph also addresses the levels of supply (appendix J) as they apply to the operation.

5. Paragraph 5. Command and Control. This paragraph specifies command relationships, succession of command, and the overall plan for communications and control.

5.3.6.2 Prepare Supporting Annexes and Appendixes

To keep the directive as simple and understandable as possible, details and amplifying information are placed in annexes. Typical annexes and their supporting appendixes and tabs include detailed plans for intelligence and operations (such as search and rescue/combat search and rescue, movement, undersea warfare, etc.), instructions necessary for C2 (such as the communications plan), and information too complex to be covered completely in the basic plan (such as the detailed CONOPS, the logistics plan, and the complete task organization). Annexes may be referenced in the appropriate part of the body and should not include matters covered in SOPs. Annexes also allow for the selective distribution of certain information. The intelligence annex is annex B. A template for producing an annex B is found in appendix J.

5.3.7 Transition

The purpose of transition is to ensure a successful shift from planning to execution. A good transition enhances the situational awareness of those who will execute the order, maintains the intent of the CONOPS, promotes unity of effort, and generates tempo. Transition facilitates the synchronization of plans between higher and subordinate commands and aids in integrated planning by ensuring the synchronization of the warfighting functions. Transition requires free flow of information between commanders and staffs by all available means. At higher echelons, where the planners may not be involved in plan execution, the commander may designate a representative as proponent for the plan or order. Normally this is the current operations representative to the planning team. After orders development, the proponent takes the approved plan or order forward to the staff charged with supervising execution. As a full participant in the development of the plan, the proponent can answer questions, aid in the use of the planning support tools, and assist the staff in determining necessary adjustments to the plan or order. Transition occurs at all levels of command. A formal transition normally occurs on staffs with separate planning and execution teams. Planning time and personnel may be limited at lower levels of command, or planners may be the same personnel as the executors.

During transition, the naval intelligence team continues to update and refine IPOE products and coordinates with subordinate commands and supporting organizations to ensure orders and requirements are clearly understood.

5.3.8 The Navy Planning Process in a Time-Constrained Environment

There are often times when forces will operate in a time-constrained or crisis environment and will need to adapt the NPP to meet their planning requirements. The focus of any planning process should be to quickly develop a flexible, tactically sound, and fully integrated and synchronized plan that increases the likelihood of mission success with the fewest casualties possible. However, any operation may outrun the initial plan. The most detailed estimates cannot anticipate every possible branch or sequel, every adversary action, unexpected opportunities, or changes in mission directed from higher headquarters. Fleeting opportunities or unexpected adversary actions may require a quick decision to implement a new or modified plan.

Before a planning staff can conduct decisionmaking in a time-constrained environment, it must master the steps in the full NPP. An organization can only shorten the process if it fully understands the role of each and every step of the process and the requirements to produce the necessary products. Training on these steps must be thorough and result in a series of staff battle drills that can be tailored to the time available. Training on the NPP must be stressful and replicate realistic conditions and time lines.

Although the task is difficult, all staffs must be able to produce a simple, flexible, tactically sound plan in a time-constrained environment. Events may make it difficult to follow the entire NPP. An inflexible process used in all situations will not work. The NPP is a sound and proven process that must be modified with slightly different techniques to be effective when time is limited. There is still only one process, however, and omitting steps of the NPP is not the solution. Anticipation, organization, and prior preparation are the keys to success in a time-constrained environment. Throughout the remainder of this chapter, reference to a process that is abbreviated is for simplicity only. It does not mean a separate process, but the same process shortened. The commander or chief of staff decides how to shorten the process. What follows are suggested techniques and procedures that may save time. They are not exhaustive or the only ways to save time, but they have proved useful to planning staffs in the past. These techniques are not necessarily sequential in nature, nor are all of them useful in all situations. What works for a staff depends on its training and the given situation. The commander or chief of staff can use these or other techniques to abbreviate the process.

The process is abbreviated any time there is too little time for its thorough and comprehensive application. The most significant factor to consider is time. It is the only nonrenewable, and often the most critical, resource. There are four primary techniques to save time. The first is to increase the commander's involvement, allowing decisions to be made during the process without waiting for detailed briefings after each step. The second technique is for the commander to become more directive in the guidance, limiting options. This saves the staff time by focusing members on those things the commander feels are most important. The third technique, and the one that saves the most time, is for the commander to limit the number of COAs developed and wargamed. In extreme cases, the commander can direct that only one COA be developed. The goal is an acceptable COA that meets mission requirements in the time available, even if it is not optimal. The fourth technique is to maximize parallel planning. Although parallel planning is the norm, maximizing its use in a time-constrained environment is critical. In a time-constrained environment, the importance of WARNORDs increases as available time decreases. A verbal WARNORD now is worth more than a written order one hour from now. The same WARNORDs used in the full NPP should be issued when the process is abbreviated. In addition to WARNORDs, planning staffs must share all available information with subordinates, especially IPOE products, as early as possible.

Decisionmaking in a time-constrained environment almost always takes place after a unit has entered into the AO and has begun to execute operations. This means that the IPOE and some portion of the staff estimates should already exist. Detailed planning before operations provides the basis for information the commander and staff need to make knowledgeable decisions as operations continue. Staff members must keep their estimates up-to-date so that when planning time is limited, they can provide accurate, up-to-date assessments quickly and move directly into COA development. When time is short, the commander and staff use as much of the previously analyzed information and products from earlier decisions as possible. Although some of these products may change significantly, many, such as the IPOE that is continuously updated, remain the same or require little change. The staff must use every opportunity to conduct parallel planning with the unit's higher headquarters. Parallel planning can save significant time, but if not carefully managed, it can also waste time. As a general rule, the staff must never get ahead of the higher headquarters in the planning process. The majority of time spent conducting parallel planning should be spent developing the foundation of the plan, such as mission analysis. The staff should not develop and analyze COAs without specific guidance and approval from higher headquarters.

The IPOE process requires constant attention. Many delays during mission analysis can be traced to the IPOE. The naval intelligence team must quickly update the IPOE based on the new mission and changed situation. This is critical to allow needed reconnaissance assets to deploy early to collect information to adjust the initial plan. Adversary event templates must be as complete as possible prior to the mission analysis briefing. Because they are the basis for wargaming, they must be constantly updated as new information becomes available. Naval intelligence professionals must be prepared to provide oral intelligence estimates and synopses of IPOE products in support of the accelerated planning effort.

5.3.9 Naval Intelligence Support to the Navy Planning Process

Figure 5-5 provides an overview of the support that naval intelligence provides to the NPP.

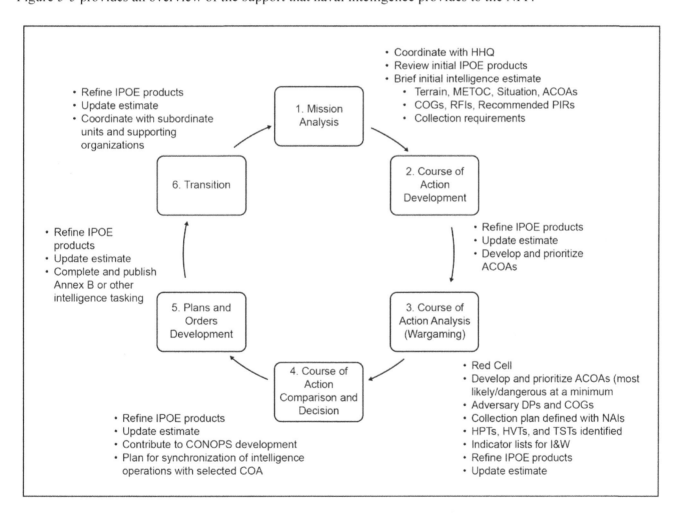

Figure 5-5. Naval Intelligence Support to the Navy Planning Process Overview

In summary, naval intelligence conducts the following activities during each step of the NPP:

1. Mission Analysis. The naval intelligence team coordinated with higher headquarters (HHQ) to obtain all relevant JIPOE/IPOE products and begins to review, tailor, and refine these products. An initial intelligence estimate is developed for the mission analysis brief and should cover terrain analysis, METOC, the situation, high-level ACOAs, and adversary COGs. RFIs and collection requirements should also be discussed as well as recommended PIRs.

2. Course of Action Development. IPOE products are refined and the intelligence estimate is updated. The naval intelligence team further develops and prioritizes ACOAs and provides them to other staff members to support friendly force COA development.

3. Course of Action Analysis (Wargaming). Red Cell plays the part of the adversary during wargaming analysis of friendly COAs. ACOAs are fully developed and prioritized with adversary most likely and most dangerous COAs developed at a minimum. Adversary DPs, COGs, HPTs, HVTs, and TSTs are identified. The collection plan is defined along with associated NAIs. Indicator lists are developed to support I&W analysis. IPOE products are refined and the intelligence estimate is updated.

4. Course of Action Comparison and Decision. The naval intelligence team contributes to the development of the CONOPS and plans for the synchronization of intelligence operations with the selected COA. IPOE products and the intelligence estimate are updated.

5. Plans and Orders Development. The naval intelligence team completes and publishes the annex B (or similar directive) as part of an OPLAN, CONPLAN, OPORD, WARNORD, or FRAGORD. IPOE products and the intelligence estimate are updated.

6. Transition. The naval intelligence team coordinates with subordinate commands and supporting organizations to ensure intelligence guidance and requirements are understood. IPOE products and the intelligence estimate are updated.

5.4 NAVAL INTELLIGENCE SUPPORT TO EXECUTION

Intelligence support is critical to all aspects of execution at all levels of war. For example, IPOE products regarding the status of foreign transportation infrastructure are vital to the successful deployment and redeployment of forces and intelligence assessments of air, sea, land, and cyber threats are crucial to ensuring force protection during sustainment operations. During execution, intelligence must not only support the current phase of the operation but must also look ahead and begin laying the groundwork to support subsequent phases and other future operations. The discussion below introduces the concept of operational phases of war and discusses the various intelligence support considerations from each phase. Given the concept of phases is more closely associated with the operational level of war, an overview of intelligence support considerations for tactical-level naval missions is also discussed in this section.

5.4.1 Operation Plan Phasing

Figure 5-6 gives an overview of the phases of an operation plan described in JP 3-0, Joint Operations and, JP 5-0, Joint Operation Planning, and the specific, high-level actions that are associated with each phase.

1. Shape (Phase 0). A prehostilities phase that focuses on building partner relationships, preventing conflict, and preparing for potential hostilities or crisis response operations. The goal of the shaping phase is to deter adversaries and assure friends.

2. Deter (Phase I). This phase focuses on deterring an adversary from conducting undesirable activities based on their perception of friendly capabilities and the will to use them. Once the crisis is defined, these actions may include mobilization, tailoring of forces, deployment into a theater; employment of ISR assets; and

development of mission-tailored C2, intelligence, force protection, and logistic requirements to support the JFC's CONOPS.

3. Seize the Initiative (Phase (II). The commander decisively uses friendly capabilities to gain access to theater infrastructure and otherwise delay, impede, or halt the adversary and deny them from accomplishing their objectives.

4. Dominate (Phase III). This phase focuses on breaking the adversary's will to fight, combat operations, and controlling the operational environment for noncombat situations.

5. Stabilize (Phase IV). During this phase, focus shifts from combat operations to stability operations and supporting the transition to a safe and secure environment where essential government services and humanitarian aid are provided.

6. Enable Civil Authority (Phase V). During this phase, the force provides support to legitimate civil governance in an effort to help the civil authority regain its ability to govern and administer to the services and other needs of its population.

Figure 5-7 provides another view of these phases, showing the approximate level of military effort that may be applied during each phase. The level of military effort, however, is not an accurate representation of the intelligence effort, which is significant in the early phases and remains high throughout all phases of an operation. Regardless of the phase, intelligence plays a major role in enabling operations by providing the commander, staff, and supporting organizations with the information on the adversary and the operational environment that is needed to adjust plans and execute orders.

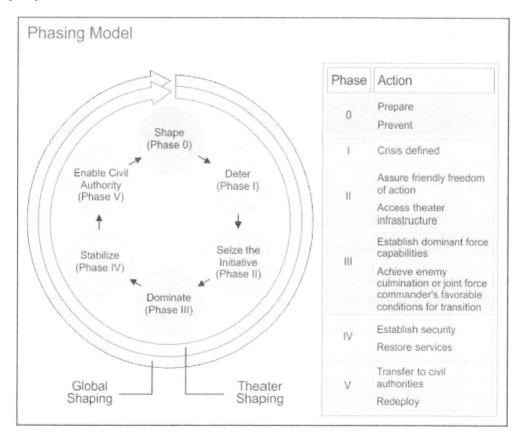

Figure 5-6. Operation Plan Phasing Model

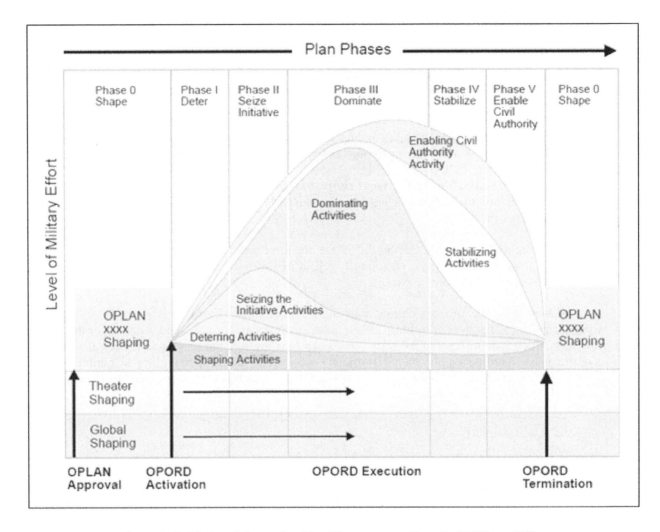

Figure 5-7. Notional Operation Plan Phases versus Level of Military Effort

5.4.2 Intelligence Support to Operation Plan Phases

JP 2-0 provides specific considerations for each phase that are discussed in the following sections.

5.4.2.1 Intelligence Support During the Shaping Phase

Intelligence activities conducted within the context of deliberate planning during the shaping phase lay the groundwork and develop the basics for intelligence operations in all subsequent operational phases of the campaign. Intelligence activities should also be conducted to support phase 0 operations, including those supporting theater campaign plans. In addition to developing IPOE products, intelligence activities during this phase may include the establishment of intelligence sharing and liaison activities with coalition and partner nations. Information operations play a significant role during the shaping phase so intelligence must support these activities throughout analysis and assessment of leadership structures and decision-making capabilities and processes. Given the lead time to implement and measure effectiveness of some phase 0 and later phase operations, intelligence must coordinate with operations personnel to determine potential targets early in this phase to support planning and execution efforts. Finally, intelligence support in understanding human sociocultural factors is critical to the planning, conduct, and assessment of civil-military operations, FHA or DSCA missions, and theater security cooperation efforts.

5.4.2.2 Intelligence Support During the Deterrence Phase

Before the initiation of hostilities, the commander must gain a clear understanding of the national and military strategic objectives; desired and undesired effects; actions likely to create those effects; COGs and decisive points; and required joint, multinational, and nonmilitary capabilities matched to available forces. The intelligence team assists the commander in visualizing and integrating relevant considerations regarding the operational environment into a plan that will lead to achievement of the objectives and accomplishment of the mission. It is therefore imperative that the IPOE effort (initiated during the shaping phase) provide the commander with an understanding of the operational environment at the outset of the deterrence phase. Since understanding of adversary perceptions of friendly actions and statements is crucial to assessing the effectiveness of deterrence operations, the decision-making processes and capabilities associated with adversary leadership are a major focus of this phase. IPOE efforts accelerate during this phase and focus not only on supporting I&W and assessing current operations but also assessing the adversary to support future phases. The naval intelligence team must concentrate on confirming adversary COGs and refine their understanding of adversary capabilities, dispositions, intentions, and probable COAs. Additionally, certain intelligence operations, such as the deployment of an ISR asset to demonstrate U.S. resolve and gain additional insights on a developing situation, may be employed as a flexible deterrent option as part of an overall effort to forestall hostilities without resorting to combat operations.

5.4.2.3 Intelligence Support During the Seize the Initiative Phase

During this phase, the decisive use of all elements of combat power are employed to seize and maintain the initiative while denying the adversary the ability to achieve their own objectives and engendering in the adversary a sense of defeat and resignation. Naval intelligence provides support to this phase through a number of capabilities. For example, target intelligence cells could potentially become more active and transition to focusing more on planning, supporting, and assessing the results of lethal operations. Intelligence support to information operations and counterintelligence support for force protection continue to be very important during this phase and must rapidly adapt and respond to changing conditions in the operational environment. The use of all intelligence disciplines and their associated collection assets and resources is also critical during this phase to provide I&W on potential threats and to support the planning and conduct of current and future operations.

5.4.2.4 Intelligence Support During the Dominate Phase

During the dominate phase, the commander engages in sustained combat operations through simultaneous employment of conventional, SOF, and information-related capabilities within the AO. These operations may be linear, with combat power directed toward the adversary in concert with adjacent units, or nonlinear, where forces are focused on objectives without geographic reference to adjacent forces. Intelligence must be equally prepared to support linear and nonlinear operations. Nonlinear operations are particularly challenging due to their emphasis on simultaneous operations along multiple lines of operations. The complexity of nonlinear operations places a premium on a continuous flow of accurate and timely intelligence to help protect individual forces. This flow of intelligence supports precise targeting, mobility advantages and freedom of action and is enabled by persistent surveillance, dynamic ISR management, and the intelligence portion of the COP. Intelligence must also anticipate requirements for the subsequent stabilization phase and set the groundwork for stability, security, transition, and reconstruction operations by providing detailed intelligence on key infrastructure, adversary governmental organizations and personnel, and anticipated humanitarian needs.

5.4.2.5 Intelligence Support During the Stabilize Phase

Stabilization may include some combat operations but as the threat diminishes and civil infrastructures are reestablished the focus shifts to returning areas back to civil authorities. Naval intelligence operations should begin to shift from support to combat operations to a focus on threats, such as insurgent groups, criminal elements, terrorist cells, or other groups, that might seek to delay, disrupt, or deny friendly forces and prevent the accomplishment of stabilization phase objectives. Counterintelligence support is vital to success during the stabilization phase to ensure force protection of naval forces engaged in stability operations. Information operations are also a significant aspect during stabilization operations and intelligence is used extensively to support the planning and assessment of information operations.

5.4.2.6 Intelligence Support During the Enable Civil Authority Phase

This phase is characterized by the establishment of a legitimate civil authority that is enabled to manage the situation without further outside military assistance. In many cases, the U.S. will transfer responsibility for the political and military affairs of the host nation to another authority. The operation is normally terminated when the stated military strategic and/or operational objectives end states have been met and redeployment of the force is accomplished. It is possible that some intelligence support may remain in place after termination of military operations to provide support to the civil authorities and to continue monitoring of the situation. During this phase, lessons learned should also be captured and recorded in the NLLS.

5.4.3 Intelligence Support to Naval Tactical Operations

The sections below address general intelligence support considerations for naval warfare missions and provide an example of some of the specific intelligence support considerations for a specific warfare area.

5.4.3.1 General Intelligence Support Considerations for Naval Warfare Missions

While a discussion of the specific intelligence support requirements for each type of naval mission are beyond the scope of this publication, naval intelligence professionals must recognize that each naval mission area is unique and each requires different approaches for the planning, conduct, and assessment of operations within each of those mission areas. NWP 2-01, Intelligence Support to Naval Operations, provides a high-level overview of the naval missions discussed throughout this publication. Naval intelligence professionals should not only educate themselves on the unique culture and capabilities associated with each of these mission areas but when assigned to a command that conducts a specific type or set of naval missions they should rapidly determine how to apply and adapt their intelligence experience to meeting the specific needs of that naval warfare area. Sources of information for conducting research on these naval warfare areas include the various naval doctrinal publications (NWPs, NTTPs, and tactical memorandums) located in the Navy Doctrine Library System Nonsecure Internet Protocol Router Network (NIPRNET)—https://ndls.nwdc.navy.mil or SIPRNET—http://ndls.nwdc.navy.smil). Additionally, certain commands have developed specialized expertise in the unique intelligence support requirements for particular naval warfare areas. For example, the NSAWC has significant expertise in the areas of strike and air warfare intelligence support while the Naval Mine and Anti-Submarine Warfare Command (NMAWC) has significant expertise in mine and antisubmarine warfare-related intelligence. Various intelligence centers and agencies also have a great depth and breadth of experience either directly or indirectly supporting analytic issues associated with naval warfare areas. For example, ONI's Surface Analysis Branch for Evaluation and Reporting, Strike Protection Evaluation and Antiair Research, and Submarine Warfare Operations Research Division divisions provide tailored operational threat assessments in the areas of surface, air, and submarine warfare and ONI has significant capabilities for merchant ship tracking and analysis as well as S&TI. Figure 5-8 discusses specific mission areas and associated commands that have specialized intelligence expertise in supporting those mission areas.

5.4.3.2 Intelligence Support Considerations for Naval Strike and Air Warfare Example

As the center of excellence for naval strike and air warfare missions, NSAWC has significant expertise in the planning, conduct, and assessment of strike and air warfare operations. These missions have very specific requirements with regard to IPOE and analytic product development, coordination with intelligence community organizations, and intelligence sources that provide the most value in supporting the mission.

With regard to IPOE products to support strike and air warfare requirements, naval intelligence analysts focus on collection requirements (to support activities from prestrike planning to poststrike BDA), availability of organic and nonorganic assets, METOC analysis, cryptologic analysis, collateral damage concerns, weaponeering, threat analysis, and all-source fusion analysis.

Recurring intelligence support products developed to support strike and air warfare missions include CONOPS matrices with OB data, prestrike and poststrike briefings, mission reports (MISREPs) and intelligence reports, joint desired point of impact graphics, targeting folders, and kneeboard cards for aircrews.

Mission Area	Associated Commands
Strike Warfare	NSAWC
Air Warfare	NSAWC
Air and Missile Defense	Navy Air and Missile Defense Command
Expeditionary	Navy Expeditionary Combat Command (NECC)
MIO	NECC
Irregular Warfare	NECC
Mine Warfare	NMAWC
Antisubmarine	NMAWC
Information Warfare	FCC
Surface Warfare	Destroyer Squadrons (DESRONs)
Antipiracy	DESRONs
Counterdrug	4th Fleet
Amphibious Warfare	ESG-2/3
FHA and DSCA	ESG-2/3
Navy Special Warfare	Naval Special Warfare Command

Figure 5-8. Naval Mission Areas and Associated Commands

Intelligence organizations that strike and air warfare intelligence analysts routinely interact with include the NGA, Defense Logistics Agency (DLA), DIA, Defense Threat Reduction Agency, combatant commands, and various joint commands. NGA and DLA are the two agencies that NSAWC most often interacts with to develop targeting intelligence and support for the strike warfare mission. Additionally, NSAWC routinely contacts cryptologic experts for additional analytic support. These agencies include but are not limited to all NIOCs, NSA, DIA, and, on occasion, the FBI.

Naval strike and air warfare intelligence analysts make use of theater in-chopper messages, air tasking orders (ATOs), special instructions, and Navy-wide/theater-specific OPTASKs to ensure intelligence support is conducted in accordance with overarching guidance. These analysts also rely on cryptologic support to analyze SIGINT and provide products at higher classification levels to support mission requirements in targeting, OB analysis, and battlefield activity.

The intelligence disciplines that are most valuable to supporting the strike and air warfare mission are GEOINT, SIGINT, HUMINT, and OSINT. GEOINT and IMINT typically provide the most value in developing target folders for strike missions with SIGINT and HUMINT augmenting the primary sources of intelligence for confirmation and clarification of target type. With GEOINT and IMINT, map data is used with the common geopositioning services (CGS) system and Falcon View mapping application to create geographically referenced funnel-down graphics, charts, images, and overlays. These elements of the target folder serve as the primary targeting products used in mission planning and briefs to the strike lead, strike warfare commander, and strike group commander. IMINT, as a precursor and requirement to GEOINT in support of strike warfare, is equally significant and plays a vital role in strike warfare missions. In accordance with the NGA Precision Point Mensuration Program and Concept of Operations (CONOPS); Training, Qualification, Certification, and Proficiency, accreditation is required for CSGs to support and conduct joint federated targeting. This is accomplished by Strike Warfare Intelligence Analysts (3923 NEC) with the CGS system by downloading imagery from the JCA.

5.5 NAVAL INTELLIGENCE SUPPORT TO ASSESSMENT

measure of performance (MOP). A criterion used to assess friendly actions that is tied to measuring task accomplishment. (JP 1-02. Source: JP 3-0)

measure of effectiveness (MOE). A criterion used to assess changes in system behavior, capability, or operational environment that is tied to measuring the attainment of an end state, achievement of an objective, or creation of an effect. (JP 1-02. Source: JP 3-0)

Continuous and timely assessments are vital to measure the progress of the naval force in accomplishing its mission. Commanders must continuously assess the operational environment and the progress of their campaigns or operations to ensure that events are unfolding in accordance with their original vision and intent. During the planning phase, assessment actions and measures are determined by the commander and staff. Assessment is conducted during the planning phase to ensure that plans are developed in accordance with good operational design practices. During execution, assessment activities are focused on measuring progress toward accomplishing mission-specific tasks and the achievement of objectives. Assessments enable the commander to adjust their operations and align resources properly to support future operations. At the strategic and operational levels, assessment efforts concentrate on broader objectives and progress toward accomplishing the end state while tactical-level assessment is focused on measuring task accomplishment. Assessment is applicable to combat and noncombat operations and often assessment for noncombat and nonlethal operations can be more complex than traditional combat assessment.

Typically, the senior naval intelligence professional works with operations and plans to coordinate assessment activities, which are led by the N-3 or Operations Department/Directorate. Intelligence supports the assessment process through tailored, flexible, and adaptive collection and analytic activities designed to measure the effectiveness of operations. The naval intelligence team monitors and assesses adversary capabilities, vulnerabilities, and intentions as well as the operational environment as part of the overall assessment effort. This includes assisting the commander with determining what aspects of the operational environment to measure and how to conduct these measurements to determine whether progress has been made in accomplishing tasks and meeting objectives.

5.5.1 Intelligence Support to the Assessment Process

The assessment process uses measures of performance (MOPs) to evaluate task performance at all levels of war, and measures of effectiveness (MOEs) to determine progress of operations toward achieving objectives. MOPs are used to measure task accomplishment, and answer questions such as, "Was the action taken, were the tasks completed to standard?" to produce the desired effect. MOEs are to assess changes in adversary behavior, capabilities, or the operational environment. MOEs help answer questions like: "Are we doing the right things, are our actions producing the desired effects, or are alternative actions required?" Well-devised measures can help the commanders and staffs understand the causal relationship between specific tasks and desired effects. MOPs and MOEs can be either quantitative or qualitative measure but, whenever possible, quantitative measures are preferred as they are less susceptible to subjective interpretations. These measures aid intelligence and operations staffs in determining whether friendly force operations are achieving their goals, when unforeseen opportunities can be exploited, or when adversary activities necessitate a change in plans. Figure 5-9 provides an overview of assessment levels and measures.

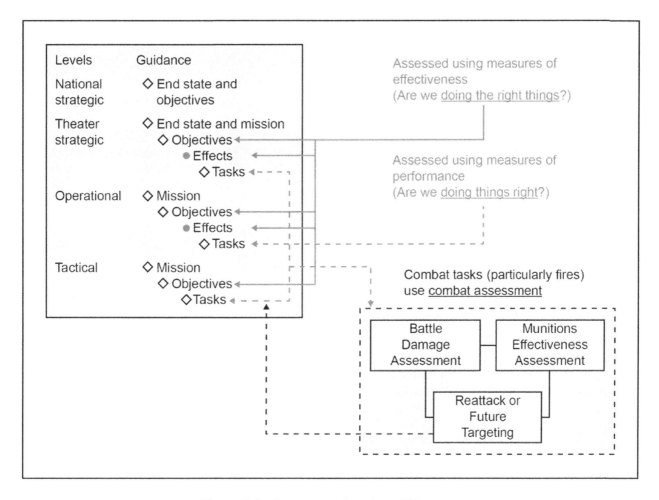

Figure 5-9. Assessment Levels and Measures

MOE assessment is an inherent component of steps one to three of the IPOE process. By continuously refining IPOE products, naval intelligence professionals can compare baseline intelligence estimates that have informed the plan with the current situation. MOE assessment is informed through collection operations that provide data and information as to whether certain conditions exist. Additionally, debriefs and friendly unit reports (MISREPs, situation reports, etc.) are extremely valuable for combat assessment given friendly units can typically provide specific, quantitative data or direct observation of an event or target to determine its status. There may be several indicators associated with a single MOE and a number of MOEs may be used to assess progress toward an objective. Favorable indicators represent progress toward accomplishing a task or objective while unfavorable indicators reflect regression that may require the execution of a branch in a plan.

5.5.2 Tactical Level Assessment

Tactical-level assessment typically uses MOPs to evaluate task accomplishment. While the results of tactical tasks are often physical in nature, they can also impact specific functions of a system through the use of nonlethal fires and other tactics. Techniques used for tactical-level assessment may include assessing progress by phase lines; neutralization of specific adversary units or forces; control of key aspects of the operational environment such as terrain, sea lanes, chokepoints, people, or resources; and completion of security or reconstruction tasks. Combat assessment determines the overall effectiveness of force employment during military operations. It can encompass many tactical-level assessment actions that typically focus on determining the results of weapons engagement (with both lethal and nonlethal capabilities), and thus is an important component of joint fires and the joint targeting process. It helps the commander understand how the joint operation is progressing and assists in shaping future operations. Combat assessment is composed of three major components: (a) battle damage assessment; (b) munitions effectiveness assessment; and (c) reattack recommendation.

5.5.2.1 Battle Damage Assessment

BDA should be a timely and accurate estimate of damage or degradation that has resulted from the use of military force, whether lethal or nonlethal, against a target. While BDA is primarily an intelligence responsibility, conducting BDA requires close coordination with operations and is often federated throughout the intelligence community. Since the BDA process is designed to answer the question, "Were the strategic, operational, and tactical objectives met in applying force against the selected targets?", the most critical factor in effective BDA is understanding the objective and how it relates to a specific target. For BDA to provide meaning, supporting MOEs must be observable, measurable, and obtainable. BDA activities must be planned and an architecture must be implemented to support these activities. BDA consists of a physical damage/change assessment phase, functional damage assessment phase, and target system assessment phase.

5.5.2.1.1 Phase 1—Physical Damage/Change Assessment

A physical damage assessment is an estimate of the quantitative extent of physical damage (through munitions blast, fragmentation, and/or fire damage effects) to a target resulting from the application of military force. A change assessment is the estimate of measurable change to the target resulting from weapons that do not create physical damage. These are is postattack target analyses that should involve a coordinated effort among combat units, Service components, JTF, combatant command, national agencies, and other organizations as required. At a national level, the Joint Staff Targeting and BDA Cell, with J-2T as lead, coordinates combatant command BDA requirements with the intelligence community. Some representative sources of data necessary to make a physical damage assessment include the ATO or master air attack.

A phase 1 BDA report provides an initial physical damage assessment of hit or miss based usually upon single source data. When appropriate, a reattack recommendation is also included.

5.5.2.1.2 Phase 2—Functional Damage Assessment

A functional damage assessment is the estimate of the effect of military force to degrade or destroy the functional or operational capability of the target to perform its intended mission and on the level of success in achieving operational objectives established against the target. This assessment is based upon all-source information, and includes an estimation of the time required for recuperation or replacement of the target function. Functional damage assessments are inferred from the phase 1 physical damage/change assessment and all-source intelligence information. These assessments must include an estimation of the time required for the recuperation or replacement of the target's function. Comparison of the original attack objective with the current target status allows BDA analysts to determine if objectives were met in striking the target. Phase 2 BDA builds upon the phase 1 report and is a fused, all-source product that provides more detail on the physical damage, an estimate of functional damage, and provides an input to phase 3 of the BDA process. Where appropriate, a reattack recommendation is provided.

5.5.2.1.3 Phase 3—Functional Assessment of the Higher-Level Target System

Functional assessment of the higher-level target system is a broad assessment of the overall impact on an adversary target system relative to the targeting objectives established. These assessments may be conducted at the combatant command or national level by fusing all phases I and II BDA reporting on targets within a target system. Phase 3 of the BDA process produces a target system assessment for a theater of operations. The functional damage assessments on individual targets within a system are compiled and applied to analyzing the impact on the whole system or adversary order of battle. It can apply whether the weapons used were lethal or nonlethal fires. For nonlethal fires, SIGINT is typically the most effective collection capability used to determine functional damage.

5.5.2.2 Munitions Effectiveness Assessment

A munitions effectiveness assessment (MEA) is conducted concurrently and interactively with battle damage assessment; the assessment of the military force applied in terms of the weapon system and munitions effectiveness to determine and recommend any required changes to the methodology, tactics, weapon system, munitions, fusing, and/or weapon delivery parameters to increase force effectiveness. Munitions effectiveness assessment is primarily the responsibility of operations with required inputs and coordination from the intelligence community. MEA targeting personnel seek to identify, through a systematic trend analysis, any deficiencies in weapon system and munitions performance or combat tactics by answering the question, "Did the systems (i.e., bomb or jamming) employed perform as expected?" Using a variety of intelligence and operations inputs, to include phase 2 functional damage assessments, operators prepare a report assessing munitions performance and tactical applications. MEA reports measure effectiveness of the weapon against specific types of targets and can have an impact on future operations and the quality of future BDA. MEA can be a long-term analytic process that incorporates observations of weapon effects at the target site after the cessation of hostilities.

5.5.2.3 Future Targeting and Reattack Recommendations

A future targeting and reattack recommendation is an assessment, derived from the results of battle damage assessment and munitions effectiveness assessment, providing the commander systematic advice on reattack of targets and further target selection to achieve objectives. The reattack recommendation considers objective achievement, target, and aimpoint selection, attack timing, tactics, and weapon system and munitions selection. The reattack recommendation is a combined operations and intelligence function. The reattack recommendation considers if the desired effect was created. That effect is then reassessed against its relative importance in the targeting effort, considering if the target is down, will it remain inoperable, or when will it be repaired.

INTENTIONALLY BLANK

APPENDIX A

Naval Intelligence Reading List

The following is a list of historical works, biographies, and scholarly material on subjects of interest to the naval intelligence professional. The list is divided into general categories to facilitate research on a specific subject. Readers desiring a more comprehensive list of resources on naval warfare in general and the U.S. Navy are encouraged to examine the CNO's reading list (http://navyreading.dodlive.mil/). Readers desiring a more comprehensive list of readings on intelligence matters should consult the CIA's Intelligence Literature: Suggested Reading List (https://www.cia.gov/library/intelligence-literature/index.html).

A.1 U.S. NAVAL INTELLIGENCE HISTORICAL WORKS

A.1.1 General Histories

A Century of U.S. Naval Intelligence	Wyman H. Packard

A.1.2 Naval Intelligence Up to World War II

Tracking the Axis Enemy: The Triumph of Anglo-American Naval Intelligence	Allan Harris Bath
The Office of Naval Intelligence: Birth of America's First Intelligence Agency, 1882–1918	Jeffery Dorwart
Conflict of Duty: U.S. Navy's Intelligence Dilemma, 1919–1945	Jeffery Dorwart
Avoiding Another Pearl Harbor The Primary Purpose of National Estimating http://www.au.af.mil/au/awc/awcgate/nic/purpose_of_estimating.pdf	Harold P. Ford
Double Edged Secrets: U.S. Naval Intelligence Operations in the Pacific During World War II	W. J. Holmes
United States Cryptologic History (NSA Monograph), World War II, Volume 6 Pearl Harbor, Revisited: United States Navy Communications Intelligence 1924–1941 http://www.nsa.gov/about/_files/cryptologic_heritage/publications/wwii/pearl_harbor_revisited.pdf	Frederick D. Parker
The Battle of Midway http://www.nsa.gov/about/cryptologic_heritage/center_crypt_history/publications/battle_midway.shtml	Patrick D. Weadon

A.1.3 Cold War

The Admiral's Advantage: U.S. Navy Operational Intelligence in World War II and the Cold War.	Christopher Ford and David Rosenberg
Adak, The Rescue of Alfa Foxtrot 586	Andrew C.A. Jampoler
The Evolution of the U.S. Navy's Maritime Strategy, 1977–1986	John B. Hattendorf
A Dangerous Business: The U.S. Navy and National Reconnaissance During the Cold War	John R. Schindler
SILENT WARRIORS: The Naval Security Group Reserve, 1945–2005	John R. Schindler
Blind Man's Bluff, The Untold Story of American Submarine Espionage	Sherry Sontag and Christopher Drew

A.1.4 Post-Cold War

Decade of War, Vol. 1; Enduring Lessons From the Past Decade of War	Joint and Coalition Operational Analysis (JCOA) a Division of Joint Staff J7
Shield And Sword The United States Navy and the Persian Gulf War	Edward J. Morolda and Robert J. Schneller, Jr
Intelligence Lessons Learned from Recent Expeditionary Operations	Robert David Steele

A.2 NAVAL INTELLIGENCE BIOGRAPHICAL OR AUTOBIOGRAPHICAL WORKS

Joe Rochefort's War: The Odyssey of the Codebreaker Who Outwitted Yamamoto at Midway	Eliot Carlson
A Woman's War: The Professional and Personal Journey of the Navy's First African American Female Intelligence Officer	Gail Harris with Pam McLaughlin
Special Agent, Vietnam: A Naval Intelligence Memoir	Douglass Hubbard
And I Was There, Pearl Harbor and Midway—Breaking The Secrets	Rear Admiral Edwin T. Layton, USN (Ret.)
Secret Missions: The Story of an Intelligence Officer	Rear Admiral Ellis M. Zacharias, USN (Ret.)

A.3 HISTORICAL OR SCHOLARLY WORKS ON THE NAVY AND WARFARE

On War	Carl von Clausewitz
Principles of Maritime Strategy	Julian S. Corbett
Fleet Tactics and Coastal Combat	Wayne P. Hughes
The Two-Ocean War: A Short History of the United States Navy in the Second World War	Samuel Eliot Morison
The Influence of Sea Power Upon History	Alfred Thayer Mahan
Sea Power: A Naval History	E. B. Potter
The History of the Peloponnesian War	Thucydides
The Art of War	Sun Tzu

A.4 GENERAL INTELLIGENCE WORKS

A.4.1 Scholarly Works on Intelligence and Intelligence Tradecraft

The Psychology of Intelligence Analysis	Richards J. Heuer, Jr.
Military Intelligence Blunders	Colonel John Hughes-Wilson
Intelligence: From Secrets to Policy	Mark M. Lowenthal
Communicating with Intelligence	James S. Major
Structured Analytic Techniques For Intelligence Analysis	Randolph H. Pherson and Richards J. Heuer, Jr.
Challenges in Intelligence Analysis	Timothy Walton
Pearl Harbor: Warning and Decision	Roberta Wohlstetter

A.4.2 Historical Works

Room 40: British Naval Intelligence 1914–1918	Patrick Beesly
Very Special Intelligence	Patrick Beesly
Very Special Admiral: The Life of Admiral J. H. Godfrey	Patrick Beesly
Winning with Intelligence	Gregory Elder
A Crucial Estimate Relived (Cuban Missile Crisis) http://www.au.af.mil/au/awc/awcgate/nic/cru_est_1992.pdf	Sherman Kent
The Final Months of the War with Japan: Signals Intelligence, U.S. Invasion Planning, and the A-Bomb Decision. https://www.cia.gov/library/center-for-the-study-of-intelligence/csi-publications/books-and-monographs/the-final-months-of-the-war-with-japan-signals-intelligence-u-s-invasion-planning-and-the-a-bomb-decision/csi9810001.html	Douglas J. MacEachin
The Wizards of Langley	Jeffrey T. Richelson

INTENTIONALLY BLANK

NWP 2-0

APPENDIX B
Naval Intelligence Case Study

B.1 INTELLIGENCE CASE STUDY—OPERATION ICEBERG (THE INVASION OF OKINAWA)

The following intelligence case study details the contributions of naval intelligence to the successful planning and execution of the U.S. invasion of Okinawa. It is excerpted from the U.S. Navy Intelligence School's training manual, Naval Intelligence, which was published in 1948. While lengthy, the case study is reprinted in full as it highlights many of the major themes and concepts that are discussed throughout this publication to include:

1. The application of operational art to synchronize complex intelligence operations in time and space.

2. The principles of naval intelligence and the attributes of naval intelligence excellence are all in evidence throughout this case study.

3. The need for centralized planning and direction and decentralized execution of intelligence operations.

4. Collaboration as one of the fundamentals of intelligence practice. This case study provides excellent examples of collaboration up, down, and across various naval echelons and shows the benefits of collaboration with Marine Corps, Coast Guard, joint Service partners, coalition forces, other government agencies, academia, and industry.

5. Demonstrates that intelligence will always face resource constraints that can only be overcome by farsighted planning and direction and innovative solutions that can come from all levels of the naval intelligence team.

6. Provides excellent examples of understanding and anticipating the needs of the intelligence user and then thinking through the end-to-end details that need to be synchronized throughout the intelligence process to meet those needs. Due to the long lead time for some collection, analysis, production, or dissemination steps, this requires careful planning of intelligence operations to ensure that the right intelligence gets to the right intelligence user in time to support their planning efforts and operations.

7. Demonstrates the value of tailoring intelligence products to meet the needs of specific intelligence users.

8. Highlights the value of an all-source, multimethod approach to producing intelligence. GEOINT, HUMINT, OSINT, and SIGINT all provided valuable information that was incorporated into fused intelligence products to support Operation ICEBERG, situational awareness plots, and to provide warning of Kamikaze attacks and other threats. Although SIGINT is not mentioned explicitly in this case study, due to classification concerns at the time of its publication, it is subtly referenced as "special dispatches."

9. Provides excellent examples of IPOE product development and the value of maintaining current intelligence plots and products to support plans and operations.

10. Demonstrates the utility of designating operations areas, areas of influence, and AOIs to focus the intelligence effort.

11. Provides excellent examples of the need to balance competing demands (support to future operations versus support to current operations, security versus the need to disseminate, etc.) and the value of taking risks to achieve the objective.

B.2 INTRODUCTION

In this appendix, the role of Naval Intelligence from the planning phase of the Okinawa campaign up to the beach landing of U.S. forces on the island is described. In the space available, a detailed account of the extremely complex intelligence functions during the operation in question is impossible. Therefore, the subject is necessarily considered only in its broadest perspective.

As the invasion of Okinawa was an amphibious operation, the part played by Naval Intelligence must necessarily be described from the standpoint of the intelligence section of ComPhibsPac, Commander of the Joint Expeditionary Force to land on Okinawa. Figure B-1 provides a map of the Operation ICEBERG invasion plan that can be used to reference some of the locations discussed in this case study.

B.3 INTELLIGENCE DURING THE PLANNING PHASE

In August 1944, the Joint Chiefs of Staff directed the occupation of certain portions of the Nansei Shoto for the development of bases there and eventual further operations against Japan. Their decision was derived from first phase strategic intelligence studies provided by ONI and Military Intelligence Service (MIS). These studies were the result of months of research and included information collected over a period of years from a wide variety of sources and by many diverse methods. However, as the basis for an amphibious operation, they naturally left much to be desired; and before the actual landings took place, first on 24 March 1945 at Kerama Retto, and next on 1 April at Okinawa proper, the facilities and personnel of Naval Intelligence were taxed to their utmost to provide the participating forces with the necessary effective strategic and tactical information.

As ComPhibsPac was assigned to be Commander of the Joint Expeditionary Force, The ComPhibsPac Intelligence Section was, of course, the highest echelon of intelligence for the amphibious operation. Announcement of Okinawa as the specific objective in the Nansei Shoto immediately set the ComPhibsPac Intelligence Section to the task of collecting all available information on the area to be attacked.

ONI supplied the basic geographic data, terrain studies, and all basic monograph material on Okinawa proper and all the islands of the Ryuku Retto. Such additional material as was subsequently collected by the district intelligence officers (DIOs), the naval attaches, and the research units of ONI itself was forwarded to Joint Intelligence Center Pacific Ocean Areas (JICPOA). In addition, ONI furnished a counterintelligence study in two parts; one on the Ryuku Retto and the other on the Satsunan Shoto. They were so compiled as to serve the needs of Combat Intelligence Corps detachments and other security forces involved in the combat phases as well as those of the military government teams charged with public safety and security. The texts covered local government organization and functions (stressing police and gendarmerie), naval installations of counterintelligence interest, organizations believed to be engaged in subversive activity, and biographical data on over 900 officials.

Thus, ONI supplied all accumulated information (primarily strategic intelligence) and in this manner assisted the Intelligence Section of ComPhibsPac in the basic planning stage. It remained for other intelligence units and sources to provide more recent and detailed strategic and tactical data from reconnaissance and local contacts.

After the preliminary survey of accumulated intelligence, conferences were held by representatives of the Intelligence Section of ComPhibsPac with the officer in charge, JICPOA, the DIO of the 14th Naval District, the Intelligence Officers of the Fifth Fleet, Tenth Army, Fleet Marine Forces (Pacific), XXIV Corps, XXX Amphibious Corps, ComAirPac, Fast Carrier Force, and all amphibious groups. These were necessary in order to apprise all planners of material available and to prepare coordinated plans and schedules for collecting new material. Here, also, the overall intelligence plan was developed and adopted. Of special importance was the agreement on the types and scales of maps and charts to be used and the photographic reconnaissance plan.

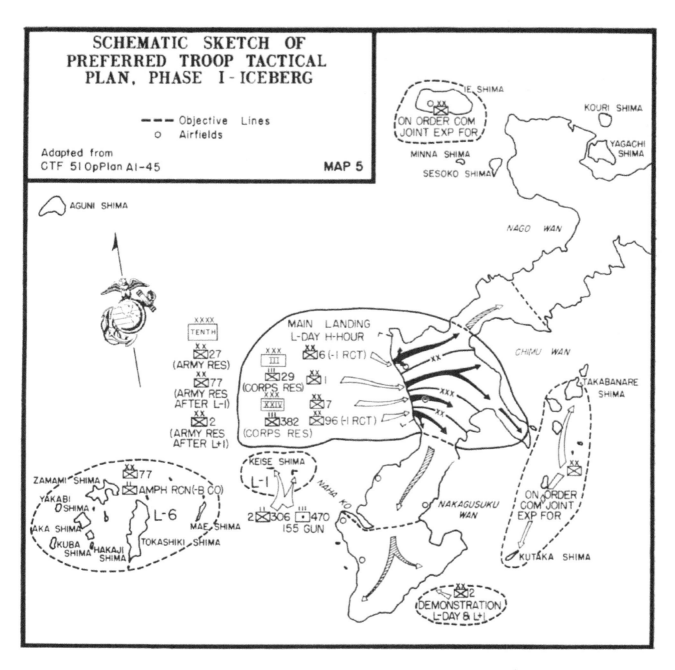

Figure B-1. Operation ICEBERG—The Invasion of Okinawa[1]

---
[1] Retrieved from http://www.ibiblio.org/hyperwar/USMC/USMC-M-Okinawa/maps/USMC-M-Okinawa-5.jpg.

The next step taken was thorough canvassing of all agencies in the Hawaiian area to obtain additional data on Okinawa and, where security permitted, to advise them of the special interest in that area so that they could intelligently coordinate and present their available information. Among these activities were JICPOA, the DIO of the 14th Naval District, the local branch of the Office of Strategic Services (O.S.S.), the Bishop Museum, the United States Geological Survey, the Hawaiian Pineapple Co., Pan-American Airways, and the Commercial Pacific Cable Co. The contact with commercial organizations and individuals was, of course, accomplished through the DIO.

Constant and close liaison was necessary from this stage on. It had already been learned that liaison was vital to prevent duplication of effort and insure receipt and use of the best possible information. Thus, liaison was established between the ComPhibsPac Intelligence Section on the one hand and on the other with JICPOA, the DIO of the 14th Naval District, the Fleet Marine Forces Intelligence Section, and numerous other cognizant commands and units. In addition, of very considerable importance was efficient liaison with the Tenth Army, composed of five United States Army and three United States Marine Divisions, which had been designated as the invading force. In this connection it must be pointed out here that the Military Intelligence Service performed a most important function throughout the operation, but one that is not discussed herein since the subject under consideration is Naval Intelligence. Also, the MIS was primarily concerned with the land operations following the initial invasion.

Early steps were taken to interrogate all possible informants including prisoners of war (POWs) who might have information on the area in question, and other categories of informants who might have visited or worked in Okinawa or its neighboring islands.

In the general category of valuable informants was a conchologist who was living in Arizona when his existence and background were discovered as a result of investigations undertaken by the DIO of the 14th Naval District. When, upon interrogating him, it was found that he had collected shells on the reefs of Okinawa and possessed much information on the reefs, surf, and channels there, he was placed under contract to the Navy, flown from the United States, and attached to the ComPhibsPac Intelligence Section.

Another excellent source of information was the United States Geological Survey Office in Honolulu in which two geologists were contacted who were of much value in analyzing terrain from aerial photographs. Other informants were Hawaiian residents of oriental extraction who had either visited Okinawa or fished in its waters.

The interrogation of POWs was carried out chiefly by graduates of the Naval School of Oriental Languages, which was established under the cognizance of ONI early in 1941 to train competent interpreters and translators of the Japanese language. In spite of the fact that only a fraction of these graduates had previously lived in the Far East, all of them were competent upon graduation in the use of spoken and written Japanese. After a brief training in intelligence methods, they were found invaluable in the field, particularly for quick exploitation of information derived from interrogation of prisoners and scanning and translation of captured documents.

As in all other Pacific operations, a great part of the information on Okinawa came from aerial reconnaissance. Since it was impossible to obtain all necessary coverage of Okinawa with land-based planes, much of the coverage had to be obtained during carrier strikes, one of the most successful of which was undertaken on 10 October 1944, when 90 percent of Okinawa was photographed and both large-scale verticals and obliques were taken. A definite improvement in the planning of photo reconnaissance was effected for the Okinawa operation by initially obtaining adequate vertical coverage for early preparation of maps. Subsequently, the large-scale coverage for detailed study of installations was obtained, and by that time interpreters had a satisfactory map on which to plot their findings. The frequent small-scale coverage by B-29s, although not always suitable for detailed study, served as an excellent means of detecting changes made in major enemy installations.

The coverage effectively closed the gap between the coverages obtained at the time of fast carrier strikes, and again emphasized the necessity for continuous photo coverage of an area. Earlier photos revealed installations that later coverage often failed to disclose because of camouflage. Likewise, the number of installations and changes in the relative strength of defense positions could be determined only by careful study of successive sorties. Some Sonne photography was obtained by planes of Task Force 58. This covered positions of the main landing beaches of Okinawa and was useful in making accurate depth determinations. Although such photography was too late for planning purposes or wide dissemination, it confirmed depth estimates made previously by the ComPhibsPac Intelligence Section.

The type of photography needed for amphibious operations is unique. The publication of a letter prepared by the ComPhibsPac Intelligence Section setting forth the requirements of the amphibious forces and illustrating good and bad coverage, notably increased the quality of photo reconnaissance. Through the air combat intelligence (ACI) officers attached to Task Forces, these requirements and requests were clarified and emphasized for pilots prior to reconnaissance. This careful briefing was found to be of great value in obtaining desired coverage.

In accordance with standard practice, ComPhibsPac requested submarine reconnaissance of the objective and this was conducted. The results were compiled in the form of annotated mosaics which included both the written reports and the periscope photos. These were especially valuable for briefing personnel operating assault craft and in familiarizing them with approaches to the beaches.

Early in 1944, a publication entitled "Information Bulletin—Nansei Shoto" had been prepared and distributed by JICPOA. On 15 November 1944, Information Bulletin—Okinawa Gunto was completed by the same activity. This latter study included all the latest general information the target area. However, as more detailed information was needed by the various units to be involved in the operation, additional material was necessary. Also it was desirable to disseminate much new information collected after the publication of the JICPOA bulletin on Okinawa.

A general study of Okinawa similar in content to the JICPOA survey was prepared by USAAFPOA, primarily for use by the Army Air Forces. A target analysis bulletin was prepared by JICPOA primarily for the use of fast carriers. The ComPhibsPac Intelligence Section and the G-2 section of the Tenth Army jointly prepared a graphic and tabular study of the hydrography, reef, and beach conditions for use of both the naval and ground forces. The Fleet Marine Forces, Pacific, published a graphic terrain study of the area assigned for Marine landings. In December 1944, the ComPhibsPac Intelligence Section prepared *Supplementary Information on Okinawa* which included detailed studies of reefs and beaches, unloading conditions, surf and swells, weather, tide tables, and daylight-dark tables. During November and December 1944, the Intelligence Section was also engaged in preparing *Preliminary Beach Sketches* which showed the assault beaches and perspective, and familiarized salient features of each area by brief paragraphs and annotations. Even though there had been very little aerial reconnaissance up to this time, these annotated beach sketches showing the depth of water, the offshore approaches, coral reefs, and many other details, were remarkably accurate. In January and February 1945, the ComPhibsPac Intelligence Section annotated oblique aerial photographs of the assault beaches and, upon request, JICPOA laid a controlled mosaic of the southern part of Okinawa from B-29 photos taken on 3 January. This mosaic was printed on 13 large sheets with the landing beaches annotated, and copies were distributed to naval units and troop commands of the Tenth Army. The engineering terrain intelligence team, a group of civilian geographers attached to the Army, prepared a detailed graphic study of the terrain behind the assault beaches at the request of ComPhibsPac. This was completed in January 1945 and proved to be very accurate. Also in January 1945, the ComPhibsPac Intelligence Section completed the Intelligence Annex for the ComPhibsPac Operation Plan and assisted in the preparation of the Annex for the Operation Plan of Amphibious Group 1. The Intelligence Section also assisted ACI officers of the commanders, air support control plan, in preparing the Photographic Plan to be executed at the objective.

The preparation and distribution of standard Hydrographic Office charts were not normally the responsibility of the ComPhibsPac Intelligence Section. However, the section assumed the duty of distributing new H.O. charts as they were printed. Sufficient copies for all commands in the operation were obtained from the Fleet Hydrographic Office in Pearl Harbor. The Okinawa approach chart was prepared and printed in Washington. The bombardment chart was prepared and printed by two Army units, located at Pearl Harbor and Guam. The information on installations and defenses was obtained from photographic interpretation by the ComPhibsPac Intelligence Section itself and reviewed and approved by the G-2 section of the Tenth Army. Anchorage charts of staging areas were supplied by the Fleet Hydrographic Office and eventually distributed by the ComPhibsPac Intelligence Section, as were a road map of Okinawa and various briefing charts of Nansei Shoto prepared by JICPOA.

The air and gunnery target map, scale 1/25,000, was the standard ground map for the operation and the largest scale map of the entire area which was produced. It was drafted and printed by Army engineers in the Pacific for use by all naval, air, and ground units. Gridded air support charts were prepared from this map, were bound in booklets, and disseminated for the use of pilots, spotters, air liaison officers, and observers. A gridded chart was printed on one side of a page and a gridded photo of the same areas was printed on the back of the succeeding page, so that both were visible to the user at the same time. The size, format, bright coloring, and grid were especially helpful to pilots who were able to use these effectively in their small cockpits.

A two-sheet plotting map of Okinawa, scale 1/100,000, was also made and reproduced by Army engineers, and was especially valuable for briefing and for maintaining generalized situation plots. A special intelligence map, scale 1/36,000, was drafted and printed by the ComPhibsPac Intelligence Section. This map showed all the defense installations observed in photographs taken up to 3 January 1945. It was brought up to date again in March. During November and December of 1944, the ComPhibsPac Intelligence Section hydrographer prepared tentative anchorage charts for five areas at the objective. These were reproduced and prepared for distribution to all ships and units immediately involved in the invasion, and to port directors for issue to ships that would later arrive at Okinawa. Since no agency had been designated to prepare Virtual Plan Position Indicator Reflectoscope and Radar Prediction Determination charts, the ComPhibsPac Intelligence Section made the necessary arrangements to have them produced and distributed to those units that would require them.

A large number of rubber relief maps of the southern part of Okinawa, scale 1/10,000 in nine panels, and a scale 1/25,000 in two panels, were produced by the Atlantic Amphibious Training Command the Photographic Intelligence Center, Navy Yard, Washington. JICPOA produced a number of plastic models, mostly of the assault beach area. The relief maps were delivered to the rear echelon intelligence section in Pearl Harbor and distribution was made to the amphibious forces from there. There were enough of these maps to supply all major commands, all TransRons, TransDivs, fire support ships, CVEs, and many of the APAs. As always, these models proved useful in briefing.

Reproduction was always a major intelligence problem in the Pacific. As operations increased in size, more and more careful planning was required to insure that material would be available in sufficient quantity for distribution. So it was gain that lack of detailed maps made photo reconnaissance imperative before commencing the major preparation of final maps, charts, and other written material. The first aerial reconnaissance of Okinawa was in late September 1944, and several more took place in early October. The preliminary distribution of material by the ComPhibsPac Intelligence Section was made on 20 January 1945, and the final distribution on 28 February 1945. Four months was the time available to reproduce the necessary material, and this period included the time employed for original preparation. In addition, the Iwo Jima preparations added a tremendous load to the already heavily taxed facilities and personnel. The vital importance of careful and detailed planning, as well as maximum and efficient use of all valuable reproduction equipment, was absolutely imperative. Fortunately, by this time, experienced officers and personnel were available.

All major reproduction facilities in the Pacific were under the control of the office in charge, JICPOA. When the facilities of the amphibious forces were inadequate to handle reproduction, requests were made to him. His cartographic and hydrographic officers then formulated a plan and the work was allocated. Army engineers and USAFPOA reproduction unit handled most of the large printing orders. However, great quantities of intelligence material were reproduced by the print shops and photo laboratories of the amphibious forces. ComPhibsPac had a print shop and photo laboratory ashore at Pearl Harbor, and every AGC in the selected force also had both. It should be pointed out that these facilities were controlled and operated by the ComPhibsPac intelligence officer, and should be emphasized that all reproduction of operation plans, communication plans, boat control diagrams, etc., for the planned invasion had to be accomplished by them in these shops and laboratories.

A preliminary distribution of basic material was made to all staffs and ships to be involved in the operation by the ComPhibsPac Intelligence Section in January 1945. This material contained most of the accumulated strategic intelligence received up to that time, and was intended to assist the intelligence officers of the various staffs and ships in preparing their individual intelligence plans and to be used as background information in the acquiring of additional intelligence. This information was transmitted in the form of terrain studies, maps, charts, beach sketches, identification sheets, and information bulletins.

The final distribution of intelligence material from the ComPhibsPac Intelligence Section was forwarded from Pearl Harbor on 23 February 1945, and included all intelligence developed from the sources described previously. The distribution was a difficult problem. It was essential that every vessel engaged in the Okinawa operation, and those that might possibly become involved, receive an adequate supply of maps, charts, photos, and all other prepared written matter. These included not only the strictly amphibious vessels but all the gunfire support ships, the components of the carrier task forces, and other vessels that might be designated to assist in bombardment and transport of troops as reinforcements. The assigned vessels were staged at seven different bases through the Western Pacific and some were almost constantly at sea. Experience has clearly demonstrated that distribution by the echelon system was unsatisfactory, for many ships had complained of not receiving packages or receiving them months after they were mailed. Therefore, arrangements were made to fly the material by special R5D planes to the various staging areas. One intelligence officer from the ComPhibsPac Intelligence Section accompanied each plane and personally made the distribution at each staging area. He had complete lists of the ships and staffs involved and was in constant touch with ComPhibsPac, and thus was able to make corrections as changes took place. The officer was also supplied with a number of additional packages for distribution to ships that might be assigned at a later date. This system operated satisfactorily although there were approximately 1,340 vessels and staffs engaged in the operation. All commands received an adequate supply of intelligence material.

With proper security provisions and detailed instructions, copies were forwarded to port directors, island commanders, and a few other important commands for possible later distribution to reserve units and ships assigned at the last moment because of unforeseen emergencies. A large supply of extra copies was furnished to the amphibious group intelligence officers for emergency distribution. Previous experience had indicated that there always were last-minute changes because of unforeseen requests for distribution of intelligence material, and these emergency measures enabled all participating units to obtain the needed intelligence regardless of the date or location when they received orders to proceed to the objective.

During the overall planning phase, which may be considered as lasting up to the end of the first week of March 1945, no intelligence briefing was conducted except for the high-ranking planners in the rear echelons and the ComPhibsPac staff, all of whom were, of course, continuously briefed as new information was developed.

B.4 INTELLIGENCE EN ROUTE TO OKINAWA

On the departure of ComPhibsPac's flagship from Iwo Jima on 9 March 1944 for Guam and then Leyte to embark the staff of the commanding general of the Tenth Army, basic plans had been completed. This, however, did not mean that intelligence could relax. On the contrary, as the various forces to be engaged in the operation began converging on the objective, its work rapidly increased. From a variety of sources, late information was being received by all intelligence units attached to staff, squadrons, ships, corps, etc. The coordination and effective dissemination of this material were heavy and vital tasks.

Upon ComPhibsPac's arrival at Guam after leaving Iwo Jima, the latest photographs taken by the fast carrier task force were obtained. These included the first of the Sonne photos. Here also the latest CincPac estimates were received. Upon arrival at Leyte the final photo coverage of Okinawa by land-based planes was received, duplicated, and disseminated. Before leaving Leyte for Okinawa, final preparations were made for the receipt and dissemination of photos to be taken by a force of CVEs between L-day minus 7 and L-day minus 3. Also, plans were perfected for the receipt and dissemination of underwater demolition team (UDT) and aerial hydrographic observers' reports.

While at Leyte, lower echelon intelligence officers began a thorough briefing of all personnel, including coxswains, boat crews, shore- and beach-party personnel. At this time, the pilots of air support units were also being briefed for the first time by ACI officers. Members of ComPhibsPac's Intelligence Section briefed UDTs who were staging there and higher echelon officers in the immediate area.

During the whole period from ComPhibsPac's departure from Iwo Jima through to the attack, the ComPhibsPac Intelligence Section maintained the following plots:

1. Strategic plot of enemy forces, including sea, land, submarine, and air

2. A plot showing the movement of all allied task forces (maintained in conjunction with the Operations Section)

3. A mine field plot, kept up to date as reports and dispatches were received

4. A running intelligence plot of the objective itself, kept current by evaluation of new photos an reports

5. A plot of the European situation, for the information of ComPhibsPac and his staff.

While en route from Iwo Jima to Leyte a photo book of vertical and oblique pictures of the principal landing beaches was prepared and distributed in quantity. It consisted entirely of contact prints thoroughly annotated and proved to be most useful. During this same leg of the trip to the objective the Sonne photography was annotated and negatives were made. These negatives were immediately printed and distributed to the commands concerned. At Leyte after receipt of the most recent photos taken by land-based planes, the 1/36,000-scale intelligence map was brought up to date, reprinted in quantity, and widely distributed. In this connection copy negatives were sent by officer messenger to the forces assembling at Ulithi, as time did not permit awaiting the completion of printing at Leyte. The intelligence officer of Amphibious Group 1 at Ultithi printed and distributed the copy negatives of the map to all fire support ships while underway to the target.

The day before the invasion, the UDT reports and the CVE photography were received. Since the UDT reports had already been distributed in accordance with a previously arranged plan, no further distribution was required by the ComPhibsPac Intelligence Section. On the same day, more photographs were received from Amphibious Group 1 and were printed in quantity and distributed during the afternoon by a destroyer. Further information received on that day from several sources on suicide boats and planes was coordinated and disseminated by dispatch to the entire force.

During several days prior to the landings the reproduction departments of ComPhibsPac AGC were worked to capacity. The photo laboratory was continuously printing photos and mosaics needed for quantity distribution. They were also engaged in reproducing special prints and mosaics requested by subordinate units for individual tasks. En route from Leyte to Okinawa, a major installation map of Okinawa, scale 1/72,000, in two sheets, was prepared for the use of the fire support ships. This was distributed early in the morning of the day of the invasion. In addition to overlays, etc., for other sections, the map reproduction unit, while en route from Leyte, prepared and printed blank charts to be later used by mobile hydrographic survey units. All of the above activity involved continuous day and night work by the yeomen, draftsmen, photographers, printers, and cartographers attached to the Intelligence Section on board.

As the forces approached Okinawa, the Intelligence Section was further burdened by planning for the invasion and occupation of the western islands of Okinawa Gunto, Kaiza Shima, and Mujako Shima. Although the latter two islands were never attacked, all maps, charts, and intelligence material pertaining to them were compiled and printed before the arrival at Okinawa.

B.5 INTELLIGENCE AT OKINAWA

With the arrival of the invasion forces before the beaches of Okinawa, the responsibilities of all intelligence officers still further increased. They were working against time at this point in the operation, and the speed with which they obtained and disseminated information was a major consideration. An attempt is made here to consider some of their multiple tasks and the volume of intelligence they were required to disseminate.

On the day of the invasion, ComPhibsPac assumed control of all photographic reconnaissance planes. This, of course, meant that his Intelligence Section was called upon to provide the necessary information to direct their activities efficiently. One photographic plane was assigned to each corps, and missions were controlled by their own air support control units. Additional photographic planes were at the disposal of ComPhibsPac and the commanding general, Tenth Army, with missions coordinated and controlled by the commander, air support control units, on the flagship of ComPhibsPac. After the arrival of land-based F5s of Photographic Reconnaissance Squadron 28 operating under the tactical air force, the majority of the photo missions requested by troop commands were flown by them. The direct control ComPhibsPac which allowed for briefing of the flights by this squadron resulted in some excellent photographic coverage of specifically designated areas.

Photo coverage of all the small islands at the objective was obtained by CVE planes from the air support force. These planes also obtained excellent low oblique coverage of other islands on the Nansei Shoto which was necessary for planning purposes. The value of having intelligence officers personally brief photo planes was again demonstrated. More or less successful efforts were made to furnish prints and comments on the results of previous sorties to the intelligence officers with the various carriers and photo groups for briefing of pilots and photographers. As in former operations, two ComPhibsPac photographers were assigned to the CVEs to assist in organizing photo flights. They acted in general as troubleshooters, and moved from one CVE to another.

Photographers were also sent ashore whenever possible after the landings to obtain photographs of installations, equipment, and unloading. One photo interpreter was assigned to the commander, Underwater Demolition Team, for the operation. He was of considerable assistance in making detailed studies of the reefs and other obstacles from late photographs and in briefing the teams. Photo interpreters on temporary duty with the commander of the heavy gunfire support ships did invaluable work in locating targets and assessing damage. The ComPhibsPac photo interpreters worked closely with the target information center set up by the gunfire support commander. Location of targets and damage assessments were passed to the center in order that assignments could be more accurately and efficiently determined for fire support ships, aircraft, and artillery.

As stated previously, the underwater demolition teams had been briefed on all available intelligence at Leyte, and latest photographs had been delivered to them there. After making a reconnaissance of Kerama Retto prior to its occupation a week before the landing on Okinawa, they proceeded to Okinawa, and 2 days before the assault there, they reconnoitered the beaches. Here they were accompanied by representatives of the various troop commanders assigned to the selected areas. Their reconnaissance produced a detailed written report including large-scale Ozalid prints of beach and reef sketches, and this proved to be very accurate and comprehensive. The day before the invasion, representatives of the various troop commands were taken by destroyer to the flagships and vessels on which troop commanders were embarked. They were furnished with this written report. At the same time the material was supplemented by oral reports of representatives of the teams.

During the underwater demolition teams' reconnaissance, 3,500 obstacles in the form of wooden posts set on the reef edges were discovered. As most of these were considered to constitute a barrier to LVTs, it was deemed necessary to blow them out of the beach approaches. This was accomplished without incident the following day by the UDTs. In addition, these incredibly courageous and successful teams were later used to reconnoiter for more suitable unloading beaches in the northern part of the island.

Three aerial hydrographic observers were assigned to the CVEs, and, beginning a week from L-day, they flew over the assault beaches daily to observe conditions of the surf, height of swells, length of breakers, and other hydrographic developments. Their reports were sent to all ships and later corroborated.

The hydrographic survey units were under the control and jurisdiction of the ComPhibsPac intelligence officer, and had begun hydrographic preparations for Okinawa prior to the Iwo Jima operation. Two of the four units employed at Okinawa had been engaged on survey work at Iwo Jima; the other two had been engaged solely on patrol and escort duties. All four had arrived at Leyte in time for rehearsals, and were then assigned to the various amphibious groups with whom they proceeded to the scene of operations. On arrival in the Okinawa area they were released from all escort and patrol duties, and were used thereafter solely for survey operations. Two of the units conducted survey operations at Kerama Retto, laying buoys, observing currents and tides, and making special examinations of beaches and anchorages. All radar reflector buoys indicated in ComPhibsPac's operation plan, and numerous additional buoys, were planted. The position of all these buoys were determined or checked.

Third class navigation buoys were laid in the ship lane, and acetylene-burning navigation lights were established on nearby islands. Small lighting buoys were laid as temporary aids until vessels operating these areas became acquainted with local conditions. Radar reflector buoys were maintained throughout the operations as aids for fire support ships. While conducting hydrographic operations in the Kerama Retto waters, the two units printed and distributed temporary charts showing the new developments. On their departure to commence new surveys at Okinawa all unfinished work and records were transferred to one survey vessel. This ship later assembled all data developed by the units into a field chart of Kerama Retto.

The other two units arrived at Okinawa and began hydrographic operations on L-day. Navigation lights were established, and all known off-shore dangers were marked. The off-shore area between the low waterline and the 30-fathom curve before the Okinawa beaches was sounded. Several uncharted coral heads inside the 10-fathom curve were found and marked. On the completion of the soundings in this area an anchorage chart was prepared, printed, and distributed by ComPhibsPac. This completed the assault hydrographic activities for Okinawa.

One unit was assigned to amphibious group for the Ie Shima operation. The results of its survey were published in the form of an anchorage chart. After the occupation of Ie Shima special surveys on a large scale were conducted there for the purpose of harbor development and were completed on 10 May 1945.

Miscellaneous hydrographic surveys of special areas were requested of the units as the operation progressed. Such requests were assigned on a priority basis established by tactical needs. These special surveys were made on scales of 1/6,000 or less, and were detailed and tedious. Ozalid prints of special surveys were furnished the island commander of our military forces as needed. In addition to the special surveys, numerous buoys were laid to make ship lanes. Positions of all buoys placed were shown on the temporary field chart, which was disseminated to all units concerned.

The force aerological officer, working under the ComPhibsPac intelligence officer, made weather forecasts twice daily, at 0800 and 2000. These reports were dispatched to all commanders within the immediate area and to CincPOA and Com Fifth Fleet. Similar forecasts were made by the task force and task group commanders daily, and were immediately relayed to ComPhibsPac for further distribution. All forecasts proved to be exceptionally reliable and accurate. Additionally, all patrol aircraft and striking aircraft operating between Okinawa and the China coast, and Okinawa and the Empire, issued radio weather reports whenever security permitted the use of radio traffic. This information was used in the preparation of the above-mentioned forecasts.

Large quantities of captured material were brought directly to the ComPhibsPac Intelligence Section for study and translation. Approximately 65 percent of this matter was taken from enemy pilots and aircraft. In general, it was processed in one of two ways:

1. A preliminary screening was made upon receipt to uncover items of immediate value. These were translated at once or forwarded to interested commands as the situation warranted. Such intelligence included top secret enemy documents, orders, tactical and briefing notes for pilots, instructions for the use of certain codes, air navigation charts, and hydrographic charts.

2. A second and more thorough examination was conducted and valuable or useful intelligence then uncovered was forwarded to a JICPOA team ashore for further dissemination. Most of this type of information was strategic rather than tactical.

Liaison was effected with language officers and other intelligence officers attached to the landing forces and to JICPOA teams, so that there was instant transmission to the ComPhibsPac Intelligence Section of any information of interest that was developed ashore from the captured documents and material.

A large number of prisoners of war were brought aboard ComPhibsPac's AGC by landing force personnel and interrogated by language officers attached to ComPhibsPac Intelligence Section. Those picked up at sea by naval commands were brought directly to the flagship. Preliminary interrogations were first conducted and if it appeared that a prisoner possessed naval technical information of strategic value he was flown to the nearest

JICPOA POW camp. If he had information of immediate value to the landing forces he was rushed to the nearest Army G-2 section ashore.

To ensure the interrogation of all POWs who had information of value, naval intelligence officers, including language officers, went ashore at an early hour and searched continuously among the POWs for those individuals who might divulge useful data. Some important information was developed in this fashion and immediately transmitted from the beaches to the interested commands.

POW ships on which intelligence and language officers had previously been assigned were designed for the removal of prisoners no longer desired for interrogation in the immediate area of combat. The ship's officers of these vessels were briefed by intelligence officers on segregation of the various classes of POWs, and as a result, steps were taken to insure that the work of future interrogators would not be impaired.

Situation plots, as indicated previously, were constantly maintained by the ComPhibsPac Intelligence Section, as was the intelligence map that showed troop positions. The map was most valuable for briefing commanders of newly arrived ships, thus preventing them from firing at night on friendly forces on the many adjacent small islands.

In accordance with previously developed practice, daily situation estimates were prepared for ComPhibsPac and his staff. These were especially important at Okinawa because of the severe losses sustained by our vessels due to Kamikaze attacks. These estimates were derived from all available sources, including the highest of classified dispatches.

A major problem of reproduction developed in the photographic laboratories of the various AGCs as a result of the enormous number of negatives taken by the photographic planes. It was necessary to develop these by the hundreds of thousands. One AGC actually produced 1,250,000 prints at the scene of this operation under combat conditions. In addition, over 3,000 more copies of the 11-sheet $20 \times 22\ 1/2$ mosaic of southern Okinawa were printed for use by the troops ashore. Besides photographs of Okinawa, pictures taken of surrounding islands considered for later phases of the operation were being developed, printed, and distributed.

The map reproduction units also were continually busy. One of the major chores was the reproduction of the captured Japanese topographical maps of Okinawa. Another was the printing of charts prepared by the hydrographic survey parties, and still others were troop situation maps, special road maps for the Army Engineers, and a quantity of smaller jobs.

While the commanding general of the Tenth Army was still on board the flagship, continuous contact with his staff and intelligence officers was maintained. After his forces moved ashore, close liaison with various echelons of his army was established. Frequent visits were interchanged between officers of the ComPhibsPac Intelligence Section and the G-2 section of the Tenth Army. Liaison between superior commanders of all Army and Marine forces was maintained as much as was physically possible.

B.6 INTELLIGENCE IN SUPPORTING COMMANDS

Thus far, we have considered the role of intelligence in the Okinawa operation from the standpoint of the ComPhibsPac Intelligence Section, at the same time bringing into the picture as much as practicable of the activities and contributions of intelligence units attached to subordinate, superior, or separate commands. An attempt to describe in more detail the functioning of three of these units follows.

On 30 October, commander second carrier task force relieved commander first carrier task force at Ulithi and commander first carrier task force returned to Pearl Harbor where he set up headquarters. It was at this time that the staff of the first carrier task force began preparations for the Iwo and Okinawa operation plans, but the formal work did not actually commence until mid-December. The Iwo Jima plan was naturally the first consideration, and was especially complex since it called for diversionary raids on the Empire. This meant that the major burden of the intelligence officers was to learn all that could be known about the Tokyo area in particular, and about the Japanese homeland in general. Once the major portion of this work was done, however, it was equally essential to turn their attention to Okinawa and the problems which that operation would bring. The Iwo Jima plan was finished in the rough stage shortly after Christmas. Inasmuch as the Okinawa operation was not to commence until mid-March, with the landings scheduled for 1 April, it did not demand immediate attention. Nevertheless, commander first carrier task force and his staff were scheduled to go to sea again late in January. Consequently this was the last opportunity to secure information on Okinawa in a somewhat leisurely fashion. Therefore, as much data was collected as could be gathered at that time from intelligence centers and commands. Liaison was maintained with the staffs of CincPac, Commander Fifth Fleet, and ComPhibsPac as well as representatives of the British Pacific Fleet who were also to take part in the operation. At this time intelligence officers made their recommendations to the intelligence centers as to the types of materials most desired for the coming operation.

Conferences were held with representatives of the other commands in an effort to consolidate activities as far as possible. Such a period of planning had been tried once before by Commander Third Fleet and Commander Second Carrier Task Force before the Philippines invasion, but was not entirely successful because the commands at that time did not change simultaneously. This was the first time that the fleet commander and the task force commander began operations together after a period of sustained long-range planning in the rear areas.

The next overt action against Okinawa took place on 22 January 1944 when the fast carrier force again struck the island; this time the primary mission was to obtain photographic coverage. Despite adverse weather, all task groups flew photographic hops and almost all the desired coverage was obtained. Later, Commander First Carrier Task Force and his staff returned to Ulithi, and, on 26 January Commander Fifth Fleet and Commander First Carrier Task Force relieved Commander Third Fleet and Commander Second Carrier Task Force, respectively. Early in February the fast carrier force again sortied from Ulithi, and on February 1 and 17 they struck at Tokyo areas preliminary to the landings on Iwo Jima on 19 February. For the next month they continued to support the Iwo Jima operation.

This is significant, in that the carrier intelligence officers who had to make their plans for the Okinawa operation scheduled for late March and April were once again at sea and at work on another complicated operation. On 1 March, Task Force 58 (the fast carrier task force) conducted another strike and photographic reconnaissance of the Nansei Shoto and specifically of Okinawa and the islands to be occupied along with it, such as Kerama Retto and the eastern islands of Okinawa. This was the last offensive action against the island until the operation began. Early in March, when the Iwo Jima operation was more or less secure, even though fighting was still heavy ashore, the bulk of the carrier forces returned to Ulithi. The commander, air support control units, began briefing the carrier ground officers and aviators about mid-March. It had been decided that the ground officers would be briefed on the entire operation, that is, the prelanding strikes at Kyushu airfields and the naval base at Kure, as well as the proposed landing operation on Okinawa itself. For purposes of security, the aviators were to be briefed at this time only on simulated offensives as far as the problem of air support was concerned, and the times of L-day and H-hour were not divulged. Squadron intelligence officers were then permitted to brief the pilots on the operations of immediate importance. Consequently, the aviators aboard the first carriers were briefed first on Kyushu and on the Inland Sea areas which they struck first. After those attacks they were briefed on air support for Okinawa. On 21 March, the escort carriers that were to protect the amphibious forces and furnish the bulk of the air support sortied from Ulithi. On L-day minus 8, Kerama Retto was captured, and on L-day, which was 1 April, Okinawa was invaded.

The formal operation plans and orders for this operation were issued as follows: CincPac Op-Plan No. 14-44 of 31 December 1944; Commander Fifth Fleet Op-Plan No. 1-45 of 3 January 1945; ComPhibsPac Commander First Carrier Task Force Operation Order No. 2-45 of 1 March 1945. Naturally, supplements were issued from time to time and the words which had appeared on the Iwo Jima operation order were by now familiar to all intelligence officers. The words were "Changes are inevitable—relax and enjoy them."

Certain problems arose at this time which were, in one sense or another, unique for the naval air force. First, the large number of operational airfields in the Kyushu-Inland Sea areas designated for diversionary attacks made it necessary to obtain the very latest estimates of enemy air strength on the reserve fields as rapidly as possible. This gave rise to a problem in dissemination, and finally a system was worked out whereby most of this information was exchanged by dispatch. Such a problem is likely to occur in any area that combines both civil and military organization, and a speedy and efficient system for exchanging such information must be adopted. Second, because of the combined nature of the operation underway, the Superfortress raids on the Empire, and the activities of the British Pacific Fleet operating as Task Force 57 in our own fleet organization, it was necessary to maintain active liaison with many units. It was especially important when the B-29s were attacking the Kyushu airfields during the period from 15 April to 15 May to learn from them as soon as possible the results of their raids. And as far as the British Pacific Fleet was concerned, it was necessary to place air combat intelligence officers aboard their ships for the purpose of indoctrinating their officers in the methods that we were using to forward intelligence. As soon as the amphibious group commanders chosen to participate in the operation were notified by ComPhibsPac of their missions, their respective intelligence officers were directed to confer with the ComPhibsPac intelligence officer. At their meeting, the work to be performed in the overall intelligence plan was determined, outlined, and discussed with the group intelligence officers. They were instructed to confine the scope of their activities to the particular beaches and approaches thereto on which their respective forces had been directed to land. They were further instructed to prepare their own beach sketches and diagrams except in the cases of the two groups that were to land troops on the principal beaches of Okinawa. It was arranged that they should prepare their sketches in conjunction with the ComPhibsPac Intelligence Section, so that a complete approved set could be printed. However, in all instances, the ComPhibsPac Intelligence Section was to furnish all basic information and render all possible assistance by supplying newly received intelligence. As the commander of Amphibious Group 1 had been designated as commander fire support group, his intelligence officer was directed to supervise preparation of special air and gunnery target folders, prepare special photographs of principal targets for issue to fire support ships, and brief the photo pilots on the CVEs. This officer was also ordered to assist in the preparation of the photographic plans.

When the various group intelligence officers were first apprised of the objective, they were also given all material available at that time in the ComPhibsPac Intelligence Section, so that they could prepare preliminary reports for their commanding officers. Thereafter, their particular studies and work dealt in the main with their immediate objectives. All strategic and tactical maps and charts were supplied by ComPhibsPac; the group intelligence officer had no duty in this respect. Generally speaking, the responsibilities of his section and his administrative organization were similar to those of the ComPhibsPac Intelligence Section.

The group intelligence officers handled their own immediate distribution problems, but if an item required wide distribution, sufficient copies were sent to ComPhibsPac, reproduced in larger quantities, and distributed by his Intelligence Section to the ships and units requiring it. The groups were furnished pertinent photo coverage as received. Arrangements were made with the photo interpretation squadrons to deliver photos to them in the forward areas. This arrangement was supervised by a ComPhibsPac liaison officer who was responsible for guaranteeing to the groups coverage of their particular areas.

In addition to close liaison with ComPhibsPac intelligence officers, group intelligence officers conferred frequently with intelligence officers attached to subordinate echelons and briefed them as to what duties they were expected to discharge. These latter officers in turn depended upon the group intelligence officer as their principal source of information. During their conference with him, plans were made for the expeditious handling of captured documents and material, processing of POWs, liaison between the various units, continuous interchange of pertinent intelligence, and agreements for the briefing of all personnel involved in their particular phase of the operation.

While en route to Okinawa, the group officers received all reports and latest photos on their respective objectives, and thus revision of their material was in constant process. The Intelligence Section of Amphibious Group 1 organized its target assessment center and printed a final intelligence map for distribution by destroyer to ships and units within its command. At Ulithi, this group had arranged three briefing rooms, complete with maps, charts, terrain models, and other equipment, on board the ComPhibs Group 1 AGC for the briefing of pilots so that this part of their program could be executed more efficiently and expeditiously. The other group intelligence officers, where possible, also arranged for one or more similar rooms and found them to be a great timesaver in briefing pilots, intelligence officers from subordinate commands, commanding officers, ship's officers, staff officers, beachmasters, and shore-party personnel. In short, the ComPhibsPac Intelligence Section's crusade on briefing had produced results, and amphibious personnel were told where they were going, what to look for, and what they could expect.

After the arrival at Okinawa, the intelligence officer of ComPhibsPac Group 7, one of the first intelligence officers on the scene, realized the necessity for rapid dissemination of information on Kamikaze methods, and personally ordered an immediate survey during which he and his subordinates interviewed 50 or more commanding officers whose ships had been under attack by suicide pilots. An analytical study was quickly prepared by his Intelligence Section and distributed to the entire task force. Later it was further distributed to all commands in the Pacific. This was the first effective study of suicide attack method and drew great praise from all recipients. As the campaign progressed, further information on Kamikaze attacks was collected and distributed to various other intelligence units.

The intelligence officer of ComPhibs Group 1 and his assistants supervised all photography during the week prior to the invasion, and evaluated and selected targets for naval gunfire from the results of photo reconnaissance sorties. In conjunction with the AGC gunnery officer, he operated a target information center at which were fixed positions or targets and where the demolished targets were eliminated and new targets recorded. Ships that failed to get on a target were materially assisted by the facilities of the center, when intelligence officers who had thoroughly studied the objectives were sent aboard with the most recent gridded photos and charts. The intelligence officer of this group also prepared duplicate negatives and prints of early photography for distribution to other groups in accordance with the previously determined photographic plan.

Intelligence officers of other groups processed intelligence for their respective areas, and assisted the G-2 sections of the Army in the reproduction of photos and printed charts required for operations ashore. Some of these officers were later detached for operations against outlying islands. In such cases, they were responsible for the preparation and distribution of all the intelligence for these special operations, including the preparation of beach sketches, briefing of personnel, interrogation, and study of captured documents.

The duties and activities of the TransRon and TransDiv intelligence officers in the Okinawa campaign, as in all others, varied considerably. Some squadrons and divisions were employed in the Philippines area and some in the Iwo Jima operation prior to the invasion of Okinawa. Staging and rehearsal areas were different. Assignments varied and planning was necessarily started at different times. It is difficult to present the work of TransRon and TransDiv intelligence officers in a general way. In order to provide a typical example, rather than an indistinct general picture of the duties of an intelligence officer assigned to a TransRon or TransDiv, certain aspects of the work of the intelligence officer attached to TransDiv 47 are described. TransDiv 47 was part of TransRon 16 which moved the Twenty-seventh Infantry Division from Espiritu Santo to Okinawa. A copy of ComPhibsPac operation plan was received by the TransDiv 47 intelligence officer while at Saipan on 3 March 1945, from the ComPhibsPac intelligence officer. During a 2-day stop at Guam, 4 and 5 March 1945, the intelligence officer checked the stock at the CincPac-CincPoa issuing office and obtained the latest bulletins on Japan, Formosa, and the China coast, large-scale plotting sheets of the Western Pacific, and copies of AMS 1/25,000 maps of Okinawa Gunto. The final distribution package from the ComPhibsPac Intelligence Section was received at Espiritu on 7 March 1945. This included a two-section plaster model of the assault beaches. All material was addressed from the ComPhibsPac Intelligence Section to the TransDiv intelligence officer, and was delivered at Espiritu by an officer of the ComPhibsPac Intelligence Section. A copy of the TransRon operation plan was received at Ulithi on 3 April 1945.

A meeting of all intelligence officers of TransRon 16 was held at Espiritu on 18 March to discuss the briefing program and probable schedule of operations and to make certain that all intelligence officers and ships had received all necessary intelligence material. The TransDiv intelligence officer, at numerous times, discussed operations plans, intelligence, and the briefing schedule with the infantry S-2 officer aboard his ship. Very close and constant liaison and free interchange of material was maintained between these officers until the latter's debarkation at Okinawa. While at Espiritu, the TransDiv officer also discussed problems with, and solicited criticism and recommendations from, the intelligence officers of the five APAs and two AKAs of the division.

En route to Espiritu from 5 to 15 March, the TransDiv intelligence officer had studied all available material and prepared an *Intelligence Digest-Okinawa Gunto*. This digest included prepared sketches for boat employment, movement, and control for the TransDiv boat flotilla commander. These were used in the TransRon operation plan. En route to Ulithi, the TransDiv intelligence officer assisted the infantry S-2 officer in writing an intelligence bulletin. For the TransDiv commander and his staff, the intelligence officer wrote a summary of the seven tentative and alternate Twenty-seventh Division missions.

At Espiritu, the TransDiv intelligence officer distributed 10 copies of the *Intelligence Digest* to each ship of the division, and 35 copies to the intelligence officers of each of the other 2 divisions in the squadron. One copy of the large-scale plotting sheet of the Western Pacific was also provided each ship in the division. En route to Ulithi he furnished various maps, charts, and gazetteers to the staff communication officer, the ship's CIC officer and ship's navigator. He provided the TransDiv beachmaster, boat flotilla commander, ship's beachmaster, ship's intelligence officer, and boat group commander with all available material. At Ulithi, the following material was distributed to each ship of the division: 4 copies of photographs of the assault beaches, 25 copies of tide curve (and boat draft lines) at Okinawa, and 2 copies of the intelligence bulletin prepared by the Twenty-seventh Division. Immediately prior to and during the Trugen Shima operation (10–12 April) all available Eastern Island intelligence material, maps and charts, estimates of troops, defenses, etc., were furnished the boat flotilla and boat group commanders, and the TransDiv and ship's beachmasters. During the unloading at Hagushi Beaches (12–16 April) all photographs, maps and charts of the area were distributed to TransDiv and ship's beachmasters, the boat flotilla and boat group commanders.

En route to Ulithi, the TransDiv intelligence officer briefed staff and ship's officers in the wardroom, CPOs in the CPO's mess room, ship's crew in the mess hall. The briefing covered the overall fleet, air, and joint expeditionary force program, task organization and composition, as well as the intelligence on Okinawa Gunto. The infantry S-2 officer briefed staff and ship's officers on enemy defenses and troop dispositions, and outlined the various possible missions of the Twenty-seventh Division. En route to Kerama Retto (4–9 April) the TransDiv intelligence officer briefed the boat group officers and men on the development of Okinawa operations, general conditions of sea, surf, weather, reef, and beach conditions at Eastern Islands and Hagushi.

The bulkheads of the flag bridge plot room were covered with various maps and charts of the Northwest Pacific, the 1/72,000 approach chart, and the 1/100,000 plotting map of Okinawa. Throughout the operation, submarine contacts, enemy air bases and operational radii, all information on mines, navigation hazards, and aids, UDT and hydrographic reconnaissance information, operating areas, summaries of enemy bases, etc., were plotted on these maps and charts. Daily dispatch summaries were always plotted immediately upon receipt. Copies of the 1/100,000 map were posted in the ship's sick bay, wardroom, and general mess. The general situation was kept plotted on all of these for the information of everyone aboard.

The role of Naval Intelligence in the Okinawa operation was not perfect. Certain of the units it served complained of the receipt of excess material and others that they were not properly supplied. In the planning phase certain factors were overlooked, and, after the invasion began, liaison sometimes broke down. However, in this operation Naval Intelligence combined past experience and training to produce its most successful operational effort.

INTENTIONALLY BLANK

APPENDIX C

National Intelligence Community

Naval intelligence is part of a broader national intelligence community (IC) that employs various sources and methods to address the needs of a broad range of decision makers. What follows is a brief overview of the national intelligence community and the seven intelligence disciplines discussed in joint doctrine that provide the context for this publication's descriptions of naval intelligence operations.

Questions concerning how to best employ internal intelligence capabilities and leverage external intelligence resources occur at every point in the intelligence process. Some members of the intelligence community, such as naval intelligence, have collection capabilities that span the full range of intelligence disciplines and favor an all-source approach to intelligence analysis. Other members of the intelligence community, such as the National Security Agency (NSA), are focused on a single intelligence discipline and primarily conduct exploitation of data collected using that discipline. This appendix provides information needed to better understand planning, synchronization, and conduct of intelligence operations that employ a multiintelligence, multimethod approach to addressing the needs of the commander, staff, and supported organizations. Given the foundational nature of the concepts presented to applying operational art to intelligence operations, readers should refer to JP 2-0, Joint Intelligence; JP 2-01, Joint and National Intelligence Support to Military Operations; and NWP 2-01, Intelligence Support to Naval Operations, for a more complete discussion of specific intelligence community members, the national intelligence community framework, and the intelligence disciplines.

As shown in figure C-1, the national intelligence community consists of 17 member organizations.

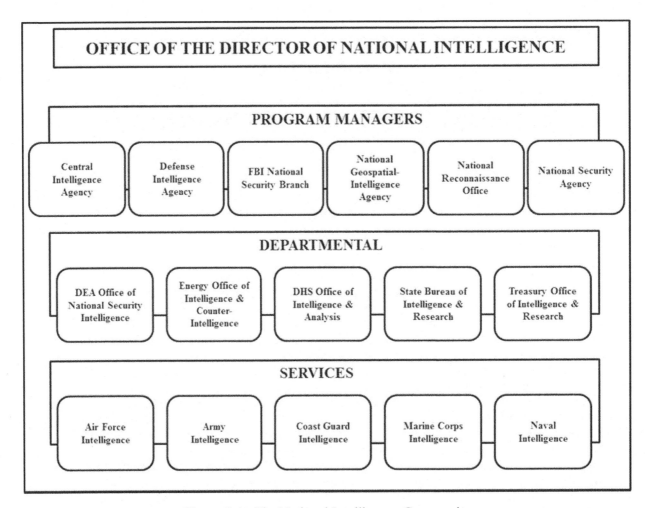

Figure C-1. The National Intelligence Community

The figure above is not meant to show a hierarchy of intelligence organizations, although there are direct lines of authority for each organization as shown in figure C-2. What is interesting to note is that there are certain organizations, such as the Central Intelligence Agency (CIA), the National Geospatial-Intelligence Agency (NGA), the Defense Intelligence Agency (DIA), and NSA that are purely intelligence agencies and have functional management and oversight responsibilities for particular intelligence disciplines and analytic areas. For example, NGA is the community's functional manager for geospatial intelligence (GEOINT) and DIA is the functional manager for defense measures and signals intelligence (MASINT). Other members of the intelligence community, like the Services and other government departments, such as the Departments of State and Homeland Security, have significant intelligence capabilities embedded in their organizations.

NWP 2-0

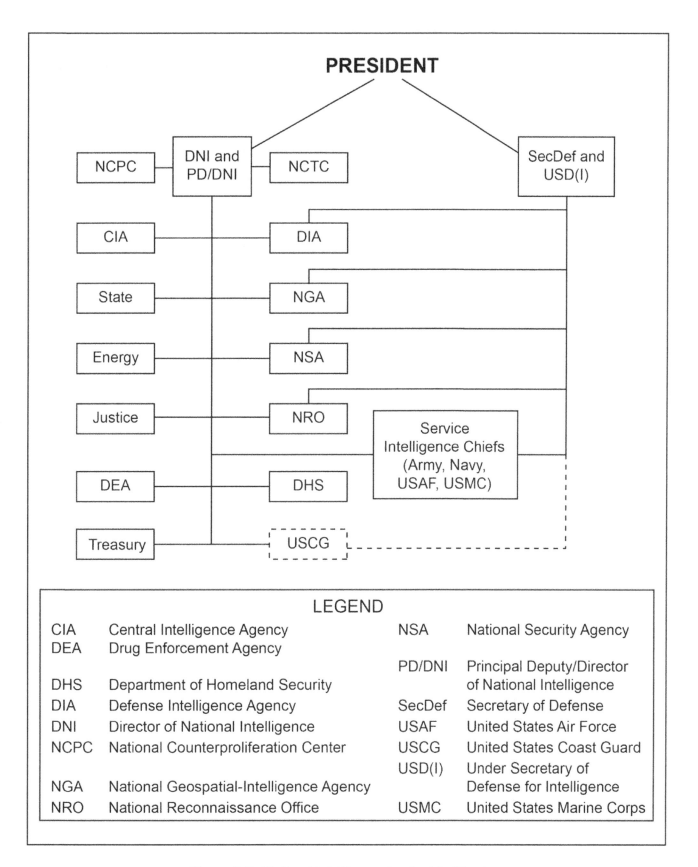

Figure C-2. National Intelligence Leadership Structure

Each member of the national intelligence community conducts extensive collection, processing, analysis, and dissemination activities. They employ specialized resources and dedicated personnel to gain information about potential adversaries and events and have specialization within specific functional and technical fields that is contributed to the entire community. For example, naval intelligence contributes significantly to the understanding of maritime threats and the maritime operational environment and this intelligence is used by organizations throughout the intelligence community. These functional "lanes in the road" are not a rigid construct as each organization conducts the collection, processing, analysis, and dissemination operations needed to support its mission. For example, there may be CIA or NGA resources dedicated to collecting and analyzing maritime intelligence issues to meet an operational need. Given the collaborative nature of intelligence analysis and production, it also likely that the maritime intelligence cells in those organizations collaborate with their counterparts throughout the intelligence community.

National intelligence organizations routinely support forces at the operational and tactical levels of war while continuing to support national level decision makers. In some cases, this support is available as a continuous service. Advances in information technologies have enabled organizations to database raw, exploited, and finished intelligence products and to make these products available via "push" broadcasts and by "pull" of needed intelligence by cleared personnel using Intelink sites on Nonsecure Internet Protocol Router Network (NIPRNET), SECRET Internet Protocol Router Network (SIPRNET), and the Joint Worldwide Intelligence Communications System (JWICS). For example, NGA makes raw and annotated imagery products available via the Web-based access and retrieval portal (WARP) and ONI permits access to its merchant ship databases via SeaLink, which is accessible via SIPRNET and JWICS Web interfaces. New requirements for leveraging external intelligence resources may require significant lead times and coordination efforts. As is discussed throughout this appendix, naval intelligence professionals must plan ahead for national intelligence capability integration and must identify new requirements for external intelligence resources as early in the planning process as possible.

In addition to the sources cited above, readers desiring a short overview of the capabilities and responsibilities of each of the intelligence community members can refer to the Office of the Director of National Intelligence (ODNI) unclassified Web site (http://www.dni.gov/index.php/intelligence-community/members-of-the-ic).

APPENDIX D

Intelligence Disciplines

D.1 INTRODUCTION

Chapter 3 defined each of the seven intelligence disciplines discussed in joint doctrine. This appendix provides a more detailed overview of each of the intelligence disciplines, as understanding their strengths and weaknesses is fundamental to naval intelligence operational art.

D.2 COUNTERINTELLIGENCE (CI)

D.2.1 Definition

CI is "information gathered and activities conducted to identify, deceive, exploit, disrupt, or protect against espionage, other intelligence activities, sabotage, or assassinations conducted for or on behalf of foreign powers, organizations or persons or their agents, or international terrorist organizations or activities."[1]

D.2.2 Discussion

CI is similar to, and often confused with, HUMINT, as CI uses many of the same techniques as HUMINT to collect information. This confusion between the two terms is exacerbated by many units or staffs being referred to as "CI/HUMINT" organizations.

CI is listed in joint doctrine as an intelligence discipline. CI, however, differs from all of the other INTs. All of the other intelligence disciplines (GEOINT, HUMINT, MASINT, OSINT, SIGINT, and TECHINT) are focused on foreign intelligence (FI). FI is defined in joint doctrine as "information relating to capabilities, intentions, and activities of foreign powers, organizations, or persons, but not including counterintelligence, except for information on international terrorist activities."[2] CI, on the other hand, involves a range of activities beyond intelligence collection and FI.

CI obtains information by or through the functions of CI operations, investigations, collection and reporting, analysis, production, dissemination, and functional services. CI organizations use that information for both offensive and defensive purposes, in coordination with other intelligence disciplines, law enforcement, and security elements.

The function of CI is to provide direct support to operational commanders, program managers, and decision makers. This support includes: CI support to force protection during all types and phases of military operations; detection identification and neutralization of espionage; antiterrorism; threat assessments; counterproliferation actions associated with CBRNE; countering illegal technology transfer; acquisitions systems protection; support to other intelligence activities; information systems protection; and treaty support.

What does make CI an intelligence discipline—even though it is separate and distinct from FI—is that many of its activities can serve as a distinct source of intelligence to the overall FI effort. For example, CI can support the FI disciplines through its contribution to the I&W function, by its collection, analysis, and production capabilities, and by maintenance of CI databases.

---

[1] JP 1-02.
[2] JP 1-02.

It is also worth noting that CI activities involve authorities that are not encompassed by fleet intelligence operations.

Note

> CI is addressed in detail in JP 2-01.2, Counterintelligence and Human Intelligence Support to Joint Operations.

D.2.3 Fleet Counterintelligence Considerations

CI capabilities extending to fleet intelligence at the operational and tactical levels of war are components of the national CI enterprise under the cognizance of the NCIX. The DOD executive agent for CI is the Director, Defense Counterintelligence and Human Intelligence Center (DCHC). The Department of the Navy authority over naval elements of this enterprise resides with the Director, Naval Criminal Investigative Service (NCIS), and the theater authority with the CCDR J-2 CI/HUMINT staff (J-2X).

As fleet commanders and intelligence staffs do not retain authorities to conduct strategic or operational-level CI activities, the NCC/numbered fleet commander (NFC) would not be in a position to expand upon this enterprise at the operational level of war without direct support from the theater NCIS regional office.

Any expansion of the CI enterprise to the tactical level of war would be limited to CI Force Protection Source Operations (CFSO) conducted by expeditionary forces. These activities do not come under the cognizance of NCIS, and would be executed under the authority of the expeditionary force commander and under the auspices of the CCDR J-2X or possibly the NCC/NFC Maritime Force intelligence directorate and human intelligence staff element (N-2X). The Navy Expeditionary Intelligence Command (NEIC) and NSW elements can assist in providing such capabilities.

Naval forces have access to a limited number of organizations capable of CI collection that are able to be tasked as assets. The principal organizations would be NCIS at the operational level of war and NEIC at the tactical level of war. There are also a limited number of joint or national organizations capable of CI collection able to be requested as resources.

D.2.4 Guidance

1. DOD Directive O-5240.02, Counterintelligence, 20 Dec 2007 (incorporating CH-1, 30 Dec 2010)

2. DOD Instruction 5240.10, Counterintelligence (CI) I in the Combatant Commands and Other DoD Components, 05 Oct 2011

3. JP 2-01.2, Counterintelligence and Human Intelligence Support in Joint Operations, 16 March 2011, CH 1 26 August 2011

4. SECNAVINST 5430.107, Mission and Functions of the Naval Criminal Investigative Service, 20 Dec 2005

5. SECNAVINST 3850.2C, Department of the Navy Counterintelligence, 20 Jul 2005

6. Fleet Intelligence, Surveillance, and Reconnaissance (ISR) Concept of Operations: Intelligence Operations Support to the Maritime Operations Center (MOC), May 2010 (Appendix C.1.7)

7. Navy-Wide OPTASK Intelligence, COMSECONDFLT 071705Z OCT 09 (Paragraphs 5.B.3, 9)

8. COMUSNAVSO Supplement to the Navy-Wide OPTASK Intelligence, COMUSNAVSO 052054Z APR 06 (Paragraph 10)

9. COMSIXTHFLT Supplement to the Navy-Wide OPTASK Intelligence, COMSIXTHFLT 211200Z APR 09 (Paragraphs 6.D, 11)

10. COMSEVENTHFLT Supplement to the Navy-Wide OPTASK Intelligence, COMSEVENTHFLT 151000Z MAR 06, 151005Z MAR 06, 151010Z MAR 06 (Paragraph D.6.F.5).

D.3 GEOSPATIAL INTELLIGENCE (GEOINT)

D.3.1 Definition

GEOINT is "the exploitation and analysis of imagery and geospatial information to describe, assess, and visually depict physical features and geographically referenced activities on the Earth. Geospatial intelligence consists of imagery, imagery intelligence, and geospatial information."[3]

D.3.2 Discussion

GEOINT encompasses a range of products from simple imagery intelligence (IMINT) reports to complex sets of layered foundation and intelligence/mission-specific data. GEOINT products are often developed through a "value added" process, in which both the producer and the user of GEOINT update a database or product with current information. Full motion video is another GEOINT intelligence collection capability that has proven key to activity-based intelligence collection by providing near-continuous or sustained collection on designated targets. The three components of GEOINT (imagery, IMINT, and geospatial information) are discussed below.

1. Imagery is a likeness or presentation of any natural or man-made feature or related object or activity and the positional data acquired at the same time the likeness or representation was acquired, including products produced by space-based national intelligence reconnaissance systems, and likenesses or presentations produced by satellites, airborne platforms, unmanned aerial vehicles, or other similar means (except that such term does not include handheld or clandestine photography taken by or on behalf of HUMINT collection organizations). Imagery is used extensively to update GEOINT foundation data and serves as GEOINT's primary source of information when exploited through IMINT. The vast majority of modern imagery products are created, processed, and disseminated in an electronic still or motion format. A few film-based systems still exist to fulfill special requirements.

2. IMINT is the technical, geographic, and intelligence information derived through the interpretation or analysis of imagery and collateral materials. It includes exploitation of imagery data derived from electro-optical (EO), radar, infrared, multispectral, and laser sensors. These sensors produce images of objects optically, electronically, or digitally on film, electronic display devices, or other media. A wide variety of platforms and sensors support IMINT operations. IMINT is a product that is the result of processing and exploiting raw imagery (information) and creating an analyzed product (intelligence). An image alone is only information in the form of pixels, digits, or other forms of graphic representation and the data behind that portrayal. Imagery source categories include commercial remote sensing, EO, ground photo, hyperspectral imagery (HSI), IR, light detection and ranging (lidar), multispectral imagery (MSI), panchromatic, polarmetric, and synthetic aperture radar.

    a. EO sensors provide digital imagery data in the infrared, visible, and ultraviolet regions of the electromagnetic spectrum. Panchromatic EO sensors detect a broad segment of the visible spectrum, while other EO sensors focus on IR energy or detect multiple narrow bands across the EO spectrum. EO sensors generally provide a high level of detail or resolution as compared to radar or other sensors, but cannot successfully image through bad weather. Panchromatic sensors provide the highest level of resolution, but cannot image at night. EO offers many advantages over nondigital (i.e., film-based) systems including improved timeliness, greater dissemination options, imagery enhancement, and additional exploitation methods.

---

[3] JP 1-02.

b. Infrared imaging sensors provide a pictorial representation of the contrasts in thermal infrared emissions between objects and their surroundings, and are effective during periods of limited visibility such as at night or in inclement weather. A unique capability available with infrared sensing is the ability to detect ongoing activity (based on heat levels) as well as past activity through residual thermal effects.

c. Spectral imagery sensors operate in discrete spectral bands, typically in the IR and visible regions of the electromagnetic spectrum. Spectral imagery is useful for characterizing the environment or detecting and locating objects with known material signatures. A multispectral image is made from a set of images taken at different intervals of continuous wavelengths, called "bands," which when viewed together produce a color image. It is similar to using a color filter when taking a black and white picture. Only the rays of the color of the filter are allowed to reach the film. Traditionally, multispectral sensors contain a red, green, and blue band, but can contain bands that image regions of the electromagnetic spectrum to which the human eye is not sensitive. The advantage of taking multispectral images (hundreds of bands) is the ability to discern different materials through their spectral signature. This information can be transferred into intelligence and aid in the analysis of targets. Some MSI sensors provide low resolution, large area coverage that may reveal details not apparent in higher resolution panchromatic imagery. Map-like products can be created from MSI data for improved area familiarization and orientation. HSI is derived from subdividing the electromagnetic spectrum into very narrow bandwidths (thousands of bands) which may be combined with, or subtracted from each other in various ways to form images useful in precise terrain or target analysis. For example, HSI can analyze electromagnetic propagation characteristics, detect industrial chemical emissions, identify atmospheric properties, improve detection of blowing sand and dust, and evaluate snow depths.

d. Radar imaging sensors provide all weather imaging capabilities and the primary night capability. Radar imagery is formed from reflected energy in the radio frequency portion of the electromagnetic spectrum. Some radar sensors provide moving target indicator capability to detect and locate moving targets such as armor and other vehicles.

e. Light detection and ranging (lidar) sensors are similar to radar, transmitting laser pulses to a target and recording the time required for the pulses to return to the sensor receiver. Lidar is used to measure shoreline and beach volume changes, conduct flood risk analysis, identify water flow issues, and augment transportation mapping applications. Lidar supports large-scale production of high-resolution digital elevation products displaying accurate, highly detailed three-dimensional models of structures and terrain invaluable for operational planning and mission rehearsal.

3. Geospatial information identifies the geographic location and characteristics of natural or constructed features and boundaries on the Earth, including: statistical data; information derived from, among other things, remote sensing, mapping, and surveying technologies; and mapping, charting, geodetic data, and related products. This information is used for military planning, training, and operations including navigation, mission planning and rehearsal, modeling and simulation, and targeting.

GEOINT includes products with significant relevance to the Navy. These include aeronautical charts, flight information publications, hydrographic products, digital nautical charts, surface/subsurface nautical charts and Notices to Mariners.

Notes

- GEOINT is addressed in detail in JP 2-03, Geospatial Intelligence Support to Joint Operations.

- This appendix treats GEOINT as a single discipline encompassing its three subdisciplines (imagery, IMINT, geospatial). However, the reality of fleet ISR programs and operations points to IMINT as the predominant GEOINT activity by fleet intelligence at the operational and tactical levels of war.

- The term "full spectrum GEOINT" (FSG) is also used in context with GEOINT, although it is not currently defined in joint doctrine. FSG represents a holistic approach to integrating data and information from a wide variety of disciplines to answer questions or needs which are geospatial in nature. FSG seeks to increase understanding by purposely incorporating a diverse set of data and information sources in such a way as to present a more inclusive geospatially related description of environments and conditions. Typical analysis will incorporate the traditional GEOINT disciplines into a larger schema exploiting such techniques as multidata fusion methods and temporal exploitation.

D.3.3 Fleet Geospatial Intelligence Considerations

The GEOINT capabilities extending to fleet intelligence at the operational and tactical levels of war are components of the national GEOINT enterprise under the cognizance of the Director, NGA as the Functional Manager for the National System for Geospatial Intelligence (NSGS). There is not a centralized Navy authority over maritime elements of this enterprise. Accordingly, GEOINT sensors on naval platforms are more closely associated with the respective platform programs offices than with a central office that coordinates the intelligence discipline within the Service. In addition, the theaters do not exercise centralized authority over GEOINT operations in the area of responsibility (AOR).

GEOINT—particularly IMINT—is an intelligence discipline that is used extensively in fleet intelligence collection operations.

Naval forces have access to a high number of platforms with sensors capable of GEOINT collection that are able to be tasked as assets. There are also a high number of joint or national platforms capable of GEOINT collection able to be requested as resources.

D.3.4 Guidance

1. National System for Geospatial Intelligence Users Guide (Intellipedia)

2. JP 2-03, Geospatial Intelligence Support in Joint Operations, 31 Oct 2012

3. USAFRICOM Standing Reconnaissance, Surveillance, and Target Acquisition (RSTA) Annex, 10 Jan 2013

4. USCENTCOM Baseline Reconnaissance, Surveillance, and Target Acquisition (RSTA) Annex, Oct 2012

5. USEUCOM Baseline Reconnaissance, Surveillance, and Target Acquisition (RSTA) Annex, 18 Apr 2012

6. Fleet Intelligence Collection Manual, ONI-1200-001-11, May 2011 (Chapter 2.D.2, 2.F.2, 2.F.3)

7. Fleet Intelligence, Surveillance, and Reconnaissance (ISR) Concept of Operations: Intelligence Operations Support to the Maritime Operations Center (MOC), May 2010 (Appendices C.1.1, C.1.2, C.1.3, C.1.4)

8. Navy-Wide OPTASK Intelligence, COMSECONDFLT 071705Z OCT 09 (Paragraphs 4.C.1, 5.B.6, 5.C)

9. COMFIFTHFLT Supplement to the Navy-Wide OPTASK Intelligence, COMUSNAVCENT 041447Z MAR 11 (Paragraphs 4.D.4, 4.D.6, 4.E.2)

10. COMSIXTHFLT Supplement to the Navy-Wide OPTASK Intelligence, COMSIXTHLT 211200Z APR 09 (Paragraphs 5.C.5, 6.C)

11. COMSEVENTHFLT Supplement to the Navy-Wide OPTASK Intelligence, COMSEVENTHFLT 151000Z MAR 06, 151005Z MAR 06, 151010Z MAR 06 (Paragraphs D.3.A.1, D.6.A).

D.4 HUMAN INTELLIGENCE (HUMINT)

D.4.1 Definition

HUMINT is "a category of intelligence derived from information collected and provided by human sources."[4]

D.4.2 Discussion

HUMINT includes all forms of information gathered by humans, from direct reconnaissance and observation to the use of recruited sources and other indirect means. This discipline also makes extensive use of biometric data (e.g., fingerprints, iris scans, voice prints, facial/physical features) collected on persons of interest.

1. Interrogation. Interrogation is the systematic effort to procure information to answer specific collection requirements by direct and indirect questioning techniques of a person who is in the custody of the forces conducting the questioning. Proper questioning of enemy combatants, enemy prisoners of war, or other detainees by trained and certified DOD interrogators may result in information provided either willingly or unwittingly. There are important legal restrictions on interrogation and source operations. Federal law and Department of Defense policy require that these operations be carried out only by specifically trained and certified personnel. Violators may be punished under the Uniform Code of Military Justice.

2. Source Operations. Designated and fully trained military HUMINT collection personnel may develop information through the elicitation of sources, to include:

   a. "Walk-in" sources, who without solicitation make the first contact with HUMINT personnel.

   b. Developed sources that are met over a period of time and provide information based on operational requirements.

   c. Unwitting persons, with access to sensitive information.

3. Debriefing. Debriefing is the process of questioning cooperating human sources to satisfy intelligence requirements, consistent with applicable law. The source usually is not in custody and usually is willing to cooperate. Debriefing may be conducted at all echelons and in all operational environments. Through debriefing, face-to-face meetings, conversations, and elicitation, information may be obtained from a variety of human sources, such as:

   a. Friendly forces personnel, who typically include high-risk mission personnel such as combat patrols, aircraft pilots and crew, long range surveillance teams, and SOF, but can include any personnel with information that can be used for intelligence analysis concerning the adversary or other relevant aspects of the operational environment. Combat intelligence, if reported immediately during an operational mission, can be used to redirect tactical assets to attack enemy forces on a time-sensitive basis.

   b. Refugees/displaced persons, particularly if they are from enemy-controlled areas of operational interest, or if their former placement or employment gave them access to information of intelligence value.

   c. Returnees, including (returned prisoners of war and defectors, freed hostages, and personnel reported as missing in action).

   d. Volunteers, who freely offer information of value to U.S. forces on their own initiative.

4. Document and Media Exploitation (DOMEX). Captured documents and media, when properly processed and exploited, may provide valuable information such as adversary plans and intentions, force locations, equipment capabilities, and logistical status. The category of "captured documents and media" includes all

---

[4] JP 1-02.

media capable of storing fixed information to include computer storage material. This operation is not a primary HUMINT function, but may be conducted by any intelligence personnel with appropriate language support.

5. Biometrics-Enabled Intelligence. Applied BEI has the potential to identify individual adversaries by revealing their personal identity attributes via BEI, digital and physical forensics, and DOMEX. Though identity can be derived from multiple intelligence disciplines, BEI provides additional layers of resolution on individuals and threat networks which enables strategic through tactical level intelligence, force protection, and targeting support.

Note

HUMINT is addressed in detail in JP 2-01.2, Counterintelligence and Human Intelligence Support to Joint Operations.

D.4.3 Fleet Human Intelligence Considerations

The HUMINT capability extending to fleet intelligence at the operational and tactical levels of war is a component of the national HUMINT architecture under the cognizance of the Director of National Intelligence (DNI) National HUMINT Manager. The Navy authority over maritime elements of this architecture resides with the Service defense human intelligence executor (DHE), the Director of Naval Intelligence (OPNAV N2/N6). Authority as the theater DHE resides with the CCDR J-2. Initiatives to expand upon this architecture at the operational level of war are coordinated between the NCC/NFC N-2X staff and the OPNAV N-2X/CCDR J-2X staffs.

Navy HUMINT operations at the operational level of war are generally conducted by a Department of the Navy (DON) HUMINT organization that cannot be identified or detailed in the NWP due to classification constraints.

Navy HUMINT operations at the tactical level of war are generally conducted by MIO-IETs, under the administrative auspices of the Navy Expeditionary Intelligence Command (NEIC).

Details on the dissemination of HUMINT information for operational decisionmaking are provided in standing theater guidance.

HUMINT is an intelligence discipline that is used routinely in fleet intelligence collection operations.

Naval forces have access to a limited number of platforms capable of HUMINT collection that are able to be tasked as assets. There are also a limited number of joint or national platforms capable of HUMINT collection able to be requested as resources.

D.4.4 Guidance

1. JP 2-01.2, Counterintelligence and Human Intelligence Support in Joint Operations, 16 March 2011, CH 1 26 August 2011

2. Defense HUMINT Manual, Vol. I, 18 Apr 2010

3. Defense HUMINT Manual, Vol. II, 23 Nov 2010

4. Fleet Intelligence Collection Manual, ONI-1200-001-11, May 2011 (Chapters 2.D.1, 2.F.1, 15, 16, 20, 21)

5. Fleet Intelligence, Surveillance, and Reconnaissance (ISR) Concept of Operations: Intelligence Operations Support to the Maritime Operations Center (MOC), May 2010 (Appendix C.1.7)

6. Navy-Wide OPTASK Intelligence, COMSECONDFLT 071705Z OCT 09 (Paragraphs 4.C.2, 4.C.6, 5.B.4, 5.F)

7. COMFIFTHFLT Supplement to the Navy-Wide OPTASK Intelligence, COMUSNAVCENT 041447Z MAR 11 (Paragraphs 4.D.4, 4.D.8, 4.E.4)

8. COMSIXTHFLT Supplement to the Navy-Wide OPTASK Intelligence, COMSIXTHLT 211200Z APR 09 (Paragraphs 5.C.6, 6.D)

9. COMSEVENTHFLT Supplement to the Navy-Wide OPTASK Intelligence, COMSEVENTHFLT 151000Z MAR 06, 151005Z MAR 06, 151010Z MAR 06 (Paragraph D.6.D).

D.5 MEASUREMENT AND SIGNATURE INTELLIGENCE (MASINT)

D.5.1 Definition

MASINT is "intelligence obtained by quantitative and qualitative analysis of data (metric, angle, spatial, wavelength, time dependence, modulation, plasma, and hydromagnetic) derived from specific technical sensors for the purpose of identifying any distinctive features associated with the emitter or sender, and to facilitate subsequent identification and/or measurement of the same. The detected feature may be either reflected or emitted."[5]

D.5.2 Discussion

The measurement aspect of MASINT refers to actual measurements of parameters of an event or object such as the demonstrated flight profile and range of a cruise missile. Signatures are typically the products of multiple measurements collected over time and under varying circumstances. These signatures are used to develop target classification profiles and discrimination and reporting algorithms for operational surveillance and weapon systems. The technical data sources related to MASINT include:

1. EO data—emitted or reflected energy across the visible/infrared portion of the electromagnetic spectrum (ultraviolet, visible, near infrared, and infrared).

2. Geophysical data—phenomena transmitted through the Earth (ground, water, and atmosphere) and manmade structures including emitted or reflected sounds, pressure waves, vibrations, and magnetic field or ionosphere disturbances. Subcategories include seismic intelligence, acoustic intelligence, and magnetic intelligence.

3. Materials data—gas, liquid, or solid samples, collected both by automatic equipment, such as air samplers, and directly by humans.

4. Nuclear radiation data—nuclear radiation and physical phenomena associated with nuclear weapons, processes, materials, devices, or facilities.

5. Radar data—radar energy reflected (reradiated) from a target or objective.

6. Radio frequency data—radio frequency/electromagnetic pulse emissions associated with nuclear testing, or other high energy events for the purpose of determining power levels, operating characteristics, and signatures of advanced technology weapons, power, and propulsion systems.

Fleet intelligence can apply many technical data sources of MASINT to intelligence collection requirements. The most common, however, is the geophysical subcategory of acoustic intelligence (ACINT).

---

[5] JP 1-02.

NWP 2-0

D.5.3 Fleet Measurement and Signature Intelligence Considerations

MASINT capabilities accessible to fleet intelligence at the operational and tactical levels of war are not components of a national enterprise. Due to the multiple subcategories and manifestations of MASINT, there is no central IC authority that directs and controls MASINT activities. The NMMO was formed as a DIA component to oversee the "U.S. MASINT System"—but that system operates in a diverse and decentralized manner. Accordingly, there is no single Navy organization with principal authority over MASINT in the maritime domain. MASINT sensors on naval platforms are more closely associated with the respective platform program office than with a central office that coordinates the intelligence discipline within the Service. In addition, the theaters do not exercise centralized authority over MASINT operations in the AOR.

Initiatives to establish or expand upon a MASINT enterprise at the OLW are coordinated between the NCC/NFC N-2 staff and the Echelon 1 or 2 force management staffs managing the platform upon which the MASINT sensor resides. Parallel coordination would be expected with the CCDR J-2 collection management staff.

Note

> The MASINT subcategory of ACINT represents the only "enterprise-like" entity in the maritime domain—as manifested in the integrated undersea surveillance system (IUSS). The IUSS operates under the authority of Commander, Undersea Surveillance (CUS) and includes theater TFs responsible for maritime patrol and reconnaissance.

MASINT is an intelligence discipline that is used routinely in fleet intelligence collection operations.

Naval forces have access to a moderate number of platforms with sensors capable of MASINT collection that are able to be tasked as assets. The principal platforms are airborne infrared sensors and multidomain ACINT systems. There are also a high number of joint or national platforms capable of MASINT collection able to be requested as resources.

D.5.4 Guidance

1. DOD Instruction 5105.58, Measurement and Signature Intelligence (MASINT), 22 Apr 2009

2. DIAM 58-8, MASINT Requirements Process Users Guide, Oct 2001

3. USAFRICOM Standing Reconnaissance, Surveillance, and Target Acquisition (RSTA) Annex, 10 Jan 2013

4. USCENTCOM Baseline Reconnaissance, Surveillance, and Target Acquisition (RSTA) Annex, Oct 2012

5. USEUCOM Baseline Reconnaissance, Surveillance, and Target Acquisition (RSTA) Annex, 18 Apr 2012

6. Fleet Intelligence Collection Manual, ONI-1200-001-11, May 2011 (Chapters 2.D.4, 2.D.5, 2.F.5, 18)

7. ONI ACINT Collection Guide, 09 Apr 2009

8. Fleet Intelligence, Surveillance, and Reconnaissance (ISR) Concept of Operations: Intelligence Operations Support to the Maritime Operations Center (MOC), May 2010 (Appendix C.1.9)

9. Navy-Wide OPTASK Intelligence, COMSECONDFLT 071705Z OCT 09 (Paragraphs 4.C.1, 4.C.4, 5.D)

10. COMFIFTHFLT Supplement to the Navy-Wide OPTASK Intelligence, COMUSNAVCENT 041447Z MAR 11 (Paragraphs 4.D.4, 4.D.9, 4.E.5)

NWP 2-0

11. COMSIXTHFLT Supplement to the Navy-Wide OPTASK Intelligence, COMSIXTHLT 211200Z APR 09 (Paragraphs 5.C.7)

12. COMSEVENTHFLT Supplement to the Navy-Wide OPTASK Intelligence, COMSEVENTHFLT 151000Z MAR 06, 151005Z MAR 06, 151010Z MAR 06 (Paragraph D.6.C).

D.6 OPEN-SOURCE INTELLIGENCE (OSINT)

D.6.1 Definition

OSINT is "information of potential intelligence value that is available to the general public."[6]

Note

> Another widely recognized definition for OSINT is "intelligence that is produced from publicly available information and is collected, exploited, and disseminated in a timely manner to an appropriate audience for the purpose of addressing a specific intelligence requirement."[7]

D.6.2 Discussion

OSINT is based on publicly available information (i.e., any member of the public could lawfully obtain the information by request or observation), as well as other unclassified information that has limited public distribution or access. Examples of OSINT include online official and draft documents, published and unpublished reference material, academic research, databases, commercial and noncommercial Web sites, chat rooms, and Web logs ("blogs"). OSINT complements the other intelligence disciplines and can be used to fill gaps and provide accuracy and fidelity in classified information databases. However, caution should be exercised when using OSINT in that open sources may be susceptible to adversary use as a mode of deception (e.g., incorrect information may be planted in public information). All-source intelligence should combine, compare, and analyze classified and open-source materiel to provide the full context and scope of the information needed to support naval forces.

Routine needs for OSINT may be satisfied by querying organization and intelligence community resources to retrieve available information. These resources include commercial online information databases and products such as Jane's Information Group yearbooks, Library of Congress country studies, and the NSA telecommunication database, libraries, organization databases containing unclassified information, Internet searches, and the Director of National Intelligence (DNI) Open-source Center products and services.

OSINT is very useful during interagency collaboration and in multinational operations where intelligence information based on OSINT sources can be easily shared. However, caution must be exercised to ensure that intelligence sharing arrangements, to include the sharing of OSINT source products, have been approved through the CCDR J-2 or MIOC foreign disclosure office. OSINT can be particularly important during peace operations that place a premium on human factors analysis and data derived from sociological, demographic, cultural, and ethnological studies. By using OSINT to supply basic information, other ISR assets are freed to be directed against priority intelligence gaps. Open-source material is useful in support of all kinds of military operations, and is particularly useful where the U.S. Government has minimal or no official presence. For example, DOD intelligence production analysts use open-source information on bridge loads, railroad schedules, electric power sources, and other logistics-related topics to support U.S. troop transport operations and noncombatant evacuation operations. Understanding the use of deception or misinformation in certain open-source media are also key to productive employment of OSINT information.

---

[6] JP 1-02.

[7] Public Law 109-163, National Defense Authorization Act for Fiscal Year 2006, Section 931.

NWP 2-0

D.6.3 Fleet Open-Source Intelligence Considerations

OSINT capabilities extending to fleet intelligence at the operational and tactical levels of war are components of the National Open Source Enterprise (NOSE) under the cognizance of the DNI. At the national level, the NOSE is comprised of the IC elements that make up the National Open Source Committee, which is chaired by the ODNI Open Source Center. The DOD executive agent for OSINT is DIA. Navy OSINT representation responsibilities have been delegated to ONI, which also operates the ONI Maritime Open Source Coordination Center (MOSCC). The theater authority for OSINT is the CCDR J-2, with oversight of OSINT activities executed within the relevant joint intelligence operations center (JIOC).

Initiatives to expand upon the OSINT enterprise at the operational level of war are coordinated between the NCC/NFC N-2 staff and the JIOC staffs.

OSINT is an intelligence discipline that is used routinely in fleet intelligence collection operations. OPSEC and tradecraft are necessary considerations for OSINT collection, which employs non- or misattribution systems for open-source research.

Naval forces have access to a limited number of organizations capable of OSINT collection that are able to be tasked as assets. The principal organizations are the JIOC or MIOC OSINT elements or CSG/ESG/ARG intelligence centers. There are also a limited number of joint or national organizations capable of OSINT collection able to be requested as resources to include the DNI OSC and ONI's MOSCC. OSINT requirements are submitted via the Open Source Collection Acquisition Requirements—Management System (OSCAR-MS).

D.6.4 Guidance

1. National Open Source Enterprise Capabilities (NOSECAP) Manual, April 2010

2. DOD Instruction 3115.12, "Open-Source Intelligence," 24 Aug 2010

3. Fleet Intelligence, Surveillance, and Reconnaissance (ISR) Concept of Operations: Intelligence Operations Support to the Maritime Operations Center (MOC), May 2010 (Appendix C.1.8)

4. Navy-Wide OPTASK Intelligence, COMSECONDFLT 071705Z OCT 09 (Paragraph 4.C.5)

5. COMFIFTHFLT Supplement to the Navy-Wide OPTASK Intelligence, COMUSNAVCENT 041447Z MAR 11 (Paragraphs 4.D.4, 4.D.10).

D.7 SIGNALS INTELLIGENCE (SIGINT)

D.7.1 Definition

SIGINT is "(1) A category of intelligence comprising either individually or in combination all communications intelligence, electronic intelligence, and foreign instrumentation signals intelligence, however transmitted; (2) Intelligence derived from communications, electronic, and foreign instrumentation signals."[8]

This definition portrays SIGINT as both an intelligence discipline and a resultant product.

D.7.2 Discussion

SIGINT is intelligence produced by exploiting foreign communications systems and noncommunications emitters. SIGINT provides unique intelligence information, complements intelligence derived from other sources and is often used for cueing other sensors to potential targets of interest. For example, SIGINT which identifies activity of interest may be used to cue GEOINT to confirm that activity. Conversely, changes detected by

---

[8] JP 1-02.

GEOINT can cue SIGINT collection against new targets. The discipline is subdivided into three subcategories: communications intelligence (COMINT), electronic intelligence (ELINT), and foreign instrumentation signals intelligence (FISINT).

1. COMINT is intelligence and technical information derived from collecting and processing intercepted foreign communications passed by radio, wire, or other electromagnetic means. COMINT also may include imagery, when pictures or diagrams are encoded by a computer network/radio frequency method for storage and transmission. The imagery can be static or streaming.

2. ELINT is intelligence derived from the interception and analysis of noncommunications emitters (e.g., radar). ELINT consists of two subcategories; operational ELINT (OPELINT) and technical ELINT (TECHELINT). OPELINT is concerned with operationally relevant information such as the location, movement, employment, tactics, and activity of foreign noncommunications emitters and their associated weapon systems. TECHELINT is concerned with the technical aspects of foreign noncommunications emitters such as signal characteristics, modes, functions, associations, capabilities, limitations, vulnerabilities, and technology levels.

3. FISINT involves the technical analysis of data intercepted from foreign equipment and control systems such as telemetry, electronic interrogators, tracking/fusing/arming/firing command systems, and video data links.

D.7.3 Fleet Signals Intelligence Considerations

The SIGINT enterprise extending to fleet intelligence at the operational and tactical levels of war is a component of the national SIGINT architecture under the cognizance of the Director, National Security Agency (NSA)/Central Security Service (CSS) as the Functional Manager for the USSS. The Navy authority over maritime elements of this architecture resides with Fleet Cyber Command (FCC). The theater SIGINT authority resides with the CCDR J-2.

FCC manages the functions of Navy Information Operations Commands (NIOCs) and fleet information operations centers (FIOCs) at NSA locations worldwide. NIOCs are force providers to both NSA sites and FIOCs under COMTENTHFLT. Personnel assigned to NSA Cryptologic Centers are focused on national and theater-specific missions. The missions of the FIOCs varies, but typically includes fleet and unit-level cryptologic and/or information operations (IO) and intelligence analysis support, as well as providing fleet direct support personnel to deploying naval units.

Initiatives to expand upon this architecture at the operational level of war are coordinated between the NCC/NFC cryptologic resource coordinator (CRC) staff and the FCC/CCDR J-2 staffs.

The TF CRC is responsible for cryptologic operations and sensor management at the tactical level of war.

SIGINT is an intelligence discipline that is used extensively in fleet intelligence collection operations.

Naval forces have access to a high number of platforms with sensors capable of SIGINT collection that are able to be tasked as assets. There are also a high number of joint or national platforms capable of SIGINT collection able to be requested as resources.

D.7.4 Guidance

1. DODD 5100.20, The National Security Agency and the Central Security Service, 26 Jan 2010

2. NTTP 3-13.4, Naval Cryptologic Operations, Aug 2009

3. USAFRICOM Standing Reconnaissance, Surveillance, and Target Acquisition (RSTA) Annex, 10 Jan 2013

4. USCENTCOM Baseline Reconnaissance, Surveillance, and Target Acquisition (RSTA) Annex, Oct 2012

5. USEUCOM Baseline Reconnaissance, Surveillance, and Target Acquisition (RSTA) Annex, 18 Apr 2012

6. Fleet Intelligence Collection Manual, ONI-1200-001-11, May 2011 (Chapters 2.D.3, 2.F.4)

7. Fleet Intelligence, Surveillance, and Reconnaissance (ISR) Concept of Operations: Intelligence Operations Support to the Maritime Operations Center (MOC), May 2010 (Appendices C.1.5, C.1.6)

8. Navy-Wide OPTASK Cryptology, COMSECONDFLT 101721Z JUL 09

9. COMTHIRDFLT OPTASK Cryptology Supplement, COMTHIRDFLT 172015Z DEC 2009

10. COMUSNAVSO OPTASK Cryptology Supplement, COMFOURTHFLT 141457Z DEC 2009

11. COMUSNAVCENT OPTASK Cryptology Supplement, COMUSNAVCENT 050929Z JUN 2011

12. COMSIXTHFLT Standing OPTASK Cryptology Supplement, COMSIXTHFLT 111205Z AUG 2011

13. COMSEVENTHFLT OPTASK Cryptology Supplement, COMSEVENTHFLT 070252Z NOV 2009

14. Navy-Wide OPTASK Intelligence, COMSECONDFLT 071705Z OCT 09 (Paragraph 4.C.3)

15. COMFIFTHFLT Supplement to the Navy-Wide OPTASK Intelligence, COMUSNAVCENT 041447Z MAR 11 (Paragraphs 4.D.4, 4.D.7, 4.E.3)

16. COMSIXTHFLT Supplement to the Navy-Wide OPTASK Intelligence, COMSIXTHLT 211200Z APR 09 (Paragraphs 5.C.4, 6.B)

17. COMSEVENTHFLT Supplement to the Navy-Wide OPTASK Intelligence, COMSEVENTHFLT 151000Z MAR 06, 151005Z MAR 06, 151010Z MAR 06 (Paragraphs D.3.A.2, D.6.B).

D.8 TECHNICAL INTELLIGENCE (TECHINT)

D.8.1 Definition

TECHINT is "intelligence derived from the collection, processing, analysis, and exploitation of data and information pertaining to foreign equipment and materiel for the purposes of preventing technological surprise, assessing foreign scientific and technical capabilities, and developing countermeasures designed to neutralize an adversary's technological advantages."[9]

D.8.2 Discussion

TECHINT is derived from the exploitation of foreign materiel and scientific information. TECHINT begins with the acquisition of a foreign piece of equipment or foreign scientific/technological information. The item or information is then exploited by specialized teams. These TECHINT teams assess the capabilities and vulnerabilities of captured military materiel and provide detailed assessments of foreign technological threat capabilities, limitations, and vulnerabilities.

TECHINT products are used by U.S. weapons developers, countermeasure designers, tacticians, and operational forces to prevent technological surprise, neutralize an adversary's technological advantages, enhance force protection, and support the development and employment of effective countermeasures to newly identified adversary equipment.

---

[9] JP 1-02.

At the strategic level, the exploitation and interpretation of foreign weapon systems, materiel, and technologies is referred to as scientific and technical intelligence (S&TI). At the operational and tactical levels of war, TECHINT applications are generally limited to the exploitation of captured enemy materiel as well as the retrieval of flotsam/jetsam in the maritime domain.

TECHINT, especially S&TI, is an intelligence discipline that often requires significant lead times for collection and exploitation.

D.8.3 Fleet Technical Intelligence Considerations

There is no national TECHINT architecture under the cognizance of a single IC agency. The TECHINT enterprise extending to fleet intelligence at the operational and tactical levels of war is a component of the DOD foreign materiel program (FMP) administered by the Defense Intelligence Agency (DIA). DIA manages joint and Service S&TI through the Joint Foreign Material Program Office (JFMPO). The Navy authority over maritime TECHINT and S&TI is the Office of Naval Intelligence (ONI), with the Navy FMP being managed by the ONI Collections Office.

There are no naval TECHINT personnel permanently deployed to forward theaters, nor are such personnel attached to rotating forces. Such expertise, however, may be requested in emergent operational scenarios. Initiatives to establish a standing capability at the OLW would be proposed by the NCC/NFC N-2 to ONI.

TECHINT is an intelligence discipline that is used infrequently in fleet intelligence collection operations.

Naval forces have access to a limited number of platforms and organizations capable of TECHINT collection that are able to be tasked as assets. The principal organizations are ONI collection and exploitation teams, with explosive ordnance disposal (EOD) assistance. There are also a limited number of joint/national platforms or organizations capable of TECHINT collection able to be requested as resources.

D.8.4 Guidance

1. DOD Directive C-3325.01E, Foreign Materiel Program (FMP) (U), 10 Oct 2006

2. DIA Manual 58-4, Foreign Materiel Program

3. OPNAV Instruction 3882.2B, Navy Foreign Materiel Program (NFMP) (C), 27 Dec 2005

4. Fleet Intelligence Collection Manual, ONI-1200-001-11, May 2011 (Chapters 2.D.1.B, 19)

5. Navy-Wide OPTASK Intelligence, COMSECONDFLT 071705Z OCT 09 (Paragraph 5.E)

6. COMUSNAVSO Supplement to the Navy-Wide OPTASK Intelligence, COMUSNAVSO 052054Z APR 06 (Paragraph 9).

NWP 2-0

APPENDIX E

Understanding Tasking, Collection, Processing, Exploitation, and Dissemination

E.1 OVERVIEW

By merging three major tenets of joint and naval intelligence doctrine described in chapter 1—the responsibilities of intelligence, the intelligence process, and the relationship between data, information, and intelligence—a structure becomes evident upon which TCPED concepts may be developed. That structure is shown in figure E-1.

Note that the middle three of the five components of TCPED—collection, processing and exploitation—align to this doctrinal construct in two key aspects: (1) these components equate to two sequential categories of the intelligence process; (2) those two categories also represent the transition of data to information. This conceptual arrangement serves as the cornerstone for defining TCPED.

Note

While collection, processing, and exploitation align to the intelligence process, the tasking and dissemination components of TCPED must be defined to fully understand TCPED.

E.2 TASKING

tasking. In the context of TCPED, the direction of intelligence discipline collection as part of a multidiscipline plan to address information or intelligence requirements.

Tasking is the output of collection planning. Specifically, it is a function of collection requirements management that translates intelligence and information requirements into a collection plan of various intelligence disciplines that are best postured to acquire corroborating data to address those requirements. Figure E-2 provides a general depiction of this planning function.

This planning process requires an understanding of the capabilities and limitations of each intelligence discipline, as well as the platforms, sensors, architectures, and organization of the elements available for tasking within that discipline. A representation of this tasking function is depicted in figure E-3.

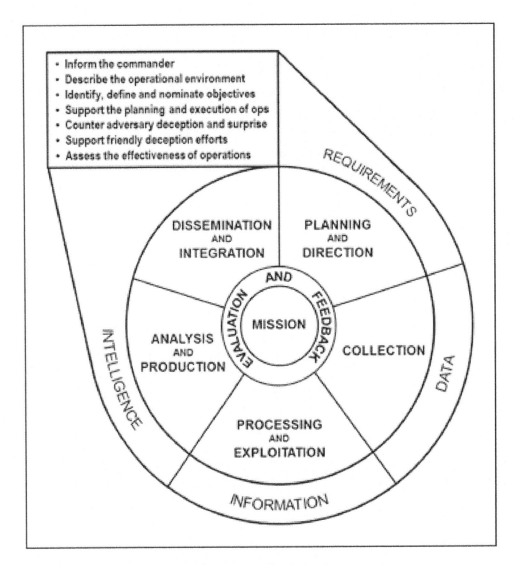

Figure E-1. The Responsibilities of Intelligence, the Intelligence Process, and the Relationship of Data, Information, and Intelligence

NWP 2-0

Information/Intelligence Requirements
Multi-INT Collection Plan

GEOINT	HUMINT	SIGINT	MASINT	OSINT	TECHINT	CI
Tasking	Tasking	Tasking	Tasking	Tasking	Tasking	Tasking

Figure E-2. Tasking of Intelligence Disciplines

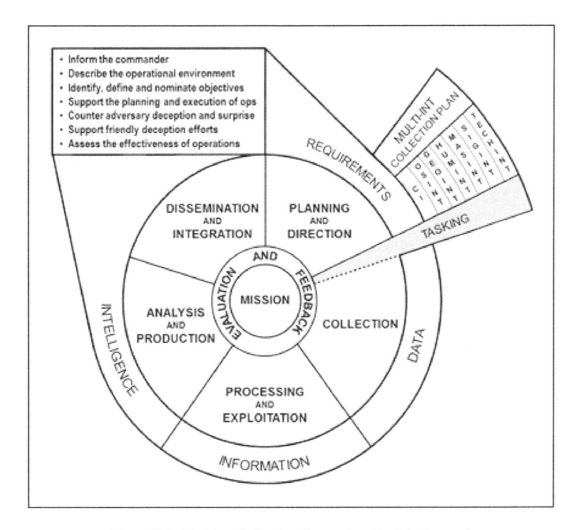

Figure E-3. Tasking, Collection, Processing, Exploitation, and Dissemination Tasking

E-3   MAR 2014

E.3 DISSEMINATION

dissemination. In the context of TCPED, the provision of an intelligence discipline information report for awareness or analysis.

The dissemination component of TCPED is easily assumed to be the same as the dissemination and integration category of the intelligence process—but it is not. Further examination of one key discriminator between TCPED and the intelligence process underscores the necessary role of postexploitation dissemination.

While JP 2-0 states that "[i]n many situations, the various intelligence operations occur nearly simultaneous with one another or may be bypassed altogether," the analysis and production category, as depicted in figure 3-2, remains a key doctrinal category in that process—even though it may not be a required step in every situation. In contrast, TCPED, as evidenced by the terms that make up the acronym itself, does not include analysis and production at any point. Instead, information derived from exploitation is directly disseminated, and the information that is disseminated is best characterized as an exploitation report.

The nature of the information will dictate its dissemination path:

1. Information considered valid with evident context and immediate relevance will be disseminated to appropriate decision makers.

2. Information requiring additional validation or context will be disseminated to appropriate intelligence analysis elements.

Note

These dissemination options are not mutually exclusive.

Joint doctrine (JP 2-0) also states "[w]hereas collection, processing, and exploitation are primarily performed by specialists from one of the major intelligence disciplines, analysis and production is done primarily by all-source analysts that fuse together information from all intelligence disciplines. The product resulting from this multidiscipline fusion effort is known as all-source intelligence."

As a result, TCPED dissemination may be characterized as exploited information from a single intelligence discipline that is an input to operational decisionmaking or all-source analysis and production. Only intelligence that is a multiintelligence output of analysis and production meets joint doctrine criteria for the category of dissemination and integration. A depiction of the nature of TCPED dissemination is provided in figure E-4.

E.4 THE COMPOSITE PROCESS

These related doctrinal concepts result in a consolidated description of TCPED as outlined in figure E-5.

Figure E-6 provides a graphic depiction of these concepts, and figure E-7 provides the simplified portrayal of TCPED and its placement within the intelligence process.

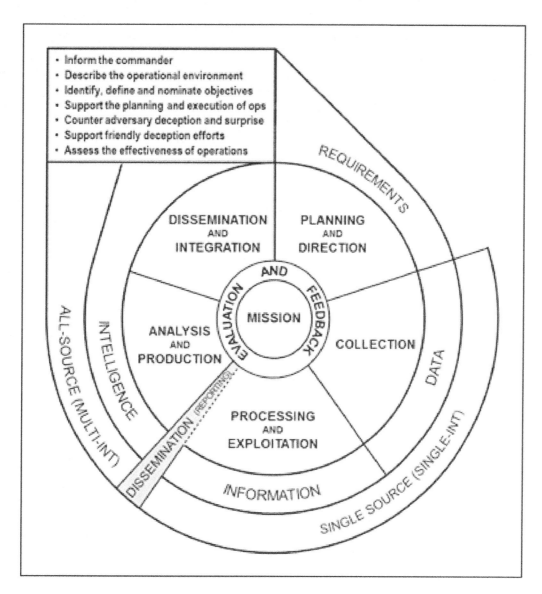

Figure E-4. Tasking, Collection, Processing, Exploitation, and Dissemination Dissemination

TCPED Component	Associated with	Intelligence Disciplines	Definition
Tasking	Requirements	Multi-INT	The direction of intelligence discipline collection as part of a multi-discipline plan to address information or intelligence requirements
Collection	Data	Single-INT	The acquisition of data and the provision of that data to processing elements
Processing	Data-Information	Single-INT	A system of operations designed to convert raw data into an exploitable format
Exploitation	Information	Single-INT	The assessment of processed data and reporting of derived information
Dissemination	Information	Single-INT	The provision of intelligence discipline information reporting for awareness or analysis

Figure E-5. Tasking, Collection, Processing, Exploitation, and Dissemination Component Characteristics and Definitions

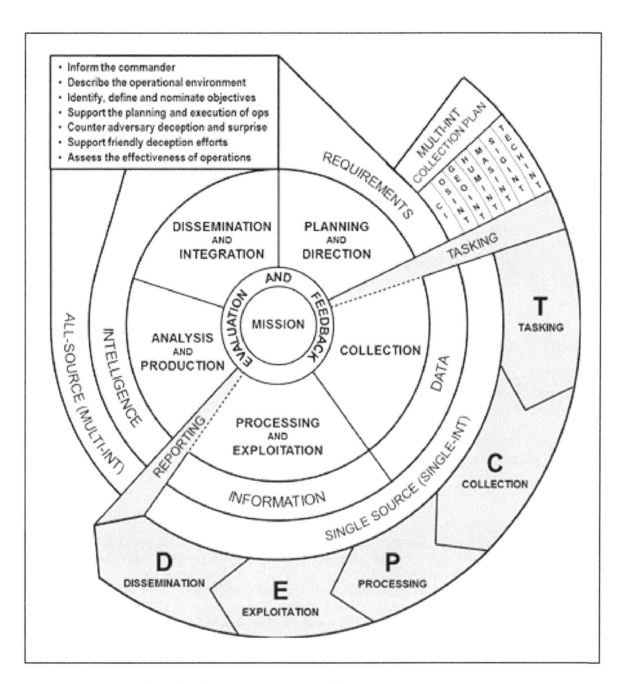

Figure E-6. The Doctrinal Depiction of Tasking, Collection, Processing, Exploitation, and Dissemination

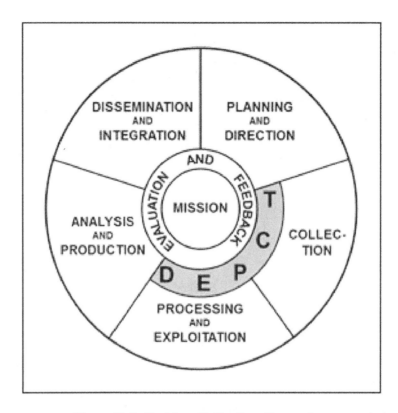

Figure E-7. Tasking, Collection, Processing, Exploitation, and Dissemination Within the Intelligence Process

Note

Figure E-6 is arranged so that each interior feature of the graphic is the foundation of any exterior features. For example, the intelligence process is an extension of (and contributes to) the mission; TCPED is an extension of (and contributes to) the intelligence process. This graphical arrangement was used to facilitate the description of TCPED doctrinal principles in this document. The positioning of TCPED in figure E-6 does not imply that TCPED is outside of or detached from the intelligence cycle

E.5 TCPED—WHAT IT IS NOT

Figure E-8 underscores that TCPED activities are encompassed by the intelligence process. However, TCPED is not the intelligence process and the intelligence process is not TCPED. Unfortunately many consumers of fleet intelligence believe that the two terms are synonymous or that advances in TCPED-related persistence and responsiveness have rendered the intelligence cycle obsolete. Not only does this viewpoint discount the critical human-centric functions of planning and direction, analysis and production, and dissemination and integration within the intelligence process, it demonstrates a lack of understanding of the relationship between data-information-intelligence described in chapter 1 and how that sequence interacts with the intelligence process shown in figure E-1.

Intelligence is developed to provide key insight to the commander. Joint doctrine JP 2-0 states that ultimately, intelligence has two critical features that make it different from information:

1. It allows anticipation or prediction of future situations and circumstances.

2. It informs decisions by illuminating the differences in available courses of action.

Additionally, intelligence provides the commander with a threat assessment based on an analysis of the full range of adversary capabilities and a prediction of the adversary's likely intentions that allows the commander to formulate plans based on this knowledge and thus decrease the risks inherent in military operations and increase the likelihood of success.

Numerous definitions highlight that intelligence is intended to be predictive in nature. It informs decision makers of potential adversary action as context for planning own force action. Information, on the other hand, is inherently more reactive. Information—like intelligence—may lead to decisions; however, decisions triggered by information are generally reacting to real-time developments while decisions triggered by intelligence are generally preparing for longer-term outcomes.

Figure E-9 underscores the essential distinctions between information and intelligence that need to be made to understand the role of TCPED within the intelligence process. Various sources of single-source information (portrayed vertically), acquired through different TCPED processes, can be integrated over a period of time into multisource intelligence assessment (portrayed horizontally) that, in turn, allows the fleet to be given indications and warning (I&W) about a potential threat. With that forewarning, a single-source effort could then allow an individual ship or aircraft to potentially detect and engage that emerging threat. The resulting operational action does not mean that TCPED has replaced intelligence. It only means that a broader intelligence effort provided the context, prediction, or warning for a single intelligence discipline to independently inform a decision maker to take appropriate action.

The final event in the above scenario is referred to as actionable intelligence ("... information that is directly useful to customers for immediate exploitation without having to go through the full intelligence production process." (JP 1-02))

Other examples of information-triggered decisions include "sensor to shooter" or select instances of time-sensitive target operations. These scenarios may be well-suited for highly automated TCPED processes to expedite tactical engagement decisions, but such instances should not be interpreted by commanders as signifying that TCPED processes should or will replace the intelligence process.

Another depiction of the differences between TCPED and intelligence processes is provided in Figure E-9.

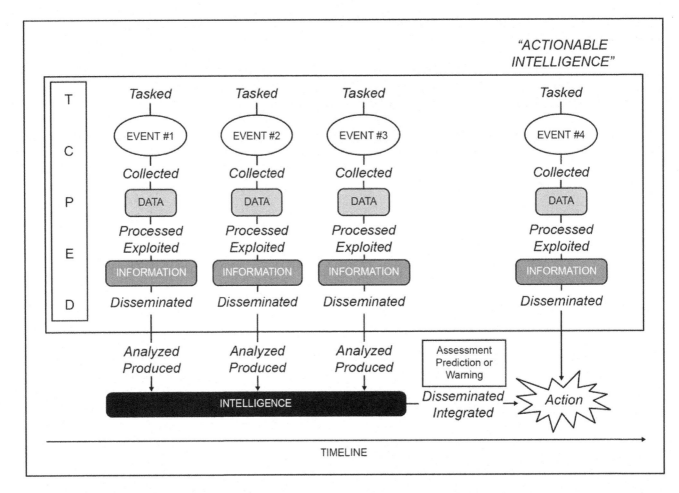

Figure E-8. The Difference Between Tasking, Collection, Processing, Exploitation, and Dissemination-Based Information and Analysis-Based Intelligence

	TCPED Process	Intelligence Process
End State	Information	Intelligence
Provides	Awareness	Understanding and Prediction
Result of	Exploitation	Fusion and Analysis
Discipline(s)	Single-INT	Multi-INT
Perspective	Short-term	Short-term to Long-term
Levels of War	(Mostly) Tactical	Strategic, Operational and Tactical

Figure E-9. A Comparison of Tasking, Collection, Processing, Exploitation, and Dissemination and Intelligence Processes

E.6 SUMMARIZING TCPED

To summarize, TCPED is identified as a process that:

1. Involves a single intelligence discipline

2. Is the output of a multiintelligence discipline planning and direction effort

3. Begins with a single intelligence discipline tasking order

4. Ends with single intelligence discipline dissemination to intelligence analysts or operational decision makers

5. Is an input to a multiintelligence discipline evaluation and feedback effort.

TCPED is differentiated from the intelligence process by:

1. The output of TCPED is information. It feeds into, but does not include, the analysis and production necessary to generate intelligence.

2. TCPED information provides awareness allowing decision makers to react to real-time developments, whereas intelligence is predictive in nature and is focused on longer-term trends.

INTENTIONALLY BLANK

APPENDIX F

Analytic Tradecraft

Major intelligence failures are usually caused by failures of analysis, not failures of collection. Relevant information is discounted, misinterpreted, ignored, rejected, or overlooked because it fails to fit a prevailing model or mind-set. The "signals" are lost in the "noise."

Richards J. Heuer, Jr.
Psychology of Intelligence Analysis

[intelligence analysts should be] . . . masters of the subject matter, impartial in the presence of new evidence, ingenious in the development of research techniques, imaginative in their use of hypotheses, sharp in the analysis of their own predilections or prejudices, and skilled in the presentation of their conclusions.

Sherman Kent[1]

As discussed in chapter 3, analysis is a core competency of the naval intelligence professional. This appendix provides additional detail on sound analytic practice by discussing some concepts related to analysis, providing a process for analysis, describing various tools and techniques used in analysis, and discussing some of the analytic pitfalls, such as cultural and cognitive biases, that are inimical to good analysis. It concludes with a reprinting of a classic text on intelligence analysis, Sherman Kent's Estimates and Influence, as that essay ties together a number of the concepts discussed in this appendix and throughout this publication.

F.1 ANALYTIC CONCEPTS

Merriam-Webster's online dictionary offers two definitions for analysis that are relevant to the following discussion. First, analysis is defined as "a separation of a whole into its component parts" or as "an examination of a complex, its elements, and their relations."[2] Intelligence analysis takes into account both definitions, which are mutually supporting. When faced with a complex problem, like determining the capability and intent of an adversary who is actively trying to deceive, analysis is needed to reduce uncertainty and provide an answer to the problem. In some cases, the assessment may be wrong. Intelligence analysis is not foolproof, but by following a sound and rigorous analytic process, one that also attempts to mitigate the factors that work against good analysis, the odds are significantly increased that analysis will at the least, reduce uncertainty, and, at best, produce a forecast that is accurate in many respects.

When faced with a complex problem, there are typically two ways to address it, both of which correspond to the two definitions of analysis given above. First, one can decompose a problem, which means to break the problem into smaller parts, examine them, and recombine them to achieve some fresh insight. For example, in assessing

---

[1] Walton, T., Challenges in Intelligence Analysis, p. 128.
[2] Merriam-Webster, http://www.merriam-webster.com/dictionary/analysis

the overall readiness and capability of an adversary naval task force, an analyst could looks at various elements such as:

1. What ships are in the naval task force?

2. What weapons systems are on those ships?

3. Who well does this adversary's Navy maintain its naval weapons systems and platforms?

What is the adversary's capability to resupply this task force at sea?

1. Do the ships of this naval task force have a history of deploying together?

2. How does the C2 of this naval task force operate?

3. What has been the observed performance of these units in previous exercises?

By determining the answers to these individual questions, an analyst could combine the results to develop their estimate of how ready and capable the adversary's naval task force is to conduct combat operations at sea.

Second, one can externalize the problem. Externalizing a problem can take many forms but, in essence, it is the process of putting the problem and the known, supposed, and unknown facts or factors related to the problem on some medium (paper, white board, computer, etc.). Externalizing a problem, which in its early stages could be as simple as brainstorming the problem, contributes to analysis as it enables one to see facts, variables, parameters, and other elements of the problem. Through the use of matrices, tables, outlines, diagrams, flow charts, and other tools that are discussed in section F.3, these various facts and factors can be combined and recombined to show relationships and achieve a new understanding of the problem. Externalizing has the additional benefits of highlighting gaps and facilitating collaboration.[3]

F.1.1 Analytic Strategies

Richards J. Heuer, Jr., in his book Psychology of Intelligence Analysis, states that there are four main analytic strategies that are used by intelligence professionals.[4] None of these strategies is better than the other but all analysts must be aware of their own cognitive and cultural biases when applying them to ensure they are used in an objective way. While these strategies can also be combined, usually one takes precedence of over the other and choices about which strategy to employ are usually driven by the analyst's preferences, training, and education; the nature of the data; and the specific intelligence problem being addressed. The four analytic strategies are:

F.1.1.1 Situational Logic

This analytic strategy is the one most often used in intelligence analysis. The analyst generates and analyzes hypotheses starting with specific details of the current situation rather than comparing the situation to some other model. The situation is seen as unique and the analyst attempts to build a narrative, one typically based on cause and effect, means and ends, or some similar construct that will form the basis of their hypothesis. The advantage of applying situational logic is that it facilitates the integration of large volumes of data. It is widely applied as any situation can be analyzed using this strategy.

Situational logic has two main weaknesses. First, it is difficult for the analyst to think like the adversary so there is a strong tendency for the analyst to "mirror-image" or project themselves into the analysis, especially if the adversary appears to be acting irrationally. Second, situational logic is somewhat limited as it does not consider other models, such as nationalism, that could be used to explain the data presented to the analyst.

---

[3] Heuer, R. J., Psychology of Intelligence Analysis, pp. 86–89.
[4] Heuer, pp. 32–48.

F.1.1.2 Applying Theory

In this strategy, the analyst accepts a set of generalizations that are based on a study of many cases of similar situations or phenomenon. The generalized model of the situation provides the analyst with a guide that can be compared with the current situation and which may prove helpful in determining indicators and in assessing how the situation is likely to unfold. Examples of applying theory would include using models the describe nationalism or tribalism to assess how groups will respond.

The main advantage to this strategy is its efficiency. It can help to winnow away unnecessary details or rule out models that lack data or information to support their validity. In cases where there is little available data, applying theory may be the best available option as it facilitates a potential understanding of the situation that can be confirmed or refuted through the collection and analysis of more data and information.

The major weakness of this strategy is that every situation is unique and so comparison with a generalization may lead analysts to ignore important data and information that may not conform to the model but which are more truly representative of the reality of the situation. Filtering of data and information based on unfounded assumptions is a key source of analytic failure so analysts must always be mindful that they are using a model to better understand aspects of the situation vice a tool that reveals adversary capabilities and intentions with certainty.

F.1.1.3 Historical Comparison

Like the strategy above, historical comparison rests on the use of analogy or modeling to better understand a current situation. This strategy looks at the country in question, or at a country in which a similar event has occurred in history, and then extrapolates how that situation unfolded to garner insights about how the current situation will develop. The difference between historical comparison and the method above (applying theory) is that historical comparison relies on comparison with only one or a few models while applying theory is a generalization based on many data points.

The advantage of historical comparison is that it is easy to apply when lacking data or a specific model. While it has less utility in reaching an analytic judgment, since situations are rarely completely analogous, historical comparison is a very good method for developing hypotheses.

The major weakness of this strategy is that the analyst is taking knowns from the past and making assumptions that these knowns are the same as present unknowns. The success of this strategy is also incumbent on the analyst knowing and understanding the historical precedent well enough to draw valid conclusions. Superficial understanding of historical example results in superficial and inaccurate analysis.

F.1.1.4 Data Immersion

In a theoretical sense, this strategy relies on the analyst to absorb all the data on a given subject and through that "immersion" in the data the analyst will begin to see patterns emerge. There are some cases where this is the only strategy available to the analyst but, in practice, this strategy is different from the theory. The significance of information rests not only in the nature of the information itself but also the context in which that information is understood. Since the mind of the analyst is used in creating this understanding, the analyst's understanding is always an interpretation that is subject to bias. If the analyst is explicit about their biases in the selecting, sorting, and organizing of data, then this strategy can be effectively applied. The goal is an objective analysis. Since there is no way to suppress biases, the analyst can only achieve this goal by stating these biases up front so that others can evaluate their validity and their impact on the analytic process.

F.1.2 Analytic Methodologies

Richards J. Heuer, Jr. and Randolph H. Pherson, in their book, Structured Analytic Techniques for Intelligence Analysis, state that there are four distinct methodologies for intelligence analysis.[5] These are:

F.1.2.1 Expert Judgment

This methodology is used by all analysts at some time. While it may use many of the same techniques discussed below in Structured Analysis, the term is used to describe the mental processes that the lone analyst undertakes in analyzing a problem. Evidentiary reasoning, comparison of the current situation to case studies, application of the historical method or other social science method, analogizing the current problem to some more general problem—these are all ways in which an analyst may think through the problem. Typically, in the case of expert judgment, the processes and assumptions used by the analyst to address the intelligence problem are not articulated until the analyst writes the intelligence report. Expert judgment is largely a function of experience, education, and practice and is not a focus of this appendix.

F.1.2.2 Structured Analysis

These are step by step analytic processes and procedures that serve to externalize the analyst's thinking about the intelligence problem and enable collaboration on that problem. The advantage to using structured analysis is it allows a review and critique of the data, steps, and assumptions used to address the intelligence problem. The techniques used in structured analysis are discussed in greater detail in section F.3. This is the type of analysis most broadly applicable to naval intelligence and is the focus of the discussion about specific analytic techniques.

F.1.2.3 Quantitative Methods using Expert-Generated Data

When lacking quantifiable data, intelligence analysts may turn to expert opinion which, like intelligence analysis, is subject to a probability determination regarding its potential accuracy. Since specialized techniques are often used to capture and analyze this type of data, this form of analysis is not discussed in this appendix.

F.1.2.4 Quantitative Methods Using Empirical Data

Empirical data are collected by various types of sensors and analysis of this type of data is typically conducted by S&TI analysts with advanced degrees in statistics, hard sciences, engineering, or economics. This method of analysis is introduced here, as it is one of the methods used by ONI S&TI analysts, but it is not discussed further in this appendix.

While structured analytic techniques are discussed in section F.3, a more general discussion of the analytic process is needed before discussing techniques. Use of specific techniques and tools is driven by the specific intelligence problem that the naval intelligence professional is called upon to address. Having a general framework for the applying those techniques and tools is the subject of the next section.

F.1.3 Analytic Principles

JP 2-01.3, Joint and National Intelligence Support to Military Operations, lists seven analytic principles that the analyst should follow in forming an intelligence assessment. The analytic principles are:[6]

1. Be precise about what is known. Decision makers and military leaders need to be informed precisely about what analysts know and the source reliability of that information. Analysts should never exaggerate what is known. They should report any important gaps in information bearing on decisionmaking and potential COAs, as well as relevant information that may contradict the analyst's leading hypothesis. Analysts

---

[5] Heuer, R. J. and Pherson, R. H., Structured Analytic Techniques for Intelligence Analysis, pp. 22–23.
[6] This list is taken directly from JP 2-01.3, Joint and National Intelligence Support to Military Operations, pp. D-7–D-8.

should be precise as well in sourcing information. The phrase, according to the U.S. embassy, for example, does not inform the reader whether the information is direct or indirect.

2. Distinguish carefully between information and fact. Analysts may have direct information on what a foreign leader said, for example, and thereby report this as factual. But what a foreign leader believes, intends to do, and will do cannot be known to be true on the basis of a report on what he or she said. From the analytic standpoint, the intelligence should be reported as such.

3. Distinguish carefully between information and estimative judgment. Analysts' estimative judgments are an important element in the process of supporting decision makers, but they must be formulated using the entire body of available information and sound inferential reasoning. Also, care should be taken to avoid confusion over whether the analyst is stating a fact or an estimative judgment.

4. Take account of substantive complexity. The more complicated an issue, the greater the demand to determine what is factual. For example, the burden of proof in determining what a terrorist group intends to do is much greater than that required for determining what they have done or said. Analysts may properly make a conditional judgment about what an entity intends to do, but this should not be stated as verified or factual information.

5. Take account of controversial sensitivities. As with substantively complex matters, the burden of proof is high on matters that are controversial among decision makers. For controversial issues, analysts should place emphasis on the relevant information, and not on estimative conclusions.

6. Take account of the possibility of deception. Deception is the manipulation of information by a foreign government, group, or individual to get intelligence analysts to reach an erroneous conclusion. Deception often works because it gives busy analysts what they are seeking—seemingly reliable information on which to base a conclusion. One test for detecting and countering deception is to determine whether all the sources and collection platforms that should be reporting on a matter have indeed done so. Databases and organized information in general help detect the possibility of deception, as does critical thinking.

7. Use the term "evidence" sparingly. Many times the term information is used synonymously with the term evidence. Both are used to refer to the content of reports and research that helps reduce the uncertainty surrounding intelligence questions. Evidence is more appropriate for law enforcement collectors and analysts. DOD analysts should avoid using the term when 'information' serves their purposes just as well. At times, characterization of the information is sufficient to make the analysts' point for the benefit of consumer.

Figure F-1 provides an overview of the analytic principles.

Analytic Principles
1. Be Precise about What is Known
2. Distinguish Carefully Between Information and Fact
3. Distinguish Carefully Between Information and Estimative Judgment
4. Take Account of Substantive Complexity
5. Take Account of Controversial Sensitivities
6. Take Account of the Possibility of Deception
7. Use the Term "Evidence" Sparingly

Figure F-1. Analytic Principles[7]

---

[7] JP 2-01.3, pp. D-7–D-8.

F.2 THE ANALYTIC PROCESS

Figure F-2 provides an overview of the steps in the analytic process. The analytic process is driven by the requirements of the intelligence problem faced by the analyst. So, each of these steps may be done to a greater or lesser degree depending on the specific intelligence problem being addressed. For example, a newly tasked assessment to determine an adversary's probable response to a naval cruise missile strike on a key C2 node may necessitate a detailed step-by-step approach to addressing that intelligence question. Updating a standing current intelligence plot that is familiar to the naval intelligence analyst may only require in-depth application of a few steps but, as is discussed in section F.4, analysts must avoid complacency with their assessments and should periodically have them reviewed in detail to determine if they still remain valid.

While variations on this process may be used, the point is that naval intelligence professionals must employ a structured and rigorous process in making analytic judgments to ensure the commander, staff, and supported organizations are provided a sound estimate based on all available sources. The discussion below provides considerations for each step in the process. In some cases, aspects of these steps have been discussed in detail in previous sections of this publication. In those cases, a brief description of the step is provided and the reader is referred to the appropriate section of the publication to obtain amplifying detail.

F.2.1 Defining the Analytic Problem

The most important step in the analytic process is to define the intelligence problem. For this reason, the discussion on this is lengthier than that provided for other steps in the process.

The analytic problem stems from the requirements, PIRs and IRs, generated by the commander, staff, and supported organizations. The type of analysis done and the tools employed to organize data and reach sound analytic conclusions should be driven by these intelligence requirements. Naval intelligence professionals must have mastery over a variety of analytic techniques and must understand, or be able to quickly determine, the depth of extant intelligence that already exists on many different topics. Intelligence analysis functions best when the requirements levied upon it are clear, specific, and unambiguous. This enables the intelligence analyst to define the problem he or she will address and enable the gap analysis needed to determine if additional collection of data and information is needed to address the question.[8]

The Analytic Process
1. Define the Problem
2. Generate Hypotheses
3. Determine Information Needs and Gather Information
4. Evaluate Sources
5. Evaluate Hypotheses and Select the Most Probable Hypothesis
6. Production and Packaging
7. Evaluation and Feedback
8. Ongoing Monitoring

Figure F-2. The Analytic Process[9]

---

[8] JP 2-01.3, p. D-3.
[9] JP 2-01.3, p. D-3; Heuer, pp. 174–179. This list provides a modification to the chart found in JP 2-01.3, adding "Select the Most Probable Hypothesis" to the "Evaluate Hypotheses" step and adding a step—"Ongoing Monitoring," since these activities are implied within the process outlined in the joint publication.

As discussed in section 3.5.1.2, the development of PIRs and IRs must result from dialog between the naval intelligence professional and the intelligence user. It is not acceptable to have ambiguous requirements, so it is incumbent upon naval intelligence professionals to be forthright in obtaining clarification of requirements.[10] As the expert on the intelligence process and the adversary, this may require that the analyst reframe the question or ask a different set of questions that provides intelligence users with the answers they really need. Figure F-3 provides examples of ways to restate an intelligence problem that may help the analyst to better understand the real requirement of the intelligence user.

In mature organizations, such as those which develop when a staff N-2 and their commander have served through a workup and deployment cycle together, the naval intelligence professional is better able to anticipate requirements. Even in these circumstances, though, active dialog between the naval intelligence professional and the intelligence user must continue to confirm the problem that will be addressed.

F.2.2 Generating Hypotheses

Once the intelligence problem has been defined, the naval intelligence analyst or team should generate reasonable hypotheses based on the problem. Collaborative brainstorming should be used, if possible, to generate as many possible hypotheses as time allows. This is a more general form of the process used to generate potential ACOAs in the IPOE process. As an example, a recent intelligence report indicates a Redland surface action group (SAG) has recently left port and the intelligence problem given to the analyst is to determine why this deployment of naval forces occurred. Potential hypotheses could include:

1. The Redland SAG is conducting training.

2. The Redland SAG is on routine patrol operations.

Initial question:	Is Redland selling the SA-37 SAM system to Orangeland?
Rephrase:	Is Orangeland buying the SA-37 system from Redland?
Ask why:	Why would Redland sell the SA-37 system to Orangeland? *Because Redland wants influence with Orangeland.*
	Why does Redland want influence with Orangeland? *Because Redland wants to reduce U.S. influence in the region.*
	Why does Redland want to reduce U.S. influence in the region? *Because Redland wants exclusive access to natural resources off Orangeland's coast.*
	Final Question: Is Redland's sale of weapons to Orangeland part of a larger effort to minimize U.S. influence in the region?
Broaden the focus:	Is there a partnership between Redland and Orangeland?
Narrow the focus:	What components, training, and support for the SA-37 system is Redland selling to Orangeland?
Redirect the focus:	Why does Orangeland want SA-37s? How will Orangeland integrate the SA-37s with their existing IADS?
Turn 180 degrees:	Is Redland buying SAM technology from Orangeland?

Figure F-3. Issue Redefinition Example[11]

---

[10] Heuer, p. 174.
[11] Heuer and Pherson, p. 51.

3. The Redland SAG is deploying for weather avoidance.

4. The ships of the Redland SAG are being resubordinated to a different fleet and are deploying to a new homeport.

5. The Redland SAG will conduct a blockade of Azureland's main port.

6. The Redland SAG will conduct a freedom of navigation (FON) operation through the Straits of Azure.

7. The Redland SAG will conduct offensive operations against the U.S. naval force in the Gulf of Azure.

8. The Redland SAG is deploying on a "show the flag" mission to other countries in the region.

9. The Redland SAG is deploying to render FHA to the people of Orangeland.

10. The Redland SAG is defecting from Redland.

At this stage, no hypothesis should be eliminated, even if data is lacking to confirm it. For example, a successful adversary deception effort would not produce the indicators expected within the context of the intelligence problem and adversary use of deception should always be considered a possibility.[12]

F.2.3 Gather Information and Determine Information Needs

As discussed in section 3.8.2.1, research is a key skill set for the analyst as it is essential to determining what data, information, and intelligence already exists to address the intelligence problem. As stated in joint doctrine, "[t]he information needed by the analyst is many times either already available or is already being sought by assets."[13] For intelligence topics the naval intelligence professional is already monitoring, gathering the information and identifying gaps to address new intelligence problems in that area may be a fairly rapid process given the analyst most likely has a level of mastery over the subject matter, well developed contacts with other analysts working the same topic, and good knowledge of the collection resources allocated to that topic and the regular collection focus of those resources. Ramping up for a new topic, such as when a crisis erupts, may take longer but very few crises are absolute surprises. Many have some build up time. If there is a possibility the naval intelligence analyst's command will respond to the crisis on some level, the analyst should begin surveying the available materials, developing contacts, and building understanding of the collection posture against that crisis in anticipation of the order to participate in the response. In doing this, the naval intelligence professional can provide their commander with initial products to build situational awareness, rapidly respond to requests for intelligence, and quickly determine where new collection or analytic effort is needed.

It cannot be overstressed that requests for "more collection" may not always be the best course of action for the analyst. Collection systems collect what they can, they do not necessarily collect what is needed by the analyst to make a sound judgment, unless the analyst has carefully considered how the data and information from a particular capability could provide a fresh insight that would confirm or alter their assessment.

Similarly, the call for "more information" has no applicability in the analytic process. What is needed is the "right" information, which can only be obtained if the analyst truly understands the specific question they are attempting to answer. In many cases, analysts today have far more data, information, and finished intelligence available to them then they can effectively manage or assimilate.[14] As Heuer states, "[t]he reaction of the Intelligence Community to many problems is to collect more information, even though analysts in many cases already have more information than they can digest. What analysts need is more truly useful information—mostly reliable HUMINT from knowledgeable insiders—to help them make good decisions. Or they need a more

---

[12] JP 2-01.3; Heuer, p. 174.
[13] JP 2-01.3.
[14] Walton, pp. 7–8.

accurate mental model and better analytical tools to help them sort through, make sense of, and get the most out of the available ambiguous and conflicting information."[15]

U.S. collection capabilities collect vast amounts of data that goes unexploited due to lack of analytic capacity, yet the proliferation of more and better collection systems has done little to improve the overall quality or accuracy of intelligence analytic judgments. Research has demonstrated two things with regard to the amount of information available to the analyst and the accuracy of their assessments. First, once an experienced analyst has the minimum amount of information that he or she needs to make a judgment, the addition of new information often does nothing to improve the quality or accuracy of their assessment. Additional information can help make the analyst more confident about the assessment but this confidence often evolves into a dangerous overconfidence and complacency. Second, experienced analysts often have insufficient understanding concerning the information that they actually use in making their assessments. While analysts may perceive they are systematically integrating and synthesizing a number of different sources to produce an estimate, in reality, analysts tend to rely on only a few key pieces of information in making their assessments.[16]

As a final note, if the naval intelligence analyst is collecting data and information to verify or invalidate a hypothesis, then they need to avoid making early judgments at this stage. The goal here is to gather data and information that can address gaps and that must be done objectively. An analyst can reach a conclusion based on very little data and information but by doing so they may filter out a significant amount of information that could aid in confirming or refuting a hypothesis.

Some theorize that intelligence analysis is like a puzzle. Known as the "Mosaic Theory," this idea states that if the analyst is just given enough of the pieces of the intelligence "puzzle" then they will be able to sort all the pieces, put them in the proper arrangement, and discern the picture made by the component parts. In reality, this is an incorrect model for intelligence analysis as one piece of data or information could fit many different possible models. The analyst's dilemma is determining which model they should choose out of the number of hypotheses that could potentially fit the data and information they have available to them. A better analogy for the analytic process is that of a doctor diagnosing a patient. In medical practice, the doctor gathers data and makes observations concerning the patient's condition (indicators) and using their knowledge of medical science they develop hypotheses that could potentially explain the pattern they are observing. The doctor then orders tests (collection) to confirm or refute their hypotheses and then makes a diagnosis on the basis of the test results.[17]

F.2.4 Evaluate Sources

Evaluation of sources was discussed in detail in section 3.8.2.3. The naval intelligence professional should always evaluate new information received for reliability and credibility. Analysts should always be conscious of the fact that the information may be the result of a deception operation.

F.2.5 Evaluate Hypotheses and Select the Most Probable Hypothesis

Once data, information, and intelligence has been gathered and evaluated, the naval intelligence analyst must objectively evaluate each hypothesis to determine which best addresses the intelligence problem. There is no one technique for evaluating hypotheses and is likely that one or a combination of analytic techniques and tools discussed in section F.3 are used to winnow down the list of probable hypotheses to one or two that appear most valid.[18]

While there is no specific technique for evaluating hypotheses, there are some general guidelines that must be followed in conducting this step. First, the emphasis should always be on trying to reject, rather than confirm, a given hypothesis. A general rule of thumb is that "[t]he most likely hypothesis is usually the one with the least evidence against it, not the most evidence for it."[19] The focus should be on data and assumptions that suggest one

---

[15] Heuer, p. 6.
[16] Heuer,, pp. 51–52.
[17] Heuer, pp. 61–62, 174–175.
[18] JP 2-01.
[19] Heuer, p.176.

hypothesis is more or less likely than another. Naval intelligence analysts must identify probabilities and areas of uncertainty when weighing the data they use and must clearly state the rationale used for accepting or rejecting a particular hypothesis.

Naval intelligence professionals must avoid satisficing in evaluating hypotheses. Satisficing is the practice of choosing the first hypothesis that fits all the information available to the analyst. The danger of this practice is that it does not allow full development of the other hypotheses being evaluated and it acts as a filter on all incoming information, which is used primarily to confirm, vice refute, the prevailing hypothesis. It is human nature to reject data and information that does not conform to a particular mindset or paradigm and selecting a hypothesis too early may result in the rejection of valuable data and information that could be used to refute the prevailing view and provide the insights needed to select a hypothesis that better answers the intelligence problem.[20]

In evaluating hypotheses, naval intelligence professionals must guard against the biases that are discussed in more detail in section F.4. Gaps in information and unknowns are acceptable—some pieces of information may just be beyond knowing in the time allowed to develop the analysis. Some pieces of information are ultimately unknowable. What is unacceptable is to fill in gaps with unfounded assumptions. A common technique when confronted with a gap is to use mirror-imaging, to assume that the adversary will act or react as someone raised in the United States would do. This is an example of an unfounded assumption as people from other cultures fundamentally perceive and respond to stimuli in completely different ways, although these differences may be very subtle. When analytic judgments are wrong, there is a strong body of evidence which shows that these judgments were wrong due to "faulty assumptions that went unchallenged" vice some fault in the information that was available to the analyst.[21]

Based on the above discussion, a key factor in ensuring an objective evaluation of hypotheses is the questioning of assumptions. While there is never sufficient time to question every aspect of the analytic process and the assumptions made in it, there are a couple of areas where analysts should focus their attention to ensure the quality of their evaluation. First, assessments should undergo an informal sensitivity analysis. To do this, the naval intelligence professional should ask how sensitive the assessment is "to changes in any of the major variables or driving forces in the analysis"[22] Those assumptions that drive the analysis are the ones that need to be questioned and the focus should be on disproving, rather than confirming, these assumptions. Second, analysts should determine if there are alternate models or interpretations that also plausibly explain the data and information that is available to the analyst. Both the questioning of assumptions and the suggestion of alternate models are best made by other analysts who disagree with the conclusions reached in evaluating the hypotheses. Naval intelligence professionals should encourage an environment where there is a respectful and free flow of knowledgeable and relevant discussion concerning analytic judgments, which ensures that assumptions are questioned and alternative assessments are considered so that the commander can receive the best analytic judgment possible.[23]

F.2.6 Production and Packaging

Production was discussed in detail in section 3.8.3. As discussed in that section, once the analysis is completed the naval intelligence professional creates a product based on the needs of the intelligence user they are supporting. Intelligence products can range from informal verbal reports to lengthy written and graphically intensive products. In all cases, though, naval intelligence analysts must provide their analytic judgment in the front of the product in a clear and concise way. They must also carefully articulate a confidence factor for their assessment and clearly articulate assumptions that were made in developing the product.[24]

---

[20] Heuer, p. 44–46.
[21] Heuer, p. 69, 176.
[22] Heuer, p. 69.
[23] Heuer, p. 69–78.
[24] JP 2-01.3.

NWP 2-0

F.2.7 Evaluation and Feedback

Evaluation and feedback of analysis and production efforts was discussed in detail in section 3.8.5. Naval intelligence professionals must engage the intelligence user to determine the utility of the assessment and to receive feedback that enables improvements in the analysis and production of intelligence. Additionally, naval intelligence leaders must foster an open environment where personnel are free to offer recommendations of ways to improve intelligence analytic and production efforts.[25]

F.2.8 Ongoing Monitoring

Given the adversary is always adapting and the operational environment is always changing, analytic judgments are always tentative. Indicators should be identified which would trigger a reassessment of a given analytic judgment based on significant changes in the operational environment or adversary actions. Additionally, indicators that could confirm or refute a given assessment should continue to be monitored. Naval intelligence professionals should be especially sensitive to feelings of surprise when new data or information arrives that does not fit their conception of the situation. In these cases, the analyst must consider whether this new information is consistent with an alternative hypothesis. A surprise may be the first indicator that an established assessment is off the mark, incomplete, or wrong.[26]

F.3 ANALYTIC TECHNIQUES

There are a number of structured analytic techniques available to the analyst. The goal of this section is not to discuss every available technique but rather to touch on some of the more commonly used techniques so that the naval intelligence professional understands that there are a range of options available for analyzing intelligence problems. Readers desiring a more in-depth understanding of specific techniques should refer to JP 2-01.3, Joint and National Intelligence Support to Military Operations, and Heuer and Pherson's Structured Analytic Techniques for Intelligence Analysis. Analytic tools and techniques have a significant impact on the ability of the analyst to make a sound judgment as they facilitate the sorting and structuring of information, provide a means for challenging assumptions, and enable the exploration of alternative interpretations of the data.[27]

Effective use of analytic tools and techniques requires the naval intelligence analyst to tailor the method or methods to the intelligence problem being addressed. In many cases, analysts use the techniques that have always worked for them in the past. While this practice has benefits and may be acceptable across a range of situations, failing to adequately assess which tools or techniques would best fit the analysis could lead to serious errors in judgment. Similarly, analysts should not select a tool out of convenience. In many cases, a specific tool will suggest itself because it appears to be the tool that works best with the preponderance of the available data and information. The trap for the analyst in selecting tools in this manner is that it may prevent them from thinking about the data that is actually required to answer the intelligence question, which may require analysis using a different technique. Finally, analysts must understand that many of these techniques are not particularly time-consuming to apply but the perception persists that they can somehow slow the process of "real" analysis. While there are some techniques, such as the analysis of competing hypotheses, which are best practiced by analysts who have specific training in that particular tool, the use of timelines, chronologies, matrices, link analysis charts, flow charts, and a host of other techniques can be performed expeditiously by very junior analysts with a high degree of rigor and accuracy. In many cases the tools and techniques discussed below can be done manually, although there are automated tools, such as the network analysis capabilities of Analyst Notebook, which can be used to facilitate analysis and enable collaboration efforts across commands that are geographically separated.

F.3.1 Brainstorming

Brainstorming is a process that is typically used at the beginning of an analytic effort to "identify a list of relevant variables, driving forces, a full range of hypotheses, key players or stakeholders, available evidence or sources of

---
[25] JP 2-01.3.
[26] Heuer, p. 177.
[27] Heuer and Pherson, p. xviii.

information, potential solutions to problems, [and] potential outcomes or scenarios."[28] Brainstorming can be done in person and virtually but the main point is that it is a structured group process that is conducted to stimulate creativity and to benefit from the insights and experiences of all team members. For that reason, naval intelligence leadership needs to foster environments where all analysts, even the most junior, feel free to express ideas, to have those ideas respectfully considered, and then to have those ideas accepted or rejected on their merits. Brainstorming may also be used at any point to stimulate further creative thought and is especially useful at times when little progress is being made or if the analytic team is caught in a rut. Brainstorming sessions are often followed by tasking of analytic and collection efforts and typically the output of the brainstorming session is used to feed other analytic products, such as a cross-impact matrix, which shows the relationship between variables and other factors. Figure F-4 provides the seven general rules that should be followed in brainstorming sessions.

F.3.2 Chronologies and Timelines

Chronologies and timelines provide a graphic means of showing the order of events that is useful for determining the relationships among events. A chronology consists of a simple list of events in the order they occurred, which may be different from the order in which the events were reported to the analyst. A timeline, also known as a time-event chart or matrix, shows not only the events in the order of occurrence but also shows when those events occurred and the time between events. The scale of the timeline can be whatever is useful in addressing the analytic problem. In the case of a cyberspace analytic problem, the timeline scale may be fractions of a second or minutes while tracing the activity of a political movement or terrorist group might be on the scale of months or years.[29]

While a simple technique, chronologies and timelines can be a powerful tool for the analyst and they can be used to convey a wealth of information to the intelligence user. Figure F-5 shows an example of a simple chronology for an adversary naval patrol. Although limited, this tool could be used to determine simple cause and effect relationships or to develop a list of indicators useful for assessing future patrol operations. While not shown here, the source of the data, information, or intelligence used to populate the chronology or timeline should always be annotated in some way so that the original report can be quickly referred to if needed. Additionally, timelines and chronologies, while they may be single use products, are typically maintained and updated by the analyst whenever new data is received.

Brainstorming Rules

1. Articulate a specific purpose for the session and have personnel come prepared with ideas, if time allows.
2. Encourage new ideas. Treat all ideas with respect and try to determine if an idea can be applied to the situation before rejecting it outright.
3. Only one person should talk at a time and everyone should have an opportunity to speak.
4. Allow sufficient time to set the stage, define rules for the session, allow discussion, and capture results.
5. Avoid "groupthink." If possible, invite an outside party with some familiarity with the topic to attend the session and provide feedback.
6. Capture group deliberations on some medium—whiteboard, computer, sticky notes, etc.
7. Summarize key findings.

Figure F-4. Brainstorming Rules[30]

---

[28] Heuer and Pherson, p. 92.
[29] Walton, pp. 9–12; Heuer and Pherson, pp. 52–53.
[30] Heuer and Pherson, p. 91–92.

REDLAND NAVAL PATROL CHRONOLOGY	
Date	Activity
15 Apr 2010	Quarterly patrol orders received for Second Fleet units.
03 May 2010	Fast cruise for three surface combatants in Second Fleet Destroyer Squadron Delta. Combatants collectively designated PG-2 (2nd patrol group observed this year)
08 May 2010	Increased pier activity, stores onload and maintenance, detected in vicinity of PG-2 combatants
12–14 May 2010	PG-2 vessels missing from imagery, assessed to be conducting training ops
15 May 2010	PG-2 vessels back in port
21–26 May 2010	PG-2 vessels missing from imagery, assessed to be conducting training ops
27 May 2010	PG-2 vessels back in port
31 May 2010	PG-2 vessels detected at ammunition pier
02 Jun 2010	PG-2 vessels return to home berths
05 Jun 2010	PG-2 vessels missing from imagery
08 Jun 2010	PG-2 vessels detected in vicinity of the Gulf of Azure, assessed as on patrol duty
21 Jun 2010	PG-2 vessels imaged at supply depot in port close to Gulf of Azure
22 Jun 2010	PG-2 vessels missing from supply depot
21 Jul 2010	PG-2 vessels imaged in homeport, assess patrol duty complete

Figure F-5. Chronology Example

When arrayed on a timeline this technique provides a number of options that can assist the analyst with achieving insights that might not be evident from a simple chronology. While there is no particular format—a timeline can be constructed horizontally or vertically—the important thing is to use a scale that is appropriate to the analytic problem. Once the timeline is constructed with all known data, the analyst can begin to look for patterns within the data. First, the chronology should be reviewed to determine the temporal distance between gaps, which should lead the analyst to question whether particularly "lengthy" gaps are routine, caused by some event, or represent missing data that indicates a potential gap in understanding. Similarly, if the pace of events seems too rapid or if the events seem unrelated, it is possible the analyst is actually viewing multiple sets of data and should investigate that possibility. The timeline could also be used to develop a list of critical events, which can form the basis of an indicator list that can be used to drive collection activities and warning problems and may also contribute to targeting efforts. Additionally, multiple data sets can be arrayed on the same timeline. For example, the patrol pattern provided above could be arrayed on a timeline and current intelligence reporting matched against it. This would allow the analyst to anticipate the adversary's next steps and to quickly detect any change in the pattern, such as the use of one training period vice two, which may indicate either a change in adversary operations or an adversary's response to some external stimulus. Similarly, events from two or more countries could be laid on the same timeline to determine if there are relationships between those data sets that would not be obvious unless viewed from a temporal perspective.[31]

Timeline charts can also be annotated to provide amplifying detail, enabling the analyst to display a significant amount of data in as little space as possible. For example, different shapes can be used to convey different

---

[31] Heuer and Pherson, pp. 52–53; Walton, p. 10.

meanings about the data. Triangles can be used to show the beginning and ending of processes or significant deviations from a routine process. Rectangles can be used to show significant events and diamonds used to show potential decision points that could branch in more than one direction. Colors or symbols can be used to show critical events or significant deviations. Callout boxes can be used to annotate and highlight specific details of the analysis.[32]

Figure F-6 shows an annotated timeline that arrays Redland normal patrol activity against that currently observed and projected to occur. In this scenario, neighboring Azureland is in the midst of a parliamentary election cycle that has been characterized by increased ethnic violence against people of Redland descent. The election results in an anti-Redland party capturing a majority of the seats in Azureland's parliament. While further collection and analysis would be needed to confirm this hypothesis, the fact that Redland disrupts its normal patrol preparation pattern and forgoes a training period to deploy its next Gulf of Azure patrol early may be based on Redland's concern over the Azureland political situation. This hypothesis suggests itself by looking at the potential correlation between Redland and Azureland activities on the timeline. The timeline can then be used as a tool for predictive analysis and collection planning. Has Redland just accelerated their normal patrol pattern or is this event significantly different from past patrols? Observing when, where, and how Redland conducts its next resupply operation for its patrolling vessels may offer insights into Redland's naval response to Azureland's political situation. An analyst may request increased collection focus on Redland naval patrol resupply operations to enable a better assessment of this activity.

Standardization or normalization of the data is an important step as improper sorting can effectively mask insights. The analyst should break data down in a matrix using as many fields or columns as are necessary to designate different types of data (people, places, things, events, amounts, dates, times, etc.). The specific intelligence problem drives what elements of data should be used as categories for the analysis. For example, an analyst attempting to discern the attitude of a nation's key military and political leaders on a U.S. policy may sort these leaders by various factors—ethnicity, educational background, religion, political influences, exchange training—and compare these attributes against leaders in that country whose attitudes are well known to determine if there is a pattern. As another example, analyzing the phone records of six individuals suspected of conspiring to commit some terrorist act may reveal insights about the level of cooperation among the conspirators, the frequency of their interactions, and patterns that may indicate potential heightened activity.[33]

F.3.3 Matrices

> . . . besides communicating your information as it arises . . . you might make out a table, or something in the way of columns, under which you might range, their magazines of forage, grain, and the like, the different corps and regiments, the Works, where thrown up, their connexion [sic] kind and extent, the officers commanding, with the number of guns . . . This table should comprehend in one view all that can be learned from deserters, spies, and persons who may come out from the enemy's boundaries. And tho' it will be a gradual work, and subject to frequent alteration and amendment, yet it may be, by attention and proper perseverance made a very useful one. Transcripts may be drawn occasionally from it as you advance. . .
>
> *General George Washington*
> *November 18, 1778*[34]

---

[32] JP 2-01.3, p. D-9.
[33] Heuer and Pherson, pp. 56–57.
[34] Walton, p. 57.

NWP 2-0

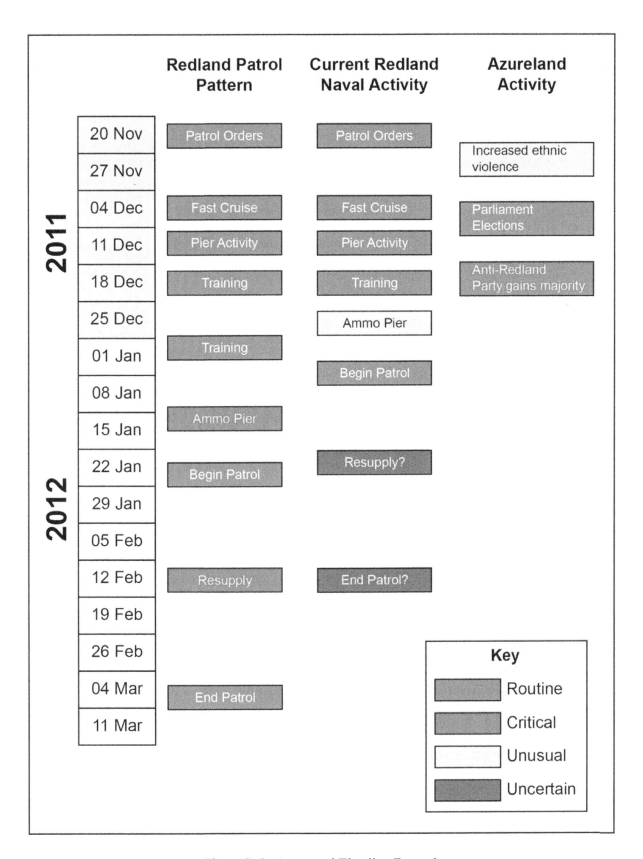

Figure F-6. Annotated Timeline Example

A matrix is one of the more versatile and powerful tools available to the analyst. It provides the ability to sort and organize data but, most importantly, it enables the analyst to show relationships between two different sets of data. While the ability to only compare two sets of data is a limitation of this technique, it is still considered one of the most fundamental and effective tools available to the analyst as it provides a means of visualizing complex relationships within and among data. In using a matrix, the analyst breaks the intelligence problem down into its component elements (represented by each cell in the matrix) and is able to analyze each element separately while still maintaining the context of the problem. An analyst may also use a series of matrices to show relationships among different, but related data sets or to show changes in data or factors over time. Matrices also help an analyst to identify gaps that may be satisfied by further collection or analytic efforts.[35]

While developing a matrix is a relatively simple and straightforward process, the naval intelligence analyst must first step back and consider the intelligence problem they are attempting to solve. This drives how the data or sets of data are broken down and categorized for comparison. The analyst must then consider how large the matrix must be, ensuring that each grid used has enough rows and columns to fit the data that will be analyzed.

Once the form of the matrix is defined, the analyst then populates the matrix with data and analyzes the data to determine patterns or obtain insights that would be difficult to ascertain if attempting to correlate and determine relationships among data through a purely mental process. Figure F-7 provides an example of a matrix used to assess the relative capabilities of four naval surface to surface missile systems operated by navies in the Gulf of Azure region. In this example, the analyst has chosen to compare each system against five different categories of information—range, speed, warhead size, failure rate, and accuracy. Other variables could be added to include more technical details, such as type of missile guidance system, or more subjective details, such as the assessed readiness of the units carrying these systems or the complexity of their detect to engage sequence. In the case below, the analyst has not only documented known and unknown variables but has also used color coding to indicate their confidence level in the data they have recorded. This provides the analyst the ability to qualify their judgment about the relative strengths and weaknesses of each system as well as indicate areas that may require more collection or analytic focus. Alternatively, the analyst could use color coding or annotations to highlight their analytic assessments, such as coloring the most capable system in green and adding a comments block to the matrix to give their rationale for this judgment.

F.3.4 Network Analysis

Network analysis, which is also known as nodal or link analysis, is a powerful tool that was introduced in section 4.4.2 and was addressed throughout the chapter 4 discussion on systems analysis and IPOE. Network analysis is defined as the "review, compilation, and interpretation of data to determine the presence of associations among individuals, groups, . . . [systems] . . . or other entities; the meaning of those associations to the people involved; and the degrees and ways in which those associations can be strengthened or weakened."[36]

Network analysis is a visually compelling tool for developing assessments and communicating linkages and influence relationships needed by targeteers and information operations planners. It can be used to aid analysts in identifying patterns of organization, authority, communications, travel, financial transactions, and other interactions that are not readily apparent from other methods of analysis. Network analysis also provides a way to visualize key leaders, information brokers, and sources of funding, ideology, or political support that may be sustaining an organization.

While network analysis may be done using manual methods, various software tools are available to the analyst to aid with this type of analysis. The use of software tools has several advantages. They allow the analyst to efficiently enter data and enable efficient analyze of that data by facilitating rapid and automated link creation. Software tools also provide the ability to filter and visualize the data from different perspectives. These tools enable the analyst to maintain the network analysis as an ongoing project that can be easily updated whenever

---

[35] Heuer and Pherson, pp. 64–65; Walton, p. 13.
[36] Heuer and Pherson, p. 68.

GULF OF AZURE NAVAL SURFACE TO SURFACE MISSILE COMPARISON					
System	Range (nautical miles)	Speed (knots)	Warhead (lbs TNT equivalent)	Failure Rate (percent)	Accuracy (feet)
Redland Hogar II	200	Unknown	400	25	100
Orangeland Mystic	65	600	200	10	5
Azureland Thumper	300	400	Unknown	35	350
Brownland Frenga IV	150	500	350	Unknown	10

Key:
- High Confidence
- Medium Confidence
- Low Confidence
- Unknown

Figure F-7. Matrix Example

new information is received. Network analysis software can aid the analytic process by providing different views of the data that may enable the analyst to achieve fresh insights. Advanced software tools can also ascertain the strengths of various links and nodes in the network and present that information to the analyst. Finally, these tools typically provide the capability to share the database which stores the network analysis data and information, enabling analysts to work collaboratively on network analysis problems despite geographic separation. As with any analytic tool, the analyst must be mindful of the accuracy of the data used to build the network analysis. Inaccurate data will significantly skew the analysis.[37]

The steps for building a network analysis product are:

1. Choose one reliable source of data to use as the starting point.

2. Identify, combine, or isolate nodes within the source data.

3. List each node on a chart or in a database, association matrix, or network analysis program.

4. Identify the interactions among nodes.

5. List interactions by type.

6. Identify each node and interaction by a specific criterion that is useful for further analysis. Examples may include frequency of contact, type of contact, type of activity (influence, financial, etc.), or source of information.

---

[37] Heuer and Pherson, p. 69.

7. Choose a central node (or one that the analyst suspects will be central) and begin with that node.

8. Draw connections (links) between nodes either through manual or software methods. Links can be annotated in many different ways. For example, dotted lines could be used to indicate suspected connections while solid lines could indicate confirmed connections; thick lines could indicate strong relationships while thin lines would indicate weak relationships; green colored lines could indicate financial transactions while red colored lines represent ideological or religious influence. Arrowheads should also be used to indicate if transactions are unidirectional or bidirectional.

9. Work from the central node or nodes and add additional links and nodes until the information from good sources of data is exhausted.

10. Add links and nodes from other sources, always ensuring the source is annotated.

11. Stop when: there is no more information to plot; all new links are dead ends; or all new links turn in on each other.[38]

Once complete, the analyst should update the network analysis product and use the following techniques to aid in analysis:

1. Try to rearrange nodes and links so that the links cross over one another as little as possible.

2. Cluster nodes by looking for "dense" areas in the chart. If possible, characterize clusters using shapes, colors, or other techniques and annotate the rationale for clustering and confidence in the assessment.

3. Cluster the clusters, using the method above, if possible.

4. Label clusters by a common theme among the nodes it contains. This aids in identifying key personnel, organizations, events, and locations. If the analyst has developed a model for what this cluster should look like, then analysis of the cluster may reveal gaps.

5. Look for "cliques" or nodes that are connected to every other node within a group but not to many nodes outside a particular group.

6. Nodes or links that connect two clusters are typically representative of information brokers. These should be annotated and watched.

7. Time stamps and directional arrows can be used to chart the flow of activity within a network. Analysis of the activity flow in a network helps the analyst to see patterns, lines of authority, network resiliency and weaknesses, and other analytic insights.[39]

Figure F-8 provides an example of a simple network analysis chart.

---

[38] Heuer and Pherson, p. 71.
[39] Heuer and Pherson, p. 72.

NWP 2-0

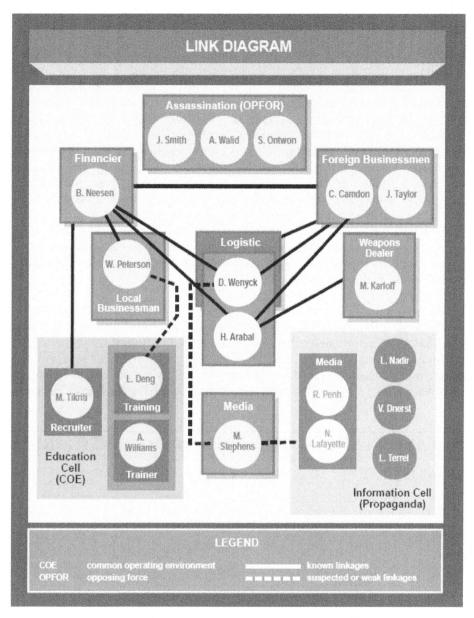

Figure F-8. Network Analysis Example[40]

---
[40] JP 2-01.3, p. D-14.

F.3.5 Flow Charting

Flow charting is a technique used to break down a process or system into its component parts and then map out the "flow" of data, information, finances, commodities, services, ideologies, or other items through the process or system. This technique is useful for determining the function of each component and how each component contributes to the process or system; visualizing and understanding how components interrelate with one another; and obtaining insights on the overall process or system, such as determining where potential weaknesses, inefficiencies, or resiliencies exist. Flow charts provide a visual means of rendering a wide array of complex data and they can be used to show internal interactions of a process or system as well as interactions with external processes and systems. These charts may be integrated with timeline charts, to show the flow of items over time, or combined with geospatial products to show the flow of items through both space and time. Analysts must tailor the flow chart to the specific intelligence problem they are addressing. A flow chart depicting the detect-to-engage sequence for an adversary naval aviation unit may only require a simplified view of how targeting information flows to the pilot from aircraft sensors and external C2 nodes. A flow chart supporting a cyberspace operation, however, may require significant levels of detail concerning personnel, organizations, networks, systems, and software versions used on specific systems. Flow charts are also an excellent tool for identifying gaps that can drive additional collection and analytic efforts.[41]

The flow charting process simply requires the naval intelligence analyst to break a process or system down to its component parts and then order those parts into a logical flow. In some cases, this may be a simple graphic, such as one that shows the flow of data or money between individuals and institutions. In other cases, a flow chart may include a significant amount of detail, not only describing the high level flow of data, services, and the like but also documenting specific systems, networks, and other supporting infrastructure used to facilitate and enable the flow. Figure F-9 provides an example of a flow chart that documents Redland's West Fleet patrol order tasking process.

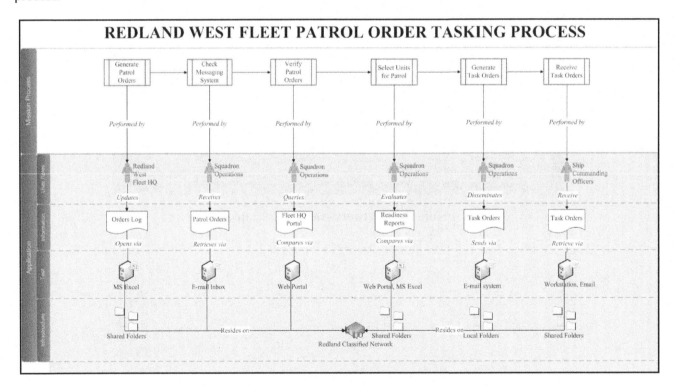

Figure F-9. Flow Chart Example

---

[41] Walton, pp. 12–13.

F.3.6 Scenario Analysis

Scenario analysis involves postulating various scenarios concerning how future events will unfold and then using these scenarios to guide collection and analytic efforts to determine which scenario is most likely to occur. Since the COA analysis process discussed in detail in section 4.6 is a type of scenario analysis, this specific technique is not discussed in depth here. It is used when analysts and decision makers are confronted with a new or highly complex situation and it serves to establish boundaries for the intelligence problem. The value of scenario analysis is that it postulates specific future end states (most likely, most dangerous, best case, status quo, etc.), which enables the naval intelligence analyst to then look at variables and drivers in the present that would result in the anticipated end states. Scenarios are useful for developing indicators, which are discussed in the next section; assist analysts and decision makers in anticipating surprise by challenging assumptions and looking for plausible "wild card" events; provide a framework for assessing costs, opportunities, and risks; enable analysts to weigh many unknown or unknowable factors yet ground these within a plausible context; and ensure that the intelligence problem is bounded and specific.[42]

F.3.7 Indicators

While previously discussed in section 3.8.3.1, indicators are often used in association with warning problems. Heuer and Pherson define indicators as "observable phenomena that can be periodically reviewed to help track events, spot emerging trends, and warn of unanticipated changes. An indicators list is a pre-established set of observable or potentially observable actions, conditions, facts, or events whose simultaneous occurrence would argue strongly that a phenomenon is present or likely to occur."[43] Indicators are closely aligned with scenario analysis since development of an indicator list often requires the analyst to develop a scenario for how an event is likely to unfold and then develop the set of observable phenomena that would indicate whether that scenario was occurring. In cases of tactical warning, indicators may be very focused and specific and are often based on the observed operational patterns of foreign navies and other forces. At the operational and strategic levels, indicators may be higher level although specific tactical actions, such as logistics and fueling activity at a ballistic missile launch site, may be part of warning scenarios developed for those levels of war.[44]

When tied with scenario analysis, each scenario should have specific and unique indicators that can be used to determine whether that particular scenario is unfolding. In the case of a specific warning problem, the development of indicators provide a collaborative means by which analysts can deconstruct a particular intelligence problem and break it into component parts that can be more easily collected on and analyzed. The other major advantage of this method is that it creates thresholds that will trigger specific actions by analysts and decision makers. A persistent danger in intelligence analysis, which is discussed in section F.4 (Analytic Pitfalls), is that the incremental addition of information often causes analysts to miss substantial change until it is too late. Once an analyst has established a particular mindset, it typically requires a very unambiguous and significant development for them to consider changing their mindset. Indicators counteract this issue by establishing the thresholds for action and preventing the analyst from rationalizing away data that does not conform to their mindset. From this discussion it should be apparent that the selection of poor indicators—ones that are too narrow, outdated, reinforce analytic biases, or cause analysts to discard relevant data—will result in intelligence failure and surprise. Indicators should be reviewed on a regular basis, to ensure their validity, and should be closely guarded since an adversary could use information about indicators to plan their deception efforts or manipulate friendly forces.[45]

Creating an indicator list can be done by a lone analyst but it is typically best done in collaboration to ensure a mix of perspectives and experiences is used. As discussed in section 3.8.3.1, indicators must be observable, valid, unique, reliable, and stable. Once the indicators are listed, the naval intelligence analyst or team must decide the method or methods they will use to monitor each indicator and how frequently each indicator will be monitored.

---

[42] Heuer and Pherson, p. 122–124.
[43] Heuer and Pherson, p. 132.
[44] Heuer and Pherson, p. 132.
[45] Heuer and Pherson, p. 133.

The process for maintaining the indicator list should also be defined as well as the method that will be used to update the list when an indicator is "tripped" and the response actions that will initiate.

Warning problem indicator lists typically use a stoplight or bubble chart to indicate whether an indicator has not been observed (green), is likely to occur or may have occurred (yellow), and has been observed (red). Arrows could also be used with an up arrow indicating an event is trending toward meeting the specific indicator, an arrow point down as indicating a decrease in activity associated with a particular indicator, and a dot or arrow pointing left and right indicating no change in the indicator's status.

As successive indicators are observed, this should drive additional collection and analytic activity, especially if events do not unfold in an expected order, which may be an indication of adversary deception efforts or a new scenario. Each indicator should be well defined and well understood by the intelligence team and levels of confidence should be applied when updating indicators.[46]

Figure F-10 provides an example of a simple indicator list associated with determining naval force mobilization efforts for Redland. Using the methodology discussed above, many of the indicators for West Fleet mobilization have been observed or may have been observed, which would indicate that the Redland West Fleet is in the process of mobilizing.

F.3.8 Analysis of Competing Hypotheses

Analysis of competing hypotheses (ACH) is an advanced analytic technique that identifies a broad range of probable hypotheses or scenarios and then subjects these hypotheses to a structured and rigorous review to determine which hypothesis or hypotheses are most probable given the data available. While ACH does not guarantee the naval intelligence team will produce the correct answer to the intelligence problem, the methodology, if properly executed, ensures the team will follow an appropriate analytic process to obtain their answer. Given it is a highly structured methodology that seeks input from a variety of viewpoints, it provides a method for mitigating the impact of many of the analytic biases discussed in section F.4. The advantages of ACH are that it begins with a wide range of alternatives, ensuring many different viewpoints are considered; it identifies and emphasizes a few key pieces of data or assumptions that are likely to have the greatest diagnostic value in assessing likely alternatives; and it is focused on seeking data to refute a given scenario vice trying to confirm a scenario as most analysts do. While a full explanation of the ACH process is beyond the scope of this publication, the steps of the process are provided in figure F-11.[47]

---

[46] Heuer and Pherson, pp. 133–137.
[47] Heuer, p. 97–109.

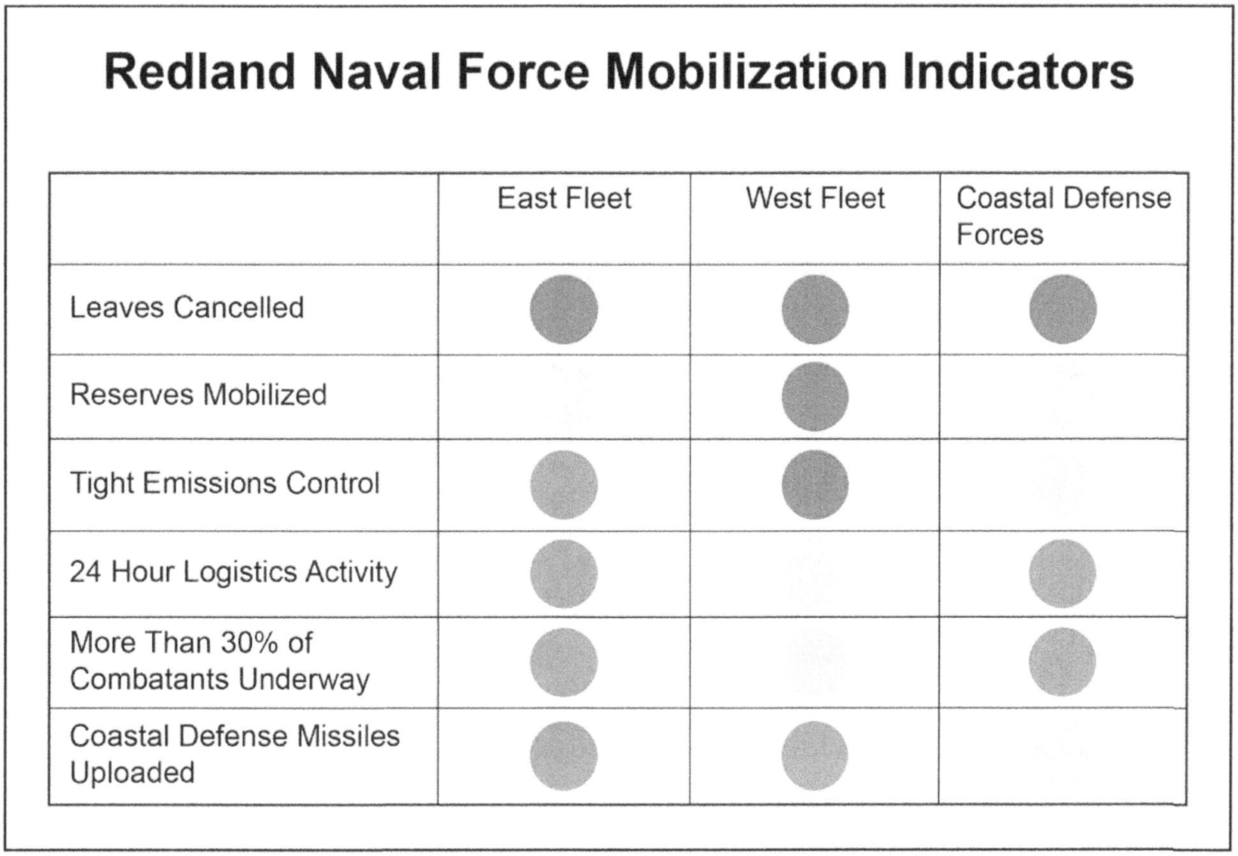

Redland Naval Force Mobilization Indicators

	East Fleet	West Fleet	Coastal Defense Forces
Leaves Cancelled	●	●	●
Reserves Mobilized		●	
Tight Emissions Control	●	●	
24 Hour Logistics Activity	●		●
More Than 30% of Combatants Underway	●		●
Coastal Defense Missiles Uploaded	●	●	

Figure F-10. Indicators Example

ANALYSIS OF COMPETING HYPOTHESES
1. Identify the possible hypotheses to be considered. Use a group of analysts with different perspectives to brainstorm the possibilities.
2. Make a list of significant evidence and arguments for and against each hypothesis.
3. Prepare a matrix with hypotheses across the top and evidence down the side. Analyze the "diagnosticity" of the evidence and arguments—that is, identify which items are most helpful in judging the relative likelihood for the hypotheses.
4. Refine the matrix. Reconsider the hypotheses and delete evidence and arguments that have no diagnostic value.
5. Draw tentative conclusions about the relative likelihood of each hypothesis. Proceed by trying to disprove the hypotheses rather than prove them.
6. Analyze how sensitive your conclusion is to a few critical items of evidence. Consider the consequences for your analysis if that evidence were wrong, misleading, or subject to a different interpretation.
7. Report conclusions. Discuss the relative likelihood of all hypotheses, not just the most likely one.
8. Identify milestones for future observation that may indicate events are taking a different course than expected.

Figure F-11. Analysis of Competing Hypotheses[48]

---
[48] Heuer, p. 97–109.

F.3.9 Red Teaming and Devil's Advocacy

Red teaming and devil's advocacy are advanced analytic techniques that are used to challenge the developed or prevailing analytic viewpoint. Both red teaming and devil's advocacy are highly effective ways of countering some of the analytic biases that is discussed in section F.4. Red teaming requires the use of experienced analysts who can adopt the mindset of the adversary and conduct an alternative analysis of the intelligence problem from the adversary's point of view. If used, the red team analysis should be considered closely by naval intelligence professionals as it can highlight important issues and alternative points of view that may require closer analysis. Devil's advocacy also requires an analyst with experience, preferably one from outside the organization, who is empowered to review an organization's analytic processes and methods, critique them, and champion an opposing viewpoint, if required. The goal of both processes is to mitigate the effect of complacency in analysis by challenging biases and perceptions that are inimical to sound analytic practice.

F.3.10 Key Assumptions Check

All intelligence analysis is founded on a combination of data and assumptions, which determine how data is analyzed and interpreted. Assumptions are a necessary part of analytic work as they allow the naval intelligence professional to work with incomplete or ambiguous data. A key assumptions check is a structured technique that should be conducted at various points in the analytic process so that assumptions are explicitly understood and questioned. Analysts should be especially sensitive to assumptions that underlie or are primary driver for a specific analytic conclusion and all efforts should be made to refute or confirm these assumptions. The process for conducting a key assumptions check is similar to brainstorming. The naval intelligence team, preferably with the assistance of knowledgeable personnel from outside the team, engage in a session to explicitly document the assumptions associated with various analytic processes and products that the team is responsible for maintaining. Each of these assumptions should be questioned to determine if they are still valid and what indicators would cause them to be invalidated. The value of this process is that it results in new perspectives about the intelligence problems faced by the naval intelligence team and establish or reset the confidence level analysts should have in their analytic judgments.

F.4 ANALYTIC PITFALLS

Good analysis depends on more than just good information. While the quality of information is very important, history is replete with examples of intelligence analysts who had access to sufficient information with which to make an accurate assessment yet their judgment proved faulty. In addition to receiving quality information, analysts must also be cognizant of the lens which they use to focus and filter the information that they receive. These lenses are known as models, mindsets, biases, or assumptions. For the purposes of the discussion below, the collection of biases and assumptions that color an analyst's thinking are referred to as a "mindset" while specific examples of these lenses are referred to as "biases."

Mindsets are often quick to form but are resistant to change. In many cases, this can be a good thing. Developing a model for perceiving the world is a useful tool for anticipating potential outcomes and navigating new situations. New information that confirms a mindset is easily accepted by an individual but ambiguous data is often rejected and, the more ambiguous the data, the easier it is to reject. People also perceive what they expect to perceive. In the case of an intelligence problem that builds incrementally, each new piece of information that confirms the prevailing hypothesis is easily accepted as "evidence" of the correctness of the assessment yet other "signals" in the "noise" are discarded. So, it typically takes more information of an unambiguous nature to overcome a person's mindset and enable them to recognize an unexpected event from an expected one. For this reason, the mind does not handle uncertainty well. It is difficult for the mind to deal with inherent uncertainty, such as the fog of war, and induced uncertainty, such as that created by adversary denial and deception efforts, yet this is exactly what the naval intelligence analyst is called to contend with on a daily basis.[49]

While no analyst can divorce themselves from their mindset or eliminate biases from their process, understanding the different forms of bias can aid the analyst in explicitly identifying them and mitigating their impacts on sound

---

[49] Heuer, pp. xviii, xix–xx, 8–15.

analytic practice. One bias that has been consistently discussed throughout this publication is "mirror-imaging," which is a type of cultural bias. The assumption that adversary's will think and act as the analyst would in a similar situation is a dangerous assumption to make as people from other cultures have a completely different mindsets, even if, superficially, their decisions and actions seem to conform to a familiar pattern. Although there are many different kinds of bias that could be discussed, the focus of the discussion below is on cognitive biases, since these are very subtle biases that are particularly inimical to sound analytic practice. Heuer says the following about cognitive biases:

> Cognitive biases are mental errors caused by our simplified information processing strategies. It is important to distinguish cognitive biases from other forms of bias, such as cultural bias, organizational bias, or bias that results from one's own self-interest. In other words, a cognitive bias does not result from any emotional or intellectual predisposition toward a certain judgment, but rather from subconscious mental procedures for processing information. A cognitive bias is a mental error that is consistent and predictable. . . . Cognitive biases are similar to optical illusions in that the error remains compelling even when one is fully aware of its nature. Awareness of the bias, by itself, does not produce a more accurate perception. Cognitive biases, therefore, are exceedingly difficult to overcome.[50]

The discussion below concentrates on some of the more common forms of cognitive bias.

F.4.1 Biases of Evaluation of Sources[51]

One of the key jobs of the analyst is to evaluate sources to assess their credibility and reliability. As discussed in section 3.8.2.3, this is a difficult task which requires analysts to develop a confidence factor for each source. Analysts should be mindful, however, that they are subject to bias in assessing sources just as they are subject to bias in conducting any other kind of analysis. Some of the more common forms of evaluation bias are discussed below:

1. Vividness of Data. As a general rule, people have a much easier time receiving vivid, concrete data and information and a much harder time absorbing and integrating data that is abstract or statistical in nature. Even though those latter forms of data may be much more meaningful, they are less likely to be accepted or well-understood by the analyst, especially the analyst under extreme time pressures. Anecdotes and case histories are easily understood and compelling to the analyst. This does not make these sources less credible or reliable but, like any source, the analyst must seek corroborating information and not accept compelling information at face value.

2. Missing Data. One of the primary characteristics of intelligence analysis, one that distinguishes it from other forms of analysis, is that key data and information is typically missing. While this publication states that analysts should determine what information is lacking from their analysis and adjust the confidence of their assessment based on this fact, in reality this happens infrequently. To deal with this bias, "analysts should identify explicitly those relevant variables on which information is lacking, consider alternative hypotheses concerning the status of these variables, and then modify their judgment and especially confidence in their judgment and act accordingly. They should also consider whether the absence of information is normal or is itself an indicator of unusual activity or inactivity."[52]

3. Oversensitivity to Consistency. Analysts may only have a small data set to work with in making their conclusions. If that data is consistent, analysts are far more likely to extrapolate a meaning that is not truly statistically valid given the size of the actual sample. Unless an analyst knows how large the representative sample size is, they should not attribute a high level of confidence on their assessment solely on the basis of the data's consistency.

---

[50] Heuer, pp. 111–112.
[51] Heuer, pp. 115–126.
[52] Heuer, p. 120.

4. Coping with Evidence of Uncertain Accuracy. When confronted with ambiguous or uncertain data the analyst's typical response is either to reject it completely or to accept it wholly. This can have serious implications for the analysis. If the analyst only has 60 percent confidence in the data yet treats it with a 100 percent confidence level throughout their analysis then it will skew the resulting analysis significantly.

5. Persistence of Impressions Based on Discredited Evidence. As discussed above, once a mindset is formed it is very resistant to change. Even if data makes it clear that the prevailing view is inaccurate, it is very difficult for analysts and decision makers to move away from that view.

F.4.2 Biases in Perception of Cause and Effect[53]

In many ways, the process of intelligence analysis shares much in common with the historical method. For many types of intelligence analysis, the analyst creates an overarching narrative to explain the connections and interrelationships among many variables and then projects how the interactions of these variables will unfold in time and space. The human mind is geared to easily receive information and analysis in a narrative form because it satisfies the need to impose order. This need also influences the way data and information is interpreted. If an analyst receives a significant amount of data that has no discernible pattern, their first thought is most likely that they lack the understanding to determine the pattern, even if the dataset they are confronting is completely random. This same attribute of the human mind is part of the reason that conspiracy theories are so prevalent—the mind wants to impose a discernible cause on events, even if one does not exist.

1. Bias Favoring Perception of Centralized Direction. There is a human tendency to attribute events to human causes and less to accident and coincidences. People also tend to overemphasize the impact of one person in making decisions, even though experience shows that the majority of organizational decisions are the result of some collaborative process. Bureaucracies, tribes, and other forms of organizational structures tend to act in similar ways. Even if access to a key decision makers thought processes and decrees are known, factional fighting, chance, and improper implementation of orders are normal and will impact the effectiveness of a leader's decisions.

2. Similarity of Cause and Effect. In the physical world, cause and effect has a very linear relationship. Big guns, for example, make loud noises. While analysts have been conditioned to accept linear cause and effect relationships as the norm, these relationships are often not linear in other spheres. It is not valid for the naval intelligence analyst to assume, for example, that political events have political causes, that large events are precipitated by some significant action, and that small events cannot change the course of history. History has many examples that prove relatively big events had small long-term consequences and that relatively small events had major consequences. While a linear cause and effect model makes a good narrative, it is not a valid assumption for the analyst to make.

3. Internal Versus External Causes of Behavior. Analysts often attribute more weight to the perceived internal motivations of persons they are analyzing than they should. Often, external causes have a significant impact on individual behavior. This bias is a form of mirror-imaging in which the analyst assumes that the adversary's perception of external events is the same as theirs. This could lead to surprise when the adversary reacts hostilely to a friendly force action that was assumed to be "nonthreatening" by an analyst who did not fully appreciate the adversary's different mindset and perceptions of events.

4. Overestimating Our Own Importance. Analysts may assume that when an adversary does respond to friendly force action in the way predicted, that this confirms the accuracy of their assessment. Drawing a direct correlation between adversary actions in response to friendly force activities is a separate intelligence problem and should not be assumed.

5. Illusory Correlation. Correlation exists when the existence of one event implies the existence of another. For example, variables are correlated when a change in one variable implies a change that will be similar in scale to the change in another variable. Correlation does not imply causation. Just because two events are

---

[53] Heuer, pp. 127–146.

correlated, there may be other reasons for the correlation such as chance or the fact that both events have the same root cause. While correlation is important to intelligence analysis, the analyst must determine the indicators that prove causation and monitor those instead of assuming that a correlation relationship exists.

F.4.3 Biases in Estimating Probabilities[54]

As a general rule, people do not have an inherent statistical sense that enables them to judge probability, particularly of complex events, with definitive accuracy. Typically, analysts use qualifiers—such as probable, possible, and likely—to introduce probability into assessments. The difficulty with the use of these terms is they are often open to interpretation on the part of the analyst and the intelligence user. Some intelligence users may interpret a report of "possible" adversary submarine activity in a particular area as confirming their preconceptions about adversary actions when, in fact, the analyst meant to communicate that there was little likelihood that there was any adversary submarine activity in the area discussed. Some biases in estimating probabilities include:

1. Availability Rule. There are two cues that analysts may unconsciously use to judge the probability of an occurrence or event. The term "availability," as used here, refers to the analyst's ability to imagine and event or the ease with which they can retrieve an example of such an event from memory. As discussed above, once a narrative is constructed and becomes accepted, that narrative seems far more likely than alternative views of the same situation. So, the availability rule is typically used to predict the likelihood or frequency of an occurrence based on past experiences. While this may be a valuable tool, especially in cases where more detailed analysis is not feasible or needed, it is still an assumption and must be characterized as such.

2. Anchoring. Anchoring occurs when an analyst chooses a starting point, such as a previous assessment, and then adjusts that assessment based on newly acquired information. The initial assessment acts as an anchor on the subsequent analysis. While it grounds the analysis, it also forces the analyst to make only incremental changes to the previous assessment vice make a more radical shift in the assessment as suggested by the data. Anchoring may lead to complacency and overconfidence in an assessment. Occasionally the analyst must examine the anchor itself to determine if the conclusions reached in that assessment require significant revision based on the weight of accumulated data and information.

3. Expressions of Uncertainty. As discussed above, if an estimate is ambiguous the intelligence user will interpret the report with a bias toward confirming what they already believe. This is the reason many intelligence users, upon reading an intelligence report, will state that it told them little that was actually new to them. When using ambiguous terms—such as probably, possibly, may, could, likely, and unlikely—the analyst should also use a percentage confidence estimate as this conveys a more concrete and objective estimate of uncertainty than a term can convey.

F.4.4 Other Biases[55]

There are other biases the analyst should be mindful of when evaluating data and creating their assessments. These include:

1. Rational Actor Model. Similar to mirror-imaging, the error in this bias is for the analyst to assume that the adversary will act in a "rational" manner, a determination which is typically anchored in one's own cultural biases. What may appear to be irrational behavior may be perfectly logical and acceptable behavior when viewed from the perspective of the adversary. Deep penetration of the adversary is needed to truly understand the adversary's perspective and mitigate the rational actor bias.

2. Confirmation Bias. As discussed, once the human mind has reached a conclusion on an issue it is very difficult to change a mindset. The confirmation bias occurs when accurate reports that do not conform to

---

[54] Heuer, pp. 147–156.
[55] Walton, pp. 127, 171.

the recipient's mindset are devalued or rejected while those pieces of information which are in alignment with the intelligence user's mindset, even if inaccurate or incomplete, are easily accepted and integrated.

3. Framing Bias. The framing bias occurs when data, information, or intelligence is presented in such a way that it leads to a preferred conclusion. This could result from an analyst's personal preconceptions or from the analyst's perceptions concerning what the intelligence user wants to see as the answer to the intelligence problem. Like all biases, the framing bias is an unconscious bias. Commanders and other users of intelligence must be conscious of the fact that their preferences regarding the outcome of the intelligence assessment can exert a subtle pressure on analysts that skews the way data is gathered, interpreted, and integrated. Analysts must guard against this bias by pressing for clear, well-defined, and objectively stated intelligence requirements; ones that do not exhibit a predisposition toward a particular answer.

4. Status Quo Thinking. When confronted with a long-standing situation, it is easy for analysts and decision makers to assume that the situation will continue into the future or, if change does occur, it will be an evolutionary or gradual change vice a dramatic or revolutionary one. The Iranian Revolution and the dissolution of the Soviet Union are two examples where status quo thinking inhibited analysts' ability to see political or social changes that would overturn the prevailing order in those countries.

F.5 ANALYTIC INTEGRITY STANDARDS

In 2007 the Director of National Intelligence issued analytic intelligence standards for the intelligence community. These standards must be adhered to when analyzing and producing intelligence. These standards are:[56]

1. Objectivity: Analysts and leadership must perform their duties from an unbiased perspective. Analysis should be free of emotional content, give due regard to alternative perspectives and contrary reporting, and acknowledge developments that necessitate adjustments to analytic judgments.

2. Independent of Political Considerations: Analysts and leadership should provide objective assessments informed by available information that are not distorted or altered with the intent of supporting or advocating a particular policy, political viewpoint, or audience.

3. Timeliness: Analytic products that arrive too later to support the work of intelligence users weaken the utility and impact of the intelligence. Analysts will strive to deliver their products in time for them to be actionable by the commander or supported organizations. Analysts have a responsibility to know the schedules and requirements of those requiring intelligence.

4. Based on All Available Sources of Intelligence: Analysis should be informed by all relevant information that is available to the analytic element. Where critical gaps exist, analytic elements should work with collectors to develop appropriate collection, dissemination, and access strategies.

5. Exhibits Proper Standards of Analytic Tradecraft:

   a. Properly describes the quality and reliability of underlying sources

   b. Properly caveats and expresses uncertainties or confidence in analytic judgments

   c. Properly distinguishes between underlying intelligence and the analyst's assumptions and judgments

   d. Incorporates alternative analysis where appropriate

   e. Demonstrates relevance to U.S. national security and the mission

   f. Uses logical argumentation

---

[56] Director of National Intelligence, Intelligence Community Directive (ICD) 203, Analytic Standards, 21 June 2007.

g. Exhibits consistency of analysis over time or highlights changes and explains rationale

h. Makes accurate judgments and assessments.

F.6 INTELLIGENCE ANALYSIS—AN ANALYST'S PERSPECTIVE

A number of issues related to analytic tradecraft were covered in this appendix. Tying together the many concepts presented here is a difficult task but one of the classic works on the subject, Sherman Kent's Estimates and Influence, is provided below to give the naval intelligence professional a greater appreciation for the craft of intelligence analysis. While written primarily for intelligence analysts serving senior policymakers, Kent's essay can be easily extrapolated for analysts at all levels.

The article below was originally published as a confidential essay in the 1968 edition of Studies in Intelligence.

Estimates and Influence by Sherman Kent[57]

There are a number of things about policymaking which the professional intelligence officer will not want to hear. For example, not all policymakers can be guaranteed to be free of policy predilections prior to the time they begin to be exposed to the product of the intelligence calling. Indeed, there will be some policymakers who could not pass a rudimentary test on the "facts of the matter" but who have the strongest views on what the policy should be and how to put it into effect. We do not need to inquire as to how these men got that way or why they stay that way, we need only realize that this kind of person is a fact of life.

Nor should we be surprised to realize that in any policy decision there are a number of issues which we who devote ourselves solely to foreign positive intelligence may almost by definition be innocent of. The bulk of them, are, of course, purely domestic ones: domestic political issues, domestic economic issues, popular attitudes, public opinion, the orientation of the congressional leadership, and so on. Even if we know in our bones of the great weight which such issues have carried in many a foreign policy decision, we do not readily and consciously acknowledge it. Our wish, is, of course, to have our knowledge and wisdom about the foreign trouble spot show itself so deep and so complete that it will perforce determine the decision. The nature of our calling requires that we pretend as hard as we are able that the wish is indeed the fact and that the policymaker will invariably defer to our findings as opposed to the cries of some domestic lobby.

But consider for a moment how people other than ourselves and our consumers view these phenomena which I have just dismissed with a mild pejorative. Look, for example, at the table of contents of any of the recent books devoted to "How Foreign Policy Is Made." Or look at the lineup of lectures and discussions in the syllabus of any of our senior service schools; look particularly at the section devoted to national security policy formulation. You will find that intelligence and what it contributes to the task, far from enjoying the overpowering importance with which we—quite understandably—like to endow it, is casually ticked off as one of a score of forces at work.

The Credibility of Intelligence

Thus a certain amount of all this worrying we do about our influence upon policy is off the mark. For in many cases, no matter what we tell the policymaker, and no matter how right we are and how convincing, he will upon occasion disregard the thrust of our findings for reasons beyond our ken. If influence cannot be our goal, what should it be? Two things. It should be relevant within the area of our competence, and above all it should be to be credible. Let things be such that if our policymaking master is to disregard our knowledge and wisdom, he will never do so because our work was inaccurate, incomplete, or patently biased. Let him disregard us only when he must pay greater heed to someone else. And let him be uncomfortable—thoroughly uncomfortable—about his decision to heed this other.

---

[57] Kent, S., Estimates and Influence, published in Studies in Intelligence (1968), https://www.cia.gov/library/center-for-the-study-of-intelligence/csi-publications/books-and-monographs/sherman-kent-and-the-board-of-national-estimates-collected-essays/4estimates.html

Being uncomfortable is surely his second choice. Before he becomes uncomfortable he is going to ask himself if it is strictly necessary. This is of course the equivalent of asking himself if he really thinks that the information he has received from his intelligence colleagues is relevant to his problem and if he has to believe it. When we in intelligence look at the matter in this light we might consider ourselves fortunate that our policymaking consumers find so much of our product relevant, credible, and hence useful. Is there any way of categorizing that which is most happily, gratefully, and attentively read and that which is least? Perhaps a start can be made by having a quick critical look at three classical families of intelligence utterances.

First, basic intelligence. No question but that credibility is highest in this area of intelligence. Time and time again our consumer has need of something comparable to the perfect World Almanac or the perfect reference service. We come close to giving him just that, and nine times out of ten he is warmly appreciative of the breadth and depth of our knowledge and the speed with which we can handle his requests.

Second, how about current intelligence? There is probably less enthusiasm among consumers for this than for basic. They have a tendency to compare it—and unfavorably—to the daily press or the weekly news magazines; or they gripe because they often find it a gloss upon something they have just read in a cable.

Lastly, in the formal estimate credibility is lowest. It was more than a decade ago that Roger Hilsman, after interrogating scores of policymaking consumers of intelligence, concluded thus. He discovered that the people with whom he talked were extremely grateful to intelligence when it came up with the facts that they felt they had to know before they went further with their policymaking and operating tasks. They seem to have gone out of their way to praise intelligence in its fact-finding role, but to be anything but grateful for intelligence utterances in the estimate category.

Why was this so? Although Hilsman does not make the point, one may safely infer from his findings: The policymaker distinguished in his own mind between things which he thought of as factual and those which he thought of as speculative. For the first he was grateful, for the second not at all.

This puts a number of questions before the house. Why should Hilsman's respondents (implicitly, at least) have questioned the credibility of intelligence estimates? Was it because the respondents had caught intelligence out in self-serving errors? Was it because they were fearful of being misled by intelligence? Had intelligence on its part ever done anything to merit this want of confidence on the part of its customers? If not, how did it come about that the very officer who besought the help of intelligence in one area eschewed intelligence in another?

The Nature of the Estimate

Let me begin with a look at estimates and the business of making them.

Let me first be quite clear as to the general and the particular meaning of the word "estimate" in the present context. In intelligence, as in other callings, estimating is what you do when you do not know. This is the general meaning. In this broad sense, scarcely an intelligence document of any sort goes out to its consuming public that does not carry some sort of estimate. Field reports are circulated only when someone has estimated that the source is sufficiently reliable and content sufficiently credible to be worthy of attention. Current intelligence items as often as not carry one of those words of likelihood—"probable," "doubtful," "highly unlikely," etc., and so forth—that indicate that someone has pondered and decided that the report should be read with something less than perfect assurance as to its accuracy. An endless number of important sentences in even the basic intelligence category carry the same evidence of this kind of speculative evaluation, i.e., estimating.

But what I have in mind in particular when I use the word "estimates" here are the formal intelligence documents which begin to examine a subject from the point of view of what is known about it, and then move on beyond the world of knowing and well into the world of speculating. When you reflect upon a whole large subject matter—the future of Greece or the armed strength of Communist China, for example—and realize that you cannot begin to know about either with the degree of certainty you know your own name, you reach for the next best thing to "knowing." You strive for some sort of useful approximation. In pursuit of this you evoke a group of techniques and ways of thinking, and with their help you endeavor logically and rationally (you hope) to unravel the unknown or at least roughly define some area of possibility by excluding a vast amount of the impossible. You know that the resultant, while still a lot better than nothing at all, will be in essence a mix of fact and judgment. Upon occasion it turns out to be almost exactly correct, but at the time you wrote it you expressed yourself with appropriate reservation.

To the extent that your judgment and the many quite subjective things which influence it are now involved, the man who reads this estimate will by no means accept it in the attitude of relaxed belief with which he reads, for example, that "not counting West Berlin, there are ten Länder in the FRG." It is this form of intelligence document that Hilsman's respondents were cool about. What follows is an attempt to explain the chill.

Let me ask you to think of one of these estimates in terms of the geometrical form called a pyramid. Think of the perfect estimate as a complete pyramid. At its base is a coagulation of all-but-indisputable fact. With an absolute minimum of manipulation on our part, the facts have arranged themselves to form what is quite clearly the base of a pyramid. They have spread out in the horizontal dimensions to the degree that we pretty well perceive its base area, and piled up in the vertical dimension generally to indicate the slope of its sides.

Knowing the nature of the base of the pyramid, to take an illustrative case, is like saying that we now have enough solid information to know that a photo image we have been wondering about is of an aircraft—not, say, a dairy ranch; more importantly, it is a bomber aircraft, not a transport. As to the other things we want to know about it— its performance characteristics—we are not at all certain. We are, however, in a good position to speculate about them.

Raising the Pyramid

Now back to the pyramid (figure F-12). Let us assume that when we know the general locus in space where the sides will converge to form the apex, we will have most of what we want. Let us assume that the exact point of the apex is exactly what we want, that if we know this with certainty we will have what we are after. For the bomber, constructing the apex would be reasoned speculations about how it will perform: How far it can fly, how high, how fast, and with what bomb load. Just as classical induction revealed the base of the pyramid, so now we call upon the other classical methodologies of deduction, and with their help we reason our way up the pyramid toward the top.

The factual stuff of the base of the pyramid is likely to be largely the fruit of our own intelligence-gathering efforts and so constitute a body of material about which we are better informed than our consumers. But we enjoy no such primacy with respect to the matter above. In fact, the talent to deduce rigorously is one which we share with any other educated and intellectually disciplined human. Furthermore, the advantage we enjoy with respect to base material can be and usually is dissipated by our habit of making it available to quite an array of nonintelligence types. The point is that the studious consumer can approach our mastery at the base and match us higher up. He can be his own estimator whenever he wishes to invest the time.

Let me not even seem to pretend that all conceptual pyramids in our area of work are constructed as described. The procedure which moves from the known to the unknown with a certain amount of tentative foraying as new hypotheses are advanced, tested, and rejected is merely the most respectable way. Its very opposite is sometimes employed, though usually with a certain amount of clandestinity.

The follower of this reverse method first decides what answer he desires to get. Once he has made this decision, he knows the exact locus of the apex of his pyramid but nothing else. There it floats, a simple assertion screaming for a rationale. This, then, is worked out from the top down. The difficulty of the maneuver comes to a climax when the last stage in the perverse downward deduction must be joined up smoothly and naturally with the reality of the base. This operation requires a very considerable skill, particularly where there is a rich supply of factual base material. Without an artfully contrived joint, the whole structure can be made to proclaim its bastardy, to the chagrin of its progenitor.

NWP 2-0

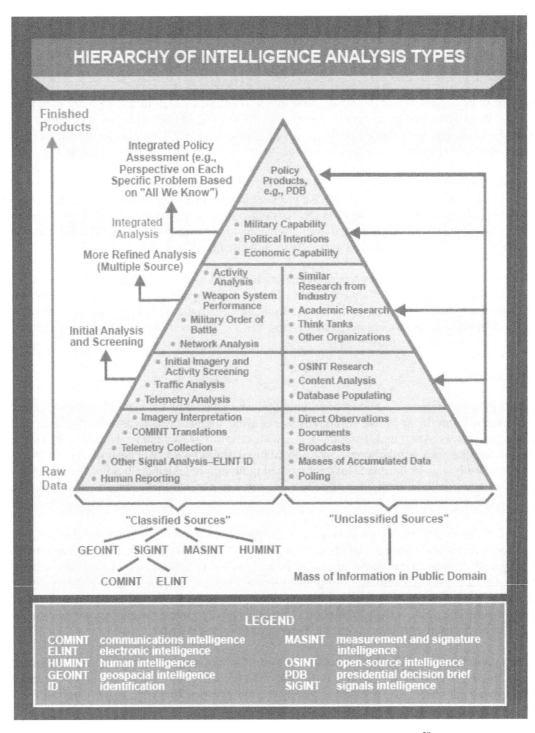

Figure F-12. Hierarchy of Intelligence Analysis Types[58]

---
[58] JP 2-01.3.

The Peak

But even under the respectable method the intelligence estimator at some moment in the construction process reaches the place where he has used his last legitimate deductive crutch and must choose one of three possible courses.

The first is to let himself be propelled by the momentum of his reasoning into a final and fairly direct extrapolation. The effect of this is to put a sharpish top on the pyramid—a measure which, in turn, has the effect of telling his audience that he is pretty sure that he has discerned the outlines of what must be the truth. For the bomber it would be like saying: "Thus we conclude that the bomber in question is almost certainly a supersonic aircraft of medium range. See table II for our estimate of its performance characteristics."

The second is not to make this final extrapolation but to leave the pyramid truncated near its apex. This has the effect of telling the reader that you have narrowed the range of possibilities down to only a few. The further down you truncate, the wider their range. Thus the most unsatisfactory kind of intelligence construction is often that which perforce has to stop where the factual stuff of the base runs out. Often it is the equivalent of issuing the most general kind of news and asking the reader to suspend judgment pending the appearance of new evidence. For example: "Thus, we are unable at this time to be more precise regarding the performance characteristics of this bomber. It is possible that it is a new supersonic medium."

The third is what I will call "the look before the leap" or the "clandestine peep ahead." It is, one may hope, less often used by the intelligence professional than by the policy officer doing his own estimating. What you do is look hard at the final extrapolation and take full stock of where you will be if you go for it. Then, having taken stock, you ask yourself if you really wish to subscribe to this conclusion.

In the case I have in mind, you recoil. It may be that by making it yours you will be depicting yourself a nonpatriot, or someone soft on Communism. It may be that by implication you can be made to seem a harsh critic of a higher authority or a scoffer at one of his policies. It may be that you will be doing the budget claims of your department or agency a grave disfavor. Or most important of all you realize that your findings may be advanced to support a policy which you oppose or that they do not support with sufficient vigor a policy which you favor.

If you have taken the peep ahead and find the prospect not to your taste, you can settle for the second course and simply not complete the estimate. Or you can back down on your argument, tearing it up as you go. Then when you have found a salubrious ground for another start, you can reargue your case upwards—perhaps using a few facts which you had dismissed as irrelevant the first time through, perhaps giving more weight to this analogy and forgetting about that, etc., etc. Thus, with a small amount of tinkering you can create a somewhat different conceptual pyramid whose base is still the same, but whose apex will lie in a zone much less dangerous to your job security or much more appropriate to the requirements of your policy preconceptions.

The Policy Welcome

Irrespective of which of the three ways of handling the problem you choose and irrespective of the substantive conclusion—or lack of it—the completed estimate will be bad news to one if not more of its important readers: it may undercut a long-held position or destroy a line of painfully developed argument; it may indicate the unwisdom of a plan or the malallocation of large sums of money. Another thing you may be sure of is that he will react as any recipient of bad news reacts—the reflex is one of "I don't believe you." Need I emphasize again that estimates are far more vulnerable to the criticism which is bound to accompany incredulity than are propositions which are stated, at least, as if they were fact.

The disappointed consumer may begin with a hard look at our pyramid's factual base. He may find some loose masonry which can be jimmied apart, and then jimmy. He may find some quite substantial building stones left off to one side, stones, which, although of the same material and cut to fit some sort of geometrical form, were not incorporated into the base structure. He will speedily perceive that if these are chiseled a bit here and there they can be made to fit into this structure, with the result that they change some important aspects of its configuration. You may be sure he will soon focus on the upper zones of our pyramid.

One thing he will be most alert to is any evidence that intelligence, having taken the "peep ahead" and found the pyramid about to peak at an unwanted place, went on to take the corrective action I have indicated. If he can find evidence of this sort of disingenuous case-making, he will attack with very weighty weaponry. Before he is done he may be able to prove to himself and a number of others that the so-called intelligence contribution is a fraud—nothing more nor less than a policy brief brazenly masquerading as an intelligence estimate.

In these terms we may readily understand why a good many of Hilsman's respondents felt as they did about the value of intelligence estimates. For purposes of fuller explanation, let us suppose that an intelligence estimate on the Banana Republics had been prepared; let us suppose that our policymaking reader Mr. "A" is his department's authority on these Republics. A tour of his psyche as he reads the paper may be illuminating.

First, let us assume that the estimate accords in very high degree with his own estimate of the present and probable future situation in Banania. His psyche will begin to purr in contentment; "What a remarkably perceptive document," it will whisper. But this may be as far as the word of praise gets. When the moment comes to articulate his comment on the estimate, he is less likely to praise it than to proclaim, "This is exactly what I have been saying all along. Why in the world do we have to have someone who knows less of the matter than I say so before anyone pays attention?" In short, as far as he is concerned, the intelligence effort that went into the study was unnecessary. "A" may not always feel this way, particularly if during the policy debate he realizes that he can make points against his opponents by citing the estimate as a dispassionate outside opinion.

Alternatively, let us assume that the estimate accords not at all with the views of Mr. "B." He will be unhappy, for he will realize that if the conclusions of the estimate are believed by his peers and superiors, the policy which he has been championing will have to be modified—perhaps drastically. If he wishes to stay in the fight, then, he must be prepared to attack the intelligence estimate as *misleading* and erect one of his own to replace it.

Lastly, let us assume that the policy issue is one of those which is going to be settled almost entirely on the basis of some purely domestic matter: The cotton lobby, the gold flow, the budget, and so on. Our policymaking consumer does not have to attack the substance of the *irrelevant* estimate. He will chuckle patronizingly to himself while his psyche warms in the feeling of superiority to those poor boobs in intelligence who have thought that what they called the "Situation and Prospects in X" could have any bearing on the way US policy towards X is being shaped today. Out loud he wonders how such naiveté can persist; he has no comment on the substance of the estimate.

These views of an estimate as unnecessary, misleading, or irrelevant may coincide with those of some of the people whom Hilsman polled and explain why they were less grateful for estimates than for what they considered factual intelligence issuances.

The Defense

How seriously should we in intelligence take the indictment which damns our estimating work as unnecessary, or misleading, or irrelevant? Take the misleading charge first. If it is made, and if it is true because the document was designed that way, then it must be taken very, very seriously indeed. For this accusation implied that the peep ahead had been taken and the necessary retracing of steps and reconstruction had followed so that the conclusion of the estimate suited the policy predispositions of the estimators. They have been caught out in their stupidity, and their credibility, at least for this estimate, is dead. It is dead not merely for the reader who found the conclusions abhorrent, but for all the others who found out by themselves or were told.

If the same group of estimators are caught out a second or third time, their credibility will probably be dead for good. Thereafter almost any intelligence pronouncement they or their associates make will be slightingly referred to as propaganda, and perhaps not even read. They have not only lost all hope of directly influencing policy, they have lost what is even more important because more attainable than direct influence. This is the indirect influence which they might have exercised through an honest contribution to the debate which ought to precede every substantial policy decision.

Suppose the charge of misleading is made simply as a function of a committed reader's general disbelief or annoyance, and suppose that, try as he may, he cannot show a trace of bad faith on the part of the estimators. The estimators are confronted with nothing more sinister than a human disagreement, perhaps from a reader whose nose is out of joint. This is just life.

What of the charge, unnecessary? The question here is—unnecessary to whom? To everyone involved in the policy decision? Already I have dealt with Mr. "A" to whom it was unnecessary because it accorded exactly with his views, and Mr. "B" to whom it was unnecessary and many times worse because he found it misleading. But are these the only two officers or two kinds of officers involved? Is there perhaps not a Mr. "C" or Messrs. "C" who have no more than a layman's knowledge of the subject but who must participate in the policy debate and decision? Of course there are the Messrs. "C," and important men they are. The President, upon many an occasion, is a Mr. "C," and so are members of his staff and his Security Council. They have found the estimate anything but unnecessary.

It does not follow, however, that the impact which the estimate may make upon the Mr. "C"s will in itself cause the defeat of the dissenting Mr. "B"s. What it will do is to force the Mr. "B"s to put forth a better effort. This will stimulate the Mr. "A"s themselves to better effort. At a minimum, the intelligence estimate will have made its contribution in the way it promoted a more thorough and enlightened debate and a higher level of discourse within the high policymaking echelon. At a maximum it may have denied a wrong-headed Mr. "B" an easy triumph.

Lastly, the charge of irrelevant. This rested upon the fact that the foreign policy decision was going to have to be made on the basis of a domestic consideration, something about which the estimate is wholly – and properly – mute. But it is just possible that the domestic consideration is not all that important and that the national interest is not really being served by this sort of deference to it. It may be that the estimate helped the policy people to reach this new appreciation of the national interest. Hence, even if the decision I am talking about gets made in conformity with the wish of the domestic pressure group, maybe the next such decision will not.

Truth Before Power

I suppose that if we in intelligence were one day given three wishes, they would be to know everything, to be believed when we spoke, and in such a way to exercise an influence to the good in the matter of policy. But absent the Good Fairy, we sometimes get the order of our unarticulated wishes mixed. Often we feel the desire to influence policy and perhaps just stop wishing here. This is too bad, because to wish simply for influence can, and upon occasion does, get intelligence to the place where it can have no influence whatever. By striving too hard in this direction, intelligence may come to seem just another policy voice, and an unwanted one at that.

On the other hand, if intelligence strives for omniscience and strives to be believed, giving a third place to influence, serendipity may take over. Unselfconscious intelligence work, even in the speculative and highly competitive area of estimates, may prove (in fact, has proved many times) a key determinant in policy decision.

INTENTIONALLY BLANK

APPENDIX G

Navy Intelligence Mission-Essential Tasks

G.1 THE INTELLIGENCE PROCESS AND MISSION-ESSENTIAL TASKS

In addition to understanding the intelligence process, the national intelligence community framework, and the intelligence disciplines, naval intelligence professionals, particularly those assigned to the fleet, should understand the linkage between the intelligence process and intelligence related mission-essential tasks (METs). The concept of a Navy mission-essential task list (NMETL) is derived from an Army effort, later adopted by the joint community and other Services, to develop a common lexicon and framework for describing the set of tasks that were necessary to accomplish a given mission. As the defining framework for intelligence operations, the intelligence process was used as the basis for developing the list of METs for naval intelligence. The discussion below provides a brief overview of the concepts behind METs, which play a key role in defining naval intelligence training standards and objectives and in assessing the readiness of the intelligence team to support fleet operations.

G.2 DEFINING TERMS

Before discussing the relationship between the intelligence process and NMETs it is important to understand some of the terms associated with Navy mission-essential tasks (NMETs). Figure G-1 provides a listing of terms important to understanding the NMET discussion. See OPNAVINST 3500.38B for a complete explanation of mission essential tasks and the Universal Naval Task List (UNTL)

G.3 NAVY MISSION-ESSENTIAL TASKS CONCEPT

Each of the intelligence operations that comprise the intelligence process can be broken down to a number of discreet tasks for all levels of warfare. The UNTL provides an expansive list that covers the entire range of naval operations and encompasses hundreds of tasks covering intelligence operations. The challenge is to choose from this list those tasks that are critical (essential) to success of an individual mission, which is a small subset of the overall task list. Once selected, each task is then broken down further. The variables within the operational environment under which that task are performed (conditions) are defined and the requirements for meeting those tasks (standards) are determined.

Standards consist of measures and criterion. A measure provides the basis for describing levels of performance. Examples would include percentage of task complete; minutes, hours, days to complete a task; level of accuracy in creating a product; or number of products completed. A criterion defines the acceptable level of performance. A specific task may have multiple measures and criteria associated with it. While the measures associated with a particular task are typically unclassified, the criterion may be classified as it is an indicator of capability and readiness. Figure G-2 provides examples of Navy tactical tasks (NTAs), with associated measures, for two tasks—one associated with collection and the other with analysis and production.

Term	Definition
UJTL	The Universal Joint Task List (UJTL) is a library of tasks, which serves as a foundation for capabilities-based planning across the range of military operations.[1] The UJTL is a common reference system for joint force commanders, combat support agencies, operational planners, combat developers, and trainers to communicate mission requirements. It is the basic language for developing joint, Service or agency mission-essential task lists (METLs) which identify required capabilities for mission success.[2]
UNTL	Universal Naval Task List (UNTL) is a combination of the Navy Tactical Task List (NTTL) and the Marine Corps Task List that contains a comprehensive hierarchical listing of the tasks that can be performed by a naval force, describes the variables in the environment that can affect the performance of a given task, and provides measures of performance that can be applied by a commander to set a standard of expected performance.
NTTL	The NTTL is the comprehensive list of Navy and Coast Guard (Department of Defense-related missions) tasks, doctrinally based, designed to support current and future mission-essential task list development. The NTTL promulgates these tasks as NTAs. These NTAs are a standardized tool for describing requirements for planning, conducting, assessing, and evaluating joint and Service training. However, because these tasks also provide a common language and reference system for addressing mission requirements, NTAs serve to identify Navy tasks at the tactical level.
MCTL	Marine Corps Task List—a comprehensive list of Marine Corps tasks, doctrinally based, designed to support current and future Marine Corps METL development.
Mission	The task, together with the purpose, that clearly indicates the action to be taken and the reason therefore.
Task	A discrete event or action, not specific to a single unit, weapon system, or individual that enables a mission or function to be accomplished.
Condition	A variable of the operational environment or situation in which a unit, system, or individual is expected to operate that may affect performance. Conditions may be physical, military, or civil. Examples would include weather conditions that limit operations or restrictions on sharing information with coalition partners.
Standard	The minimum acceptable proficiency required in the performance of a particular task under a specified set of conditions, expressed as quantitative or qualitative measures. The commander establishes standards.
MET	A task selected by a force commander from the UNTL deemed essential to mission accomplishment.
METL	A list of tasks considered essential to the accomplishment of assigned or anticipated missions. A METL includes essential tasks, conditions, standards, and associated supporting and command-linked tasks.

Figure G-1. Definitions of Navy Mission-Essential Tasks Associated Terms[3]

---

[1] CJCSI 3500.02 Series, UJTL Policy and Guidance for the Armed Forces of the United States, designates the authoritative UJTL as the online version of the UJTL Task Development Tool (UTDT), which resides on the Joint Doctrine, Education, and Training Electronic Information System (JDEIS).
[2] The structure of the UJTL is by four levels of war: strategic national (SN), strategic theater (ST), operational (OP), and tactical (TA).
[3] OPNAVINST 3500.38B/MCO3500.26A/USCG COMDTINST 3500.1, Universal Naval Task List, version 3.0, January 2007, p. 1-3-1-4.

Task Number	NTA 2.2.3
Task Name	Perform Tactical Reconnaissance and Surveillance
Task Definition	To obtain, by various detection methods, information about the activities of an enemy or potential enemy or tactical area of operations. This task uses surveillance to systematically observe the area of operations by visual, aural, electronic, photographic, or other means. This includes development and execution of search plans. (JP 2-0 Series, MCDP 2, Marine Corps warfighting publication (MCWP) 2-1, NWP 2-01, NWP 3-01 Series, NWP 3-02.12, NWP 3-15 Series, NWP 3-21 Series, NWP 3-62M)

Measure Number	Measure Scale	Measure Nomenclature
M1	Days	From receipt of tasking, unit reconnaissance/surveillance assets in place.
M2	Percent	Of collection requirements fulfilled by reconnaissance/surveillance assets.
M3	Percent	Of time able to respond to collection requirements.
M4	Hours	To respond to emergent tasking.
XM5	Hours	To produce a RECCEXREP.

Task Number	NTA 2.4.4
Task Name	Analyze and Synthesize Information
Task Definition	To assess, synthesize and fuse new information and existing intelligence from all sources to develop timely, accurate mission-focused intelligence estimates in order to provide meaningful knowledge pertinent to the supported commanders' current and future planning and decision-making needs, and to determine the significance of information in relation to the current situation. (JP 2-0 Series, MCDP 2, MCWP 2-1, NWP 2-01, 3-02.21, 3-13, 3-62M, 4-02)

Measure Number	Measure Scale	Measure Nomenclature
M1	Percent	Of enemy branches and sequels correctly identified during planning.
M2	Hours	To process new intelligence data and integrate within the targeting cycle.
M3	Percent	Of forecasted significant enemy actions, were false alarms.
M4	Percent	Of enemy targets or vulnerabilities identified within targeting cycle.

Figure G-2. Example of Navy Tactical Tasks with Associated Measures

The value of the NMETL construct is apparent with regard to training and readiness. By establishing the tasks, conditions, and standards for each MET, intelligence team leadership can develop training plans needed to ensure their personnel are proficient to the levels needed by the commander in support of a given mission or set of missions. NMETLs play a key role in establishing skills training goals for the intelligence schoolhouses and are used by Afloat Training Groups, Strike and Expeditionary Strike Groups, and Expeditionary force training organizations to certify unit and group readiness for deployments. They are a foundational component of the Navy Warfare Training System (NWTS) and created and maintained in the Navy Training Information Management System (NTIMS). NMETLs are integral to the Defense Readiness Reporting System-Navy (DRRS-N), which is used to provide fleet leadership with a tool for visualizing the readiness of forces throughout the fleet. Additionally, by breaking down intelligence tasks to a discreet level, the UNTL provides a ready reference for naval intelligence professionals when called upon to perform a new task or function. While the UNTL and UJTL do not cover every conceivable intelligence related task, the listing of tasks is quite extensive. However, tasks can be modified or new tasks can be added as advances in threats, technology, or capabilities dictate. Referring to the UNTL enables the naval intelligence professional to quickly ascertain the key components of a given task and the associated measures that are used to establish performance standards.

G.4 OPERATIONAL LEVEL INTELLIGENCE TASKS AND THE INTELLIGENCE PROCESS

The UJTL provides the basis for operational level intelligence tasks. As shown in Figure G-3, the operational level intelligence tasks (OP 2 series) are joint focused. In addition, each operational task has many subtasks, each of which comprises discreet intelligence operations that contribute to the accomplishment of a specific task and are aligned to the intelligence process discussed above.

NTTP 3-32.1 specifically identifies intelligence-related functions executed within the maritime operations center (MOC)—either by the senior intelligence officer (SIO), the intelligence staff (N-2), the MIOC or the intelligence support elements (ISEs). These functions do not equate to METs, but they do help delineate the intelligence operations doctrinally assigned to the operational level of war (e.g., MOC) from those doctrinally assigned to the tactical level of war (CSG/ESG and below). These intelligence process "lanes in the road" between the MOC and deployed forces are an important distinction with implications for manning, training, and equipping the fleet for intelligence capabilities.

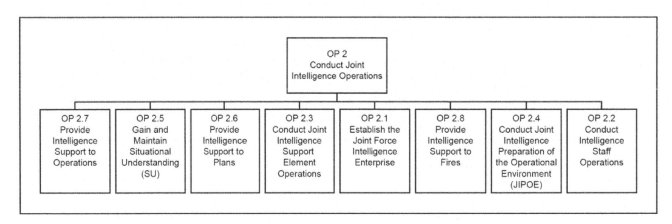

Figure G-3. Universal Joint Task List Operational Level Intelligence Tasks[4]

---

[4] UJTL Task List, January 2013, pp. 869–946.

NWP 2-0

The MOC intelligence functions, as defined in chapter 2 of NTTP 3-32.1, are:[5]

1. Intelligence Plans

2. Intelligence Preparation of the Maritime Operational Environment

3. Intelligence Operations

4. Counterintelligence/Human Intelligence

5. Collection Management

6. Production and Analysis

7. Dissemination

8. ISR Task, Collect, Process, Exploit, and Disseminate Cell

9. Interagency/Coalition/Host Nation Group.

The listing above demonstrates the importance of understanding the intelligence process as that process provides the framework around which naval intelligence operations and responsibilities are defined at the operational level of war.

G.5 TACTICAL LEVEL INTELLIGENCE TASKS AND THE INTELLIGENCE PROCESS

As discussed, the NTTL identifies tasks at the TLW. While a commander may designate a set of NMETLs based on a specific mission, naval intelligence professionals primarily encounter NMETLs as part of the workup cycle for deployment. Each class of unit has particular NMETLs assigned to it. Depending on the intelligence resources available on that unit, the list of intelligence related NMETLs can be extensive. The selection of tasks for each unit is driven by operational requirements and enables certification of an intelligence readiness to conduct intelligence operations in support of a wide range of naval missions. As shown in figure G-4, the intelligence process provides the foundation for the set of NTAs that make up the current set of naval intelligence METs. Each of the highlighted boxes aligns to a particular intelligence operation and within each intelligence operation is a collection of lower level tasks that tactical level units may be expected to perform.

Note

While there are subtasks that go beyond the third level for each of the NTAs shown in figure G-4, those subtasks were not shown for readability of the figure. Figure G-4 is provided to give an appreciation of the range of tasks that naval intelligence professionals may need to perform at the tactical level to support a wide variety of naval missions.

---

[5] NTTP 3-32.1, Maritime Operations Center.

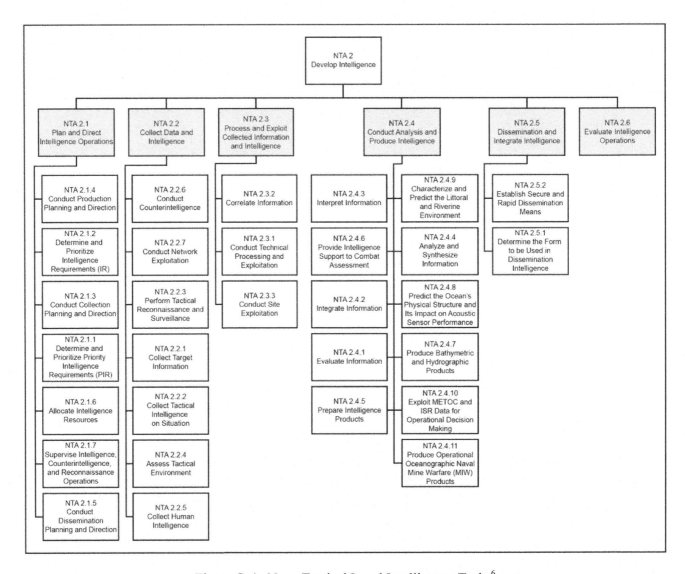

Figure G-4. Navy Tactical Level Intelligence Tasks[6]

---

[6] UNTL, Report Number: 38.1, Generated: November 2012.

APPENDIX H

Intelligence Preparation of the Maritime Operational Environment Worksheets

Chapter 4 provides a complete overview of the IPOE process as applied to the planning and execution of operations in the maritime operational environment. These worksheets provide a means for the naval intelligence professional to focus on the outputs of various phases of the IPOE effort. The worksheets provided are followed by a historical example to provide naval intelligence professionals a better sense of how these worksheets support planning and execution of naval operations.

H.1 DEFINE THE MARITIME OPERATIONAL ENVIRONMENT

Area of Operations: Defined by LAT/LONG or displayed on a map/chart for clarity and reference.

Area of Interest (AOI): Adjacent or other designated geographic area where political, military, economic, or other developments have an effect within a given theater or particular AO. Defined by LAT/LONG or displayed on a map/chart for clarity and reference.

H.2 DETERMINE THE ADVERSARY CENTER OF GRAVITY

Strategic Objective(s)

Operational Objective(s)

Tactical Objective(s) (if operating at the tactical level of war)

Critical Strengths	Critical Weaknesses

Strategic Center(s) of Gravity

Operational Center(s) of Gravity

Tactical Center(s) of Gravity (if applicable—these may be objectives or decisive points)

Critical Capabilities

NWP 2-0

Critical Requirements

Critical Vulnerabilities

Decisive Points

H.3 DETERMINE THE ADVERSARY CENTER OF GRAVITY HISTORICAL EXAMPLE

For brevity, this example only examines the single critical capability of IADS.

Strategic Objective(s)

- *Retain Kuwait as 19th province*
- *Enhance Saddam Hussein's hold on power*
- *Increase Iraq's political and military influence in the Arab world*
- *Increase Iraq's power and influence within OPEC*

Operational Objective(s)

- *Defeat or neutralize a coalition attack to liberate Kuwait*
- *Prevent coalition forces from obtaining air superiority*
- *Prevent coalition forces from obtaining sea control in the northern part of the Persian Gulf*

Tactical Objective(s) (if operating at the tactical level of war)

N/A

NWP 2-0

Critical Strengths	Critical Weaknesses
• IADS • WMD • land-based ballistic missiles (scuds) • Republican Guards in the Kuwait theater of operations (KTO) • forces are in defensive positions • Saddam and his strategic C2 • combat experienced units and commanders • missile-armed surface combatants • sea mine inventories and delivery platforms	• long and exposed land LOCs from Iraq to KTO • world opinion; Arab world outrage • combat skills and readiness of the Air Force • numerical and qualitative inferiority of naval forces • low morale and poor discipline of regular forces • class IX for weapon systems • inadequate forces to protect the Iraq-Iran border

Strategic Center(s) of Gravity

Saddam and his inner-circle security apparatus

Operational Center(s) of Gravity

Republican Guards in the KTO

Critical Capabilities

- sustain Republican Guard forces in KTO (Log)
- receive strategic direction and provide directives to subordinate units (C3)
- **protect forces from coalition airpower**
- employ conventional defensive forces as a screening force
- maintain organizational morale

Critical Requirements (per the IADS example)

- radar sites
- communication nodes
- Iraqi Air Force
- class IX for IADS
- resupply of class V
- morale of fixed site crews

Critical Vulnerabilities

- radars
- Air Force
- LOCs
- resupply

Decisive Points (not an exhaustive list)

- *APODs and SPODs in Saudi Arabia*
- *Strait of Hormuz*
- *APODs in Turkey*
- *Kuwait SPOD*

H.4 DETERMINE ADVERSARY COURSES OF ACTION

Identify the Full Set of Courses of Action Available to the Adversary	
Adversary COAs (ACOAs) must be suitable, feasible, acceptable, unique, and consistent with adversary doctrine or patterns of operation	
ACOA 1	
ACOA 2	
ACOA 3	
ACOA 4	
ACOA 5	

H.5 EVALUATE AND PRIORITIZE EACH ADVERSARY COURSE OF ACTION

1. Analyze each COA to identify its strengths and weaknesses, COGs, and decisive points.

2. Evaluate how well each COA meets the criteria of suitability, feasibility, acceptability, uniqueness, and consistency with doctrine. Avoid cultural bias by considering these criteria in the context of the adversary's culture.

3. Evaluate how well each COA takes advantage of the operational environment.

4. Compare each COA and determine which one offers the greatest advantages while minimizing risk.

5. Consider the possibility that the adversary may choose the second or third most likely COA while attempting a deception operation portraying adoption of the best COA.

6. Analyze the adversary's current dispositions and recent activity to determine if there are indications that one COA has already been adopted.

7. Guard against being "psychologically conditioned" to accept abnormal levels and types of adversary activity as normal. Identify and focus in greater detail on those adversary preparations not yet completed that are, nevertheless, mission-essential to accomplish a specific COA.

	Retained ACOAs (Prioritized)	Vulnerabilities
ACOA #		
ACOA #		
ACOA #		
ACOA #		

NWP 2-0

Example Prioritization of Retained ACOAs[1]

	Retained ACOAs (Prioritized)	**Vulnerabilities**
ACOA 3	REDLAND conducts a two pronged ground attack against the APOD with the 3rd RGB from the North and the 1st RGB from the South, with supporting air operations. (Most Likely)	• No operational Reserves remaining • Extended LOCs • Complex C3, little experience
ACOA 5	REDLAND conducts a delay and interdicts friendly APODs / SPODs	• Weak maritime interdiction capability • Limited Operational environment for delay
ACOA 1	REDLAND initially conducts joint operations to disrupt JTF Blue Sword forced entry operations, and upon establishment of the JTF Blue Sword in REDLAND, the REDLAND armed forces disperse into small-unit formations in the mountains and cities and initiate insurgency operations to defeat the JTF ground forces. (Most Dangerous)	• Limited popular support • Limited sustainment capability • Centralized C3 required, minimal capability

---

[1] Naval War College, Joint Operation Planning Process (JOPP) Workbook, 1-24.

H.6 ADVERSARY COURSES OF ACTION WORKSHEET EXAMPLE

Note

Adversary objectives, critical factors, critical requirements, and decisive points need to be reassessed by phase.

Identify Adversary Objective(s) (examples)
Strategic:
Pinkland's full and unconditional diplomatic and military support
Redland's influence and prestige in the region
Operational:
Oil and gas production and refining installations in the Blueland North Sea
Remaining ports and airfield in Pinkland

List Adversary Critical Factors (examples)

Critical Strengths:	Critical Weaknesses:
Strategic:	**Strategic:**
Country XXX's diplomatic and military support	World opinion negative towards current actions
Operational:	**Operational:**
23rd Guards Division	Unsound C2 organization
Short lines of communication	Poor discipline
Mining capabilities	Limited control over its terrorist forces
Terrorist forces	**Tactical:**
Tactical:	Poor capacity of supply lines to currently forward-deployed naval force
High-speed patrol craft	
Kilo class submarine	Lack of night-fighting capabilities

List Adversary Centers of Gravity (examples)
Strategic:
Military and diplomatic support
Operational:
23rd Guards Division

List Adversary Critical Capabilities (examples)
Operational: (Also complete for strategic and tactical levels as required)
Integrated air defense
Air and sea transportation for terrorist forces
C2
Sea denial forces

List Adversary Critical Requirements Identify Critical Vulnerabilities
C2: (just one example)
Communications link between Redland High Command and terrorist network X
Communications link between Redland High Command and 23rd Guards Division X
23rd Guards Division internal C2 X

NWP 2-0

List Decisive Points (examples)
Redland airfield
Redland port
Pinkland ISB

Identify the Full Set of ACOAs Available to the Adversary (examples)
ACOA #1
All-out attack from North Sea forces on Blueland
ACOA #2
Attack on friendly shipping using conventionally powered submarines
ACOA #3
Full withdrawal from current forward positions

Evaluate and Prioritize Each ACOA		
	Retained ACOAs (prioritized)	Vulnerabilities
ACOA #1	All-out attack from North Sea forces on Blueland	Assets would be spread very thin Lack of sustainability Would require long LOCs
ACOA #2	Attack on friendly shipping using conventionally powered submarines	Limited attack would allow coalition forces to mass effects Give continued rise in negative world opinion

INTENTIONALLY BLANK

APPENDIX I

Intelligence Estimate

I.1 INTELLIGENCE ESTIMATE TEMPLATE

CLASSIFICATION

>Issuing Headquarters
>Place of Issue
>Day, Month, Year, Hour, Zone

APPENDIX 11 TO ANNEX B TO [COMMANDER] OPLAN/CONPLAN/OPORD XXXX-YY ( )[1]

INTELLIGENCE ESTIMATE[2]

References: List documents essential to this appendix.

1. Mission. State the assigned task and its purpose. The mission of the command as a whole is taken from the commander's mission analysis, planning guidance, or other statement.

2. Enemy Situation. State the conditions that exist and indication of effects of these conditions on adversary capabilities and the assigned mission. This paragraph describes the area of operations, the enemy military situation, and the effect of these two factors on enemy capabilities.

    a. Characteristics of the Operational Environment (OE). Discuss the effect of the physical and nonphysical characteristics of the OE on the activities of actors of concern and/or the military activities of both adversary and friendly forces. If an analysis of the area has been prepared separately, this paragraph in the intelligence estimate may simply refer to it. Also discuss the effects of the existing situation on military operations in the area.

        (1) Military Geography.

            (a) Topography.

                1. Existing Situation. Describe relief and drainage, vegetation, surface materials, cultural features, and other characteristics in terms of their effects on key terrain, observation, fields of fire, obstacles, cover and concealment, avenues of approach, lines of communication, and landing areas and zones.

                2. Effect on Adversary Capabilities. Discuss the effect of topography on the activities of individual or multiple actors of concern and/or broad adversary capabilities such as attack and defense, describing generally how the topography affects each type of activity. The effect on employment of weapons of mass destruction; amphibious, airborne, or air-land forces;

---

[1] Chairman of the Joint Chiefs of Staff Manual (CJCSM) 3130.03, Adaptive Planning and Execution (APEX) Planning Formats and Guidance, 18 October 2012

[2] This Intelligence Estimate format is intended to support the development of either contingency plans for specified adversaries and actors of concern or theater or functional campaign plans addressing multiple actors.

intelligence collection capability surveillance devices and systems; communications equipment and systems; electronic warfare; psychological operations; OPSEC and military deception; logistic support; and other appropriate considerations should be included.

3. Effect on Friendly Course of Action (COA). Discuss the effects of topography on friendly forces' military operations (attack, defense, etc.), in the same fashion as for adversary capabilities in the preceding subparagraph.

(b) Hydrography.

1. Existing Situation. Describe the nature of the coastline; adjacent islands; location, extent, and capacity of landing beaches and their approaches and exits; nature of the offshore approaches, including type of bottom and gradients; natural obstacles; surf, tide, and current conditions.

2. Effect on Adversary Capabilities. Discuss the effects of the existing situation on the activities of individual or multiple actors of concern and/or broad adversary capabilities.

3. Effect on Friendly COAs. Discuss the effects of the existing situation on broad COAs for friendly forces.

(c) Climate and Weather.

1. Existing Situation. Describe temperature, cloud cover, visibility, precipitation, light data, and other climate and weather conditions and their general effects on roads, rivers, soil trafficability, and observation in addition to effects on ground, air, and sea operations and equipment.

2. Effect on Adversary Capabilities. Broadly discuss the effects of the existing climate and weather situation on the activities of individual or multiple actors of concern and/or adversary capabilities.

3. Effect on Friendly COAs. Broadly discuss the effects of the existing climate and weather situation on COAs for friendly forces.

(2) Systems Perspective of the OE. Discuss the significant interrelationships between Political, Military, Economic, Social, Information, Infrastructure (PMESII) systems of the OE without regard to geographic boundaries, but addressing the focus area of the commander. If developed, insert graphics to facilitate a more in depth understanding of key nodes and links or include them as tabs to this appendix. Use the following subparagraphs to describe relevant details of PMESII systems.

(a) Political.

1. Existing Situation. Describe the organization and operation of civil government in the area of operation. Describe how the political system relates to the military, economic, social, information, and infrastructure systems of the OE.

2. Effect on Adversary Capabilities. Discuss how the relationships between the political system and the other systems of the OE affect the activities of individual or multiple actors of concern and/or adversary capabilities.

3. Effect on Friendly COAs. Discuss how the relationships between the political system and the other systems of the OE affect COAs for friendly forces.

(b) Military.

1. Existing Situation. Describe how the military system relates to the political, economic, social, information, and infrastructure systems of the OE.

2. Effect on Adversary Capabilities. Discuss how the relationships between the military system and the other systems of the OE affect the activities of individual or multiple actors of concern and/or adversary capabilities.

3. Effect on Friendly COAs. Discuss how the relationships between the military system and the other systems of the OE affect COAs for friendly forces.

(c) Economic.

1. Existing Situation. Describe industry, public works and utilities, finance, banking, currency, commerce, agriculture, trades and professions, labor force, and other related factors. Describe how the economic system relates to the political, military, social, information, and infrastructure systems of the OE.

2. Effect on Adversary Capabilities. Discuss how the relationships between the economic system and the other systems of the OE affect the activities of individual or multiple actors of concern and/or adversary capabilities.

3. Effect on Friendly COAs. Discuss how the relationships between the economic system and the other systems of the OE affect COAs for friendly forces.

(d) Social.

1. Existing Situation. Describe language, religion, social institutions and attitudes, minority groups, population distribution, health and sanitation, and other related factors. Describe how the social system relates to the political, military, economic, information, and infrastructure systems of the OE.

2. Effect on Adversary Capabilities. Discuss how the relationships between the social system and the other systems of the OE affect the activities of individual or multiple actors of concern and/or adversary capabilities.

3. Effect on Friendly COAs. Discuss how the relationships between the social system and the other systems of the OE affect COAs for friendly forces.

(e) Information.

1. Existing Situation. Describe how the information system relates to the political, military, economic, social, and infrastructure systems of the OE.

   a. Dissemination. Discuss the aggregate of individuals, organizations and groups that disseminate or act on information. Identify key individuals having an influence on relevant populations. Describe the modes of information dissemination such as print or broadcast media and social media networks. Elaborate on the technical aspects of telecommunications systems under the infrastructure subparagraph (f) below.

   b. Electromagnetic Battlespace (EMB). Discuss frequency restrictions that may affect the employment of intelligence collection systems. Address frequency use in neighboring countries, host nation agreements, and documented interference reports within the AOR.

c. Use additional subparagraphs as required to describe other relevant aspects of the information environment.

2. Effect on Adversary Capabilities. Discuss how the relationships between the information system and the other systems of the OE affect the activities of individual or multiple actors of concern and/or adversary capabilities.

3. Effect on Friendly COAs. Discuss how the relationships between the Information system and the other systems of the OE affect COAs for friendly forces.

(f) Infrastructure.

1. Existing Situation. Describe how the infrastructure system relates to the political, military, economic, social, and information systems of the OE.

a. Transportation. Describe roads, railways, inland waterways, airfields, and other physical characteristics of the transportation system; capabilities of the transportation system in terms of rolling stock, barge capacities, and terminal facilities; and other pertinent data.

b. Telecommunications. Describe telecommunications facilities and capabilities in the area. Include Internet and cell phone penetration rates, cable and broadcast media and other telecommunications infrastructure.

c. Use additional subparagraphs as required to describe other infrastructure categories.

2. Effect on Adversary Capabilities. Discuss how the relationships between the infrastructure system and the other systems of the OE affect the activities of individual or multiple actors of concern and/or adversary capabilities.

3. Effect on Friendly COAs. Discuss how the relationships between the infrastructure system and the other systems of the OE affect COAs for friendly forces.

b. Enemy Military Situation (Ground, Naval, Air, Space, other Service).

(1) Strength. State the number and size of enemy units committed and enemy reinforcements available for use in the area of operations. Ground strength, air power, naval forces, WMD, electronic warfare, unconventional warfare, surveillance potential, and all other strengths (which might be significant) should be considered.

(2) Composition. Outline the structure of adversary forces (order of battle) and describe unusual organizational features, identity, armament, and weapon systems.

(3) Location and Disposition. Describe the geographic location of adversary forces in the area, including fire support elements, command and control facilities, ground, air, naval, and missile forces, and bases.

(4) Availability of Reinforcements. Describe adversary reinforcement capabilities in terms of ground, air, naval, missile, and weapons of mass destruction; terrain, weather, road and rail networks, transportation, replacements, labor forces, prisoner of war policy; and possible aid from sympathetic or participating neighbors.

(5) Movements and Activities. Describe the latest known adversary activities in the area.

(6) Logistics. Describe levels of supply, resupply ability, and capacity of beaches, ports, roads, railways, airfields, and other facilities to support supply and resupply. Consider hospitalization and evacuation, military construction, labor resources, and maintenance of combat equipment.

(7) Operational Capability to Launch Missiles. Describe the total missile capability (air, ground, and naval) that can be brought to bear on forces operating in the area, including characteristics of missile systems, location and capacity of launch or delivery units, initial and sustained launch rates, size and location of stockpiles, and other pertinent factors.

(8) Serviceability and Operational Rates of Aircraft. Describe the total aircraft inventory by type, performance characteristics of operational aircraft, initial and sustained sortie rates of aircraft by type, and other pertinent factors.

(9) Operational Capabilities of Combatant Vessels. Describe the number, type, and operational characteristics of ships, boats, and craft in the naval inventory; base location; and capacity for support.

(10) Technical Characteristics of Equipment. Describe the technical characteristics of major items of equipment in the adversary inventory not already considered such as missiles, aircraft, and naval vessels.

(11) Intelligence Capabilities. Describe adversary intelligence collection system capabilities and limitations with SIGINT, HUMINT, GEOINT and MASINT subparagraphs as appropriate.

(12) Weapons of Mass Destruction (WMD). Describe the types and characteristics of WMD in the adversary inventory, stockpile data, delivery capabilities, nuclear and WMD employment policies and techniques, and other pertinent factors.

(13) Information Operations. Address adversary information operations capabilities as appropriate.

(14) Significant Strengths and Weaknesses. Discuss the significant adversary strengths and weaknesses assessed from the facts presented in the preceding subparagraphs.

c. Adversary Unconventional and Psychological Warfare Situation.

(1) Guerrilla. Describe the adversary capability for, policy with regard to, and current status in the area of guerrilla or insurgent operations.

(2) Psychological. Describe adversary doctrine, techniques, methods, organization for, and conduct of psychological operations in the area of operations.

(3) Subversion. Describe adversary doctrine, techniques, methods, organization for, and conduct of subversion in the area of operations.

(4) Sabotage. Outline adversary organization and potential for and conduct of sabotage in the area of operations.

3. Adversary Capabilities. List each adversary capability in a series of subparagraphs that can affect the accomplishment of the assigned mission. Adversary capabilities may include ground, air, naval, space, WMD, nuclear, biological, chemical, unconventional warfare, and any other special weapons capabilities such as electromagnetic pulse, etc. Each adversary capability should contain an assessment including the following: what the adversary can do; where they can do it; when they can start it and get it done; what strength they can devote to the task. In describing adversary capabilities, the J-2 must also be able to inform the commander what the adversary can accomplish using its forces in a joint effort.

4. Analysis of Adversary COAs. Analyze each COA in light of the assigned mission. Consider all of the relevant factors from the preceding subparagraphs and determine their relative order probability of adoption. Utilizing separate subparagraphs, examine the adversary COAs by discussing the factors that favor or adversely affect their adoption. When applicable, the analysis of each COA should also include a discussion of adversary vulnerabilities attendant to that COA (i.e., conditions or circumstances of the adversary situation that render the adversary especially liable to damage, deception, or defeat). Finally, the analysis should include a discussion of any indications that point to possible adoption of the COA.

5. Conclusions. This paragraph contains the summary conclusions of the Intelligence Estimate with associated analytic confidence levels. In separate subparagraphs, list the adversary capabilities most likely to be adopted based on the discussion in the previous paragraph. Include a concise statement of the effects of each capability on the accomplishment of the assigned mission and where applicable list exploitable vulnerabilities.

NWP 2-0

APPENDIX J

Format for Annex B: Intelligence

J.1 ANNEX B TEMPLATE

CLASSIFICATION

Issuing Headquarters
Place of Issue
Day, Month, Year, Hour, Zone

ANNEX B TO [COMMANDER] OPLAN/CONPLAN/OPORD XXXX-YY ( )
INTELLIGENCE

References: List documents essential to this annex.

1. Situation.

   a. Characteristics of the operational environment (OE). Summarize the relevant characteristics of the OE. The OE encompasses the physical areas and factors, information environment, and systems perspective within the context of an analysis of all aspects of the mission and commander's intent that could influence the commander's decisions or affect friendly and adversary courses of action (COAs). Include sufficient analysis of the OE to provide context to the Concept of Intelligence Operations, but do not repeat basic information included in the general situation paragraph of the Base plan. Refer the reader to the complete Intelligence Estimate (Appendix 11 to Annex B) for additional analysis required to inform the development of supporting plans.

      (1) Physical Areas and Factors. Physical areas include the assigned operational area and the associated areas of influence and interest necessary for the conduct of operations within the air, land, maritime, space, and cyberspace domains.

      (2) Information Environment.

         (a) Provide a brief summary of analysis of the aggregate of individuals, and key individuals and groups having influence among the indigenous population as well as the source of their influence.

         (b) Foreign language capability needed to exploit or collect intelligence information. Identify specific language capability needed in terms of language and/or dialects for open source, signal or human intelligence collection or exploitation.

      (3) Systems Perspective. Provide a brief description of the interrelated political, military, economic, social, information, and infrastructure (PMESII) systems (to include sociocultural factors) without regard to geographic boundaries but addressing a focus area specified by the commander.

b. Enemy.

   (1) Evaluate the OE from the adversary's perspective in terms of a prioritized set of adversary military COAs, to include any related diplomatic, informational, or economic options.

   (2) Prioritize adversary COAs based on how well each is supported by the overall impact of the OE. For JSCP-tasked plans, refer to the Dynamic Threat Assessment or the Theater Intelligence Assessment from DIA for adversary threat, capabilities, and intentions.

   (3) Outline the enemy's capability to collect, communicate to intelligence centers, process, and disseminate through telecommunications or other methods.

   (4) Determine Adversary COAs.

      (a) Discern the adversary's ability to integrate air, sea, and land capabilities in combined arms operations.

      (b) Consolidate a list of all adversary COAs appropriate to the current situation and accomplishment of likely objectives as indicated by recent activities or events. Each identified COA should meet the following five criteria: suitability; feasibility; acceptability; uniqueness; and consistency with adversary doctrine or operational patterns.

      (c) Analyze each COA to identify strengths and weaknesses, centers of gravity (COGs), critical capabilities (CC), critical requirements (CR), critical vulnerabilities (CVs), decisive points, compatibility with the OE, and how well each COA satisfies the above five criteria.

c. Friendly.

   (1) Friendly Intelligence Capability. Identify available friendly intelligence capabilities to include interagency, allied and coalition capabilities when appropriate.

   (2) Employment Limitations. Provide a brief description of factors affecting the employment of friendly intelligence capabilities. These may include but are not limited to a lack of access due to legal restrictions, technical limitations, or basing rights considered in the J2 Staff Estimate.

d. Legal Considerations. Identify any legal considerations relevant to intelligence operations.

2. Mission. (Restate Base Plan mission statement). Based on the approved J-2 Staff Estimate and Intelligence Estimate (Appendix 11), confirm the purpose of intelligence operations and describe how federated intelligence operations will be integrated and synchronized to support CCDR decision points and joint operation assessments.

3. Execution.

   a. Concept of Intelligence Operations.

      (1) Summarize the means and agencies employed in managing tasks associated with, collection, processing and exploitation, analysis and production, dissemination, and integration. Specify procedures for evaluation and feedback to assess the conduct of intelligence operations.

      (2) Identify defense, non-DOD Intelligence Community, allied, and coalition intelligence support requirements. Ensure all intelligence support is properly aligned with each phase of the operation.

      (3) Identify national intelligence augmentation requirements including national intelligence support teams (NISTs), Quick Reaction Teams (QRTs), and agency liaison requirements.

b. Tasks.

   (1) Priority Intelligence Requirements (PIRs). List the PIRs required to accomplish the mission by phase. The list of PIRs should include requirements for intelligence during peacetime to confirm threat-related planning assumptions and to inform plan revisions. The list of PIRs should also include requirements associated with the execution phases to assess progress towards the achievement of objectives and inform critical decisions. Describe the coordination or collaboration process for dynamically updating and satisfying PIR. List the PIRs in the coordinating instructions of the Basic Plan if Annex B is not published. Provide amplifying information regarding PIRs in Appendix 1 to Annex B.

   (2) Identify intelligence tasks required to support the plan, and the OPR within the Defense IC for accomplishing each task. Include the responsibilities of allied nations and coalition partners (approved during IPR A/C) for support to multinational operations.

      (a) List the Priority Intelligence Requirements (PIRs) required to accomplish the mission by phase. PIRs should include requirements during peace, crisis, and war, both prior to and during execution. Describe the coordination or collaboration process for satisfying PIRs. List the PIRs in the coordinating instructions of the Base plan if Annex B is not published.

      (b) Based on the completed J-2 Staff Estimate and the PIRs required to accomplish the mission by phase develop the Intelligence Task List (ITL) by phase and include the ITL as Tab A to Appendix 1 or as an additional appendix. The ITL serves as a baseline for federated analysis and production, and will be used, as required, to coordinate the supporting capabilities of CSAs, Services, and other DOD and non-DOD intelligence organizations.

   (3) Orders to Subordinate Units. List detailed instructions for each subordinate unit performing intelligence functions in separate subparagraphs.

   (4) Requirements to Higher and Supporting Organizations. List intelligence support requirements to units not organic or attached as well as to supporting intelligence agencies, interagency intelligence organizations, allied or coalition forces, and organizations providing federated support.

c. Collection. Provide guidance for the collection of information and material. Provide guidance for managing collection activities not covered by regulation or standard operating procedure (SOP). Include operations security (OPSEC) planning guidance and guidance for the use of tactical military deceptions during the planning and conduct of intelligence collection activities.

   (1) Signals Intelligence (SIGINT). Provide guidance for assignment and coordination of communications intelligence (COMINT) and electronic intelligence (ELINT) resources. Include guidance on the interaction of SIGINT activities with imagery intelligence (IMINT), human intelligence (HUMINT), and measurement and signature intelligence (MASINT) and electronic warfare support activities.

   (2) Geospatial Intelligence (GEOINT). Provide guidance for establishing and conducting imagery and geospatial activities. Include guidance on synchronizing GEOINT collection activities with SIGINT, HUMINT, and MASINT activities.

   (3) HUMINT. Provide information and guidance pertaining to the organization, direction, and coordination of HUMINT collection operations and support activities. Include guidance, if appropriate, on interaction with GEOINT, SIGINT, and MASINT activities.

   (4) MASINT. Provide guidance on obtaining intelligence by quantitative and qualitative analysis of data derived from specific technical collection sensors other than those normally associated with

SIGINT, GEOINT, and HUMINT. Include guidance on the interaction of GEOINT, HUMINT, SIGINT, and technical intelligence.

(5) Counterintelligence (CI). Provide guidance pertaining to the assignment and coordination of operations using CI agents and sources in support of force protection efforts.

(6) Other Collection Activities. Provide guidance for collection by other specialized means such as visual, amphibious, reconnaissance, and medical collection activities to support plan requirements. Include guidance on how these activities are expected to interact with collection efforts discussed elsewhere in this plan.

d. Processing and Evaluation. Provide guidance for converting information into usable form, including required provisions for document translation; imagery, signals, and technical sensor processing and interpretation; and other pertinent processing activity.

e. Analysis and Production. Provide guidance on analyzing and reporting collected intelligence information by all collection sources employed in support of the plan. Include guidance on multidiscipline reports that fuse information from multiple sources and on development and integration of independent and alternative assessments by the Red Team. Reference appropriate regulations, directives, and SOPs specifying U.S.-only and multinational reporting procedures. Identify the production effort, including any intelligence and counterintelligence products, required to support the plan.

f. Dissemination and Integration. Provide necessary guidance for conveying intelligence to appropriate operational levels, including the forces of allied nations. Include criteria to satisfy expanded requirements for vertical and lateral dissemination of finished intelligence and spot reports. Identify alternate means of disseminating intelligence to combat units and headquarters during crises and combat operations. The following items may be covered in this subparagraph:

(1) Intelligence reports required from units with specific guidance regarding periodicity and distribution

(2) Formats for intelligence reports including additional appendices if required

(3) Distribution of intelligence studies

(4) Requirements for releasability or disclosure to allied nations

(5) Requirements for secondary imagery dissemination.

g. Coordinating Instructions. Detail instructions applicable to two or more supporting intelligence organizations. Identify which, if any, collaborative tools will be used. The following items may be covered in this subparagraph:

(1) Periodic or special conferences for intelligence officers

(2) Intelligence liaison with adjacent commanders, crisis intelligence federation partners, foreign government agencies or military forces, and host countries

(3) Coordination of protected frequencies to be nominated for the Joint Restricted Frequency List

(4) Reconnaissance and Surveillance conferences

(5) Guidance, Apportionment, and Targeting conferences

(6) Release or disclosure of intelligence information to coalition partners

(7) Sanitization of intelligence information.

4. Administration and Logistics.

   a. Shortfalls and Limiting Factors. Based on the J-2 estimate list shortfalls and limiting factors, including foreign language and regional expertise capabilities, significantly affecting intelligence support. Include an impact assessment for each shortfall. Identify resource challenges and specify key tasks that may not be accomplished.

   b. Mitigation. Identify or describe the mitigation strategy, which may include contracted capabilities, in response to shortfalls or limiting factors. Cross-reference to Tab A to Appendix 3 to Annex W.

   c. Miscellaneous. Identify OPSEC, evasion and escape, deception, disclosure of intelligence, releasability to coalition forces, public affairs, use of specialized intelligence personnel, military information support operations (MISO), exploitation of captured foreign materiel and documents, and composition of the J-2 staff.

   d. Logistics. Identify logistic requirements specific to intelligence support of the plan.

   e. Reporting. Identify any unique reporting requirements in support of the plan.

5. Command and Control.

   a. Command Relationships. Describe any unique command relationships for intelligence operations.

   b. Communications. Summarize the U.S. and non-U.S. communications systems and procedures to be used to carry out the intelligence function or reference the appropriate paragraphs of Annex K. Include comments on interoperability of these systems.

Appendixes
1—Priority Intelligence Requirements (PIR)
2—Signals Intelligence (SIGINT)
3—Counterintelligence (CI)
4—Targeting
5—Human Intelligence (HUMINT)
6—Intelligence Support to Information Operations (IO)
7—Geospatial Intelligence (GEOINT)
8—Measurement and Signature Intelligence (MASINT)
9—Captured Enemy Equipment (CEE)
10—National Intelligence Support Team (NIST)
11—Intelligence Estimate

CLASSIFICATION

INTENTIONALLY BLANK

REFERENCES

JOINT DOCTRINE

JP 1-02, Department of Defense Dictionary of Military and Associated Terms

JP 2-0, Joint Intelligence

JP 2-01, Joint and National Intelligence Support to Military Operations

JP 2-01.2, Counterintelligence and Human Intelligence Support to Joint Operations

JP 2-01.3, Joint Intelligence Preparation of the Operational Environment

JP 3-0, Joint Operations

JP 3-12, Cyberspace Operations

JP 3-32, Command and Control for Joint Maritime Operations

JP 5-0, Joint Operation Planning

NAVY DOCTRINE

NDP 1, Naval Warfare

NDP-2, Naval Intelligence (30 Sep 1994) (superseded by NDP-1 in Mar 2010)

NTTP 3-32.1, Maritime Operations Center

NWP 2-01, Intelligence Support to Naval Operations

NWP 5-01, Navy Planning

MARINE CORPS DOCTRINE

MCDP 2, Intelligence

MCWP 2-1, Intelligence Operations

OTHER

A Cooperative Strategy for 21st Century Seapower, 2007

Alfred Thayer Mahan, The Influence of Sea Power Upon History, 1660–1783

CJCSI 3500.02 Series, UJTL Policy and Guidance for the Armed Forces of the United States

CJCSM 3130.03, Adaptive Planning and Execution (APEX) Planning Formats and Guidance

Clausewitz, On War

Director of National Intelligence, Intelligence Community Directive (ICD) 203, Analytic Standards

Flynn, M.T., Juergens, R., Cantrell, T.L., Employing ISR: Special Operations Forces (SOF) Best Practices Joint Force Quarterly, Third Quarter 2008

FM 34-8, Intelligence Officer's Handbook

General Carl E. Mundy, Jr., Reflections on the Corps: Some Thoughts on Expeditionary Warfare, Marine Corps Gazette, (March 1995)

Heuer, R. J., Psychology of Intelligence Analysis

Heuer, R. J. and Pherson, R. H., Structured Analytic Techniques for Intelligence Analysts

Joint Warfighting Center, "Commander's Handbook for Persistent Surveillance," p. B-2

JP 3-28, Defense Support of Civil Authorities

JP 3-29, Foreign Humanitarian Assistance

Julian S. Corbett, Some Principles of Maritime Strategy

Kent, S., Estimates and Influence, published in Studies in Intelligence (1968), https://www.cia.gov/library/center-for-the-study-of-intelligence/csi-publications/books-and-monographs/sherman-kent-and-the-board-of-national-estimates-collected-essays/4estimates.html

Lowenthal, M. M., Intelligence: From Secrets to Policy, 2009

Major, J., Communicating with Intelligence

Maritime Component Commander Guidebook, Naval War College, July 2012

Merriam-Webster, http://www.merriam-webster.com/dictionary/analysis

Milan Vego, Joint Operational Warfare: Theory and Practice

Naval Operations Concept 2010

Naval War College, Joint Operation Planning Process (JOPP) Workbook

Navy Cyber Forces, "FID/FIAF Frequently Asked Questions," p. 1, retrieved from http://www.public.navy.mil/bupers-npc/officer/Detailing/IDC_FAO/intelligence/Documents/FIDFIAF.pdf

Navy Cyber Forces, "Recommendation for Navy Tactics, Techniques, and Procedures (TTP) for Intelligence Collection Management for Commander Task Force Level Operations," p. 2

NWP 3-29, Disaster Response Operations

OPNAV N2/N6, Information Dominance Roadmap, 2013

OPNAVINST 3500.38B/MCO3500.26A/USCG COMDTINST 3500.1, Universal Naval Task List, version 3.0, January 2007

Public Law 109-163, National Defense Authorization Act for Fiscal Year 2006, Section 931

The National Military Strategy of the United States, Joint Chiefs of Staff, 08 Feb 2010

United States Naval War College, Operational Art Primer, 2012

UJTL Task List, January 2013

UNTL, Report Number: 38.1, Generated: November 2012

U.S. Naval Intelligence School, Intelligence for Naval Officers, 1956

U.S. Naval Intelligence School, Naval Intelligence, 1948

Walton, T., Challenges in Intelligence Analysis

Wayne P. Hughes, Fleet Tactics and Coastal Combat

http://www.ibiblio.org/hyperwar/USMC/USMC-M-Okinawa/maps/USMC-M-Okinawa-5.jpg

INTENTIONALLY BLANK

NWP 2-0

GLOSSARY

actionable intelligence. Intelligence information that is directly useful to customers for immediate exploitation without having to go through the full Intelligence production process. (JP 1-02. Source: JP 2-01.2)

analysis and production. In intelligence usage, the conversion of processed information into intelligence through the integration, evaluation, analysis, and interpretation of all source data and the preparation of intelligence products in support of known or anticipated user requirements. (JP 1-02. Source: JP 2-01)

area of influence. A geographical area wherein a commander is directly capable of influencing operations by maneuver or fire support systems normally under the commander's command or control. (JP 1-02. Source: JP 3-0)

area of interest. That area of concern to the commander, including the area of influence, areas adjacent thereto, and extending into enemy territory. This area also includes areas occupied by enemy forces who could jeopardize the accomplishment of the mission. (JP 1-02. Source: JP 3-0)

area of operations. An operational area defined by the joint force commander for land and maritime forces that should be large enough to accomplish their missions and protect their forces. (JP 1-02. Source: JP 3-0)

battle damage assessment (BDA). The estimate of damage composed of physical and functional damage assessment, as well as target system assessment, resulting from the application of lethal or nonlethal military force. (JP 1-02. Source: JP 3-0)

center of gravity (COG). The source of power that provides moral or physical strength, freedom of action, or will to act. (JP 1-02. Source: JP 5-0)

collection. In intelligence usage, the acquisition of information and the provision of this information to processing elements. (JP 1-02. Source: JP 2-01)

collection asset. A collection system, platform, or capability that is supporting, assigned, or attached to a particular commander. (JP 1-02. Source: JP 2-01)

collection management. In intelligence usage, the process of converting intelligence requirements into collection requirements, establishing priorities, tasking or coordinating with appropriate collection sources or agencies, monitoring results, and retasking, as required. (JP 1-02. Source: JP 2-0)

collection operations management (COM). The authoritative direction, scheduling, and control of specific collection operations and associated processing, exploitation, and reporting resources. (JP 1-02. Source: JP 2-0)

collection requirement. A valid need to close a specific gap in intelligence holdings in direct response to a request for information. (JP 1-02. Source: JP 2-0)

collection requirements management (CRM). The authoritative development and control of collection, processing, exploitation, and/or reporting requirements that normally result in either the direct tasking of requirements to units over which the commander has authority, or the generation of tasking requests to collection management authorities at a higher, lower, or lateral echelon to accomplish the collection mission. (JP 1-02. Source: JP 2-0)

collection resource. A collection system, platform, or capability that is not assigned or attached to a specific unit or echelon which must be requested and coordinated through the chain of command. (JP 1-02. Source: JP 2-01)

commander's critical information requirement (CCIR). An information requirement identified by the commander as being critical to facilitating timely decision making. (JP 1-02. Source: JP 3-0)

counterintelligence (CI). Information gathered and activities conducted to identify, deceive, exploit, disrupt, or protect against espionage, other intelligence activities, sabotage, or assassinations conducted for or on behalf of foreign powers, organizations or persons or their agents, or international terrorist organizations or activities. (JP 1-02. Source: JP 2-01.2)

course of action (COA). 1. Any sequence of activities that an individual or unit may follow. 2. A scheme developed to accomplish a mission. 3. A product of the course-of-action development step of the joint operation planning process. (JP 1-02. Source: JP 5-0)

critical capability. A means that is considered a crucial enabler for a center of gravity to function as such and is essential to the accomplishment of the specified or assumed objective(s). (JP 1-02. Source: JP 5-0)

critical requirement. An essential condition, resource, and means for a critical capability to be fully operational. (JP 1-02. Source: JP 5-0)

critical vulnerability. An aspect of a critical requirement which is deficient or vulnerable to direct or indirect attack that will create decisive or significant effects. (JP 1-02. Source: JP 5-0)

current intelligence. Current intelligence provides updated support for ongoing operation. It involves the integration of time-sensitive, all-source intelligence and information into concise, objective reporting on the current situation in a particular area. The term "current" is relative to the time sensitivities of the decision maker and the context of the type of operation being supported. For example, in some contexts intelligence may be considered "current," whereas other circumstances may require intelligence in near real time. (JP 2-0. Source: N/A)

decisive point. A geographic place, specific key event, critical factor, or function that, when acted upon, allows commanders to gain a marked advantage over an adversary or contribute materially to achieving success. (JP 1-02. Source: JP 5-0)

dissemination and integration. In intelligence usage, the delivery of intelligence to users in a suitable form and the application of the intelligence to appropriate missions, tasks, and functions. (JP 1-02. Source: JP 2-01)

essential elements of information (EEI). The most critical information requirements regarding the adversary and the environment needed by the commander by a particular time to relate with other available information and intelligence in order to assist in reaching a logical decision. (JP 1-02. Source: JP 2-0)

estimative intelligence. Intelligence that identifies, describes, and forecasts adversary capabilities and the implications for planning and executing military operations. (JP 1-02. Source: JP 2-0)

friendly force information requirement (FFIR). Information the commander and staff need to understand the status of friendly force and supporting capabilities. (JP 1-02. Source: JP 3-0)

functional damage assessment. The estimate of the effect of military force to degrade or destroy the functional or operational capability of the target to perform its intended mission and on the level of success in achieving operational objectives established against the target. (JP 1-02. Source: JP 3-60)

geospatial intelligence (GEOINT). The exploitation and analysis of imagery and geospatial information to describe, assess, and visually depict physical features and geographically referenced activities on the Earth.

Geospatial intelligence consists of imagery, imagery intelligence, and geospatial information. (JP 1-02. Source: JP 2-03)

high-value target (HVT). A target the enemy commander requires for the successful completion of the mission. (JP 1-02. Source: JP 3-60)

human intelligence (HUMINT). A category of intelligence derived from information collected and provided by human sources. (JP 1-02. Source: JP 2-0)

indications and warning (I&W). Those intelligence activities intended to detect and report time-sensitive intelligence information on foreign developments that could involve a threat to the United States or allied and/or coalition military, political, or economic interests or to U.S. citizens abroad. It includes forewarning of hostile actions or intentions against the United States, its activities, overseas forces, or allied and/or coalition nations.

information requirements (IR). In intelligence usage, those items of information regarding the adversary and other relevant aspects of the operational environment that need to be collected and processed in order to meet the intelligence requirements of a commander. (JP 1-02. Source: JP 2-0)

intelligence. 1. The product resulting from the collection, processing, integration, evaluation, analysis, and interpretation of available information concerning foreign nations, hostile or potentially hostile forces or elements, or areas of actual or potential operations. 2. The activities that result in the product. 3. The organizations engaged in such activities. (JP 1-02. Source: JP 2-0)

intelligence operations. The variety of intelligence and counterintelligence tasks that are carried out by various intelligence organizations and activities within the intelligence process. (JP 1-02. Source: JP 2-01)

intelligence preparation of the battlespace (IPB). The analytical methodologies employed by the Services or joint force component commands to reduce uncertainties concerning the enemy, environment, time, and terrain. Intelligence preparation of the battlespace supports the individual operations of the joint force component commands. (JP 1-02. Source: JP 2-01.3)

intelligence requirement (IR). 1. Any subject, general or specific, upon which there is a need for the collection of information, or the production of intelligence. 2. A requirement for intelligence to fill a gap in the command's knowledge or understanding of the operational environment or threat forces. (JP 1-02. Source: JP 2-0)

joint intelligence preparation of the operational environment (JIPOE). The analytical process used by joint intelligence organizations to produce intelligence estimates and other intelligence products in support of the joint force commander's decision-making process. It is a continuous process that includes defining the operational environment; describing the impact of the operational environment; evaluating the adversary; and determining adversary courses of action. (JP 1-02. Source: JP 2-01.3)

link. 1. A behavioral, physical, or functional relationship between nodes. 2. In communications, a general term used to indicate the existence of communications facilities between two points. (JP 1-02. Source: JP 3-0)

littoral. The littoral comprises two segments of operational environment: 1. Seaward: the area from the open ocean to the shore, which must be controlled to support operations ashore. 2. Landward: the area inland from the shore that can be supported and defended directly from the sea. (JP 1-02. Source: JP 2-01.3)

maritime domain. The oceans, seas, bays, estuaries, islands, coastal areas, and the airspace above these, including the littorals. (JP 1-02. Source: JP 3-32)

measure of performance (MOP). A criterion used to assess friendly actions that is tied to measuring task accomplishment. (JP 1-02. Source: JP 3-0)

measure of effectiveness (MOE). A criterion used to assess changes in system behavior, capability, or operational environment that is tied to measuring the attainment of an end state, achievement of an objective, or creation of an effect. (JP 1-02. Source: JP 3-0)

measurement and signature intelligence (MASINT). Information produced by quantitative and qualitative analysis of physical attributes of targets and events to characterize, locate, and identify targets and events, and derived from specialized, technically derived measurements of physical phenomenon intrinsic to an object or event. (JP 1-02. Source: JP 2-0)

modified combined obstacle overlay (MCOO). A joint intelligence preparation of the operational environment product used to portray the militarily significant aspects of the operational environment, such as obstacles restricting military movement, key geography, and military objectives. (JP 1-02. Source: JP 2-01.3)

munitions effectiveness assessment (MEA). Conducted concurrently and interactively with battle damage assessment, the assessment of the military force applied in terms of the weapon system and munitions effectiveness to determine and recommend any required changes to the methodology, tactics, weapon system, munitions, fusing, and/or weapon delivery parameters to increase force effectiveness. Munitions effectiveness assessment is primarily the responsibility of operations with required inputs and coordination from the intelligence community. (JP 1-02. Source: JP 2-01)

naval intelligence. Intelligence used to support naval and maritime decisionmaking, the planning, execution, and assessment of naval operations, and the naval personnel and associated organizations engaged in such activity.

node. An element of a system that represents a person, place, or physical thing. (JP 1-02. Source: JP 3-0)

open-source intelligence (OSINT). Relevant information derived from the systematic collection, processing, and analysis of publicly available information in response to known or anticipated intelligence requirements. (JP 1-02. Source: JP 2-0)

operational art. The cognitive approach by commanders and staffs—supported by their skill, knowledge, experience, creativity, and judgment—to develop strategies, campaigns, and operations to organize and employ military forces by integrating ends, ways, and means. (JP 1-02. Source: JP 3-0)

operational environment (OE). A composite of the conditions, circumstances, and influences that affect the employment of capabilities and bear on the decisions of the commander. (JP 1-02. Source: JP 3-0)

operational intelligence. Intelligence that is required for planning and conducting campaigns and major operations to accomplish strategic objectives within theaters or operational areas. (JP 1-02. Source: JP 2-0) (Note: For naval intelligence professionals, the term "OPINTEL" refers to the products resulting from an all-source intelligence analysis process that can provide the commander, staff, and supported organizations the best intelligence estimate possible to support their planning, efforts, and operations.)

physical damage assessment. The estimate of the quantitative extent of physical damage to a target resulting from the application of military force. (JP 1-02. Source: JP 3-60)

planning and direction. In intelligence usage, the determination of intelligence requirements, development of appropriate intelligence architecture, preparation of a collection plan, and issuance of orders and requests to information collection agencies. (JP 1-02. Source: JP 2-01)

priority intelligence requirement (PIR). An intelligence requirement, stated as a priority for intelligence support, that the commander and staff need to understand the adversary or other aspects of the operational environment. Also called PIR. (JP 1-02. Source: JP 2-01)

processing and exploitation. In intelligence usage, the conversion of collected information into forms suitable to the production of intelligence. (JP 1-02. Source: JP 2-01)

reattack recommendation (RR). An assessment, derived from the results of battle damage assessment and munitions effectiveness assessment, providing the commander systematic advice on reattack of a target. (JP 1-02. Source: JP 3-60)

request for information (RFI). 1. Any specific time-sensitive ad hoc requirement for intelligence information or products to support an ongoing crisis or operation not necessarily related to standing requirements or scheduled intelligence production. A request for information can be initiated to respond to operational requirements and will be validated in accordance with the combatant command's procedures. 2. A term used by the National Security Agency/Central Security Service to state ad hoc signals intelligence requirements. (JP 1-02. Source: JP 2-0) 3. In Navy usage, a general term for an information request that can be used to meet an information need associated with an operation.

scientific and technical intelligence (S&TI). The product resulting from the collection, evaluation, analysis, and interpretation of foreign scientific and technical information that covers: a. foreign developments in basic and applied research and in applied engineering techniques; and b. scientific and technical characteristics, capabilities, and limitations of all foreign military systems, weapons, weapon systems, and materiel; the research and development related thereto; and the production methods employed for their manufacture. (JP 1-02. Source: JP 2-01)

signals intelligence (SIGINT). 1. A category of intelligence comprising either individually or in combination all communications intelligence, electronic intelligence, and foreign instrumentation signals intelligence, however transmitted. 2. Intelligence derived from communications, electronic, and foreign instrumentation signals. (JP 1-02. Source: JP 2-0)

source. 1. A person, thing, or activity from which information is obtained. 2. In clandestine activities, a person (agent), normally a foreign national, in the employ of an intelligence activity for intelligence purposes. 3. In interrogation activities, any person who furnishes information, either with or without the knowledge that the information is being used for intelligence purposes. (JP 1-02. Source: JP 2-01)

strategic intelligence. Intelligence required for the formation of policy and military plans at national and international levels. Strategic intelligence and tactical intelligence differ primarily in level of application, but may also vary in terms of scope and detail. (JP 1-02. Source: JP 2-01.2)

system. A functionally, physically, and/or behaviorally related group of regularly interacting or interdependent elements; that group of elements forming a unified whole. (JP 1-02. Source: JP 3-0)

tactical intelligence. Intelligence required for the planning and conduct of tactical operations. (JP 1-02. Source: JP 2-01.2)

target intelligence. Target intelligence portrays and locates the components of a target or target complex, networks, and support infrastructure, and indicates its vulnerability and relative importance to the adversary. (JP 2-0. Source: N/A)

tasking, collection, processing, exploitation, and dissemination (TCPED). The process through which a specific intelligence discipline provides information to address an information or intelligence requirement. Information derived from tasking, collection, processing, exploitation, and dissemination processes may directly inform operational decision makers or serve as an input to multi-source intelligence analysis. (Source: NWP 2-0)

technical intelligence (TECHINT). Intelligence derived from the collection, processing, analysis, and exploitation of data and information pertaining to foreign equipment and materiel for the purposes of preventing technological surprise, assessing foreign scientific and technical capabilities, and developing countermeasures designed to neutralize an adversary's technological advantages. (JP 1-02. Source: JP 2-0)

warning intelligence. Those intelligence activities intended to detect and report timesensitive intelligence information on foreign developments that forewarn of hostile actions or intention against United States entities, partners, or interests. (JP 1-02. Source: JP 2-0)

LIST OF ACRONYMS AND ABBREVIATIONS

ACH	analysis of competing hypotheses
ACI	air combat intelligence
ACINT	acoustic intelligence
ACOA	adversary course of action
AI	area of influence
AO	area of operations
AOA	amphibious objective area
AOI	area of interest
AOR	area of responsibility
APOD	aerial port of debarkation
ARG	amphibious ready group
ASAT	antisatellite weapon
ATO	air tasking order
BDA	battle damage assessment
Bde	brigade
BPT	beach party team
C2	command and control
C5I	command, control, computers, communications, combat systems, and intelligence
CAP	combat air patrol
CBRN	chemical, biological, radiological, and nuclear
CBRNE	chemical, biological, radiological, nuclear, and high-yield explosives
CCDR	combatant commander
CCIR	commander's critical information requirement
CDCM	coastal defense cruise missile

NWP 2-0

CGS	common geopositioning services
CI	counterintelligence
CIA	Central Intelligence Agency
CINCPACFLT	Commander in Chief, United States Pacific Fleet (World War II)
CMWS	collection management workstation
COA	course of action
COG	center of gravity
COM	collection operations management
COMFIFTHFLT	Commander, Fifth Fleet
COMINT	communications intelligence
COMSECONDFLT	Commander, Second Fleet
COMSEVENTHFLT	Commander, Seventh Fleet
COMSIXTHFLT	Commander, Sixth Fleet
COMTENTHFLT	Commander, Tenth Fleet
COMTHIRDFLT	Commander, Third Fleet
COMUSNAVCENT	Commander, United States Naval Forces, Central Command
COMUSNAVSO	Commander, United States Naval Forces, Southern Command
CONOPS	concept of operations
CONPLAN	contingency plan
CONREP	connected replenishment
COP	common operational picture
CP	command post
CR	critical requirement
CRC	cryptologic resource coordinator
CRM	collection requirements management
CSG	carrier strike group
CSS	Central Security Service
CTF	combined task force

CV	critical vulnerability
CVE	escort aircraft carrier (World War II)
CVOA	carrier operating area
CWC	composite warfare commander
DCHC	Defense Counterintelligence and Human Intelligence Center
DCGS-N	Distributed Common Ground System – Navy
DCO	defensive cyberspace operations
DESRON	destroyer squadron
DHE	defense human intelligence executor
DIA	Defense Intelligence Agency
DIME	diplomatic, informational, military, and economic
DIO	district intelligence officer (World War II)
DLA	Defense Logistics Agency
DMZ	demilitarized zone
DNI	Director of National Intelligence
DOD	Department of Defense
DOMEX	document and media exploitation
DP	decision point
DR	disaster relief
DRAW-D	defend, reinforce, attack, withdraw, and delay
DSCA	defense support of civil authorities
ECM	electronic countermeasures
EEI	essential element of information
ELINT	electronic intelligence
EO	electro-optical
ESG	expeditionary strike group
EW	electronic countermeasures
FBI	Federal Bureau of Investigation

FCC	Fleet Cyber Command
FFIR	frIendly force information requirement
FHA	foreign humanitarian assistance
FI	foreign intelligence
FID	fleet intelligence detachment
FIOC	fleet information operations center
FISINT	foreign instrumentation signals intelligence
FIST	fleet imagery support team
FMP	foreign materiel program
FMV	full-motion video
FOSIC	Fleet Ocean Surveillance Information Center
FOV	field of view
FRAGORD	fragmentary order
GCCS	Global Command and Control System
GCCS-M	Global Command and Control System–Maritime
GEOINT	geospatial intelligence
GMI	general military intelligence
HA	humanitarian assistance
HHQ	higher headquarters
HPT	high-payoff target
HQ	headquarters
HSI	hyperspectral imagery
HUMINT	human intelligence
HVT	high-value target
I&W	indications and warning
IADS	integrated air defense system
IBS	integrated broadcast service
IC	intelligence community

ICW	in conjunction with
IMINT	imagery intelligence
Inf	infantry
IO	information operations
IPB	intelligence preparation of the battlespace
IPOE	intelligence preparation of the operational environment
IR	information requirement
ISB	intermediate staging base
ISE	intelligence support element
ISO	in support of
ISR	intelligence, surveillance, and reconnaissance
IUSS	integrated undersea surveillance system
J-2	intelligence directorate of a joint staff
J-2X	joint force counterintelligence and human intelligence staff element
J-3	operations directorate of a joint staff
JCA	Joint Concentrator Architecture
JCMB	Joint Collection Management Board
JFACC	joint force air component commander
JFC	joint force commander
JFLCC	joint force land component commander
JFMCC	joint force maritime component commander
JFMPO	Joint Foreign Materiel Program Office
JIC	joint intelligence center
JIOC	joint intelligence operations center
JICPOA	Joint Intelligence Center Pacific Ocean Areas (World War II)
JIPOE	joint intelligence preparation of the operational environment
JISE	joint intelligence support element
JOA	joint operations area

JP	joint publication
JTENS	Joint Tactical Exploitation and National Systems
JTF	joint task force
JWICS	Joint Worldwide Intelligence Communications System
KML	Keyhole Markup Language
lidar	light detection and ranging
LTIOV	latest time intelligence is of value
LOC	line of communications
MASINT	measurement and signature intelligence
MBT	main battle tank
MCDP	Marine Corps doctrinal publication
MCM	mine countermeasures
MCOO	modified combined obstacle overlay
MCWP	Marine Corps warfighting publication
MDA	maritime domain awareness
MEA	munitions effectiveness assessment
Mech	mechanized
MET	mission-essential task
METL	mission-essential task list
METOC	meteorological and oceanographic
MEU	Marine expeditionary unit
MG	major general
MIDB	modernized integrated database
MILDEC	military deception
MIO	maritime interception operations
MIO-IET	maritime interception operation–intelligence exploitation team
MIOC	maritime intelligence operations center
MISO	military information support operations

MISREP	mission report
MOC	maritime operations center
MOE	measure of effectiveness
MOP	measure of performance
MPA	maritime patrol aircraft
MPSRON	maritime pre-positioning ships squadron
MSI	multispectral imagery
MSO	maritime security operations
N-2	Navy staff intelligence directorate
N-2X	maritime force counterintelligence and human intelligence staff element
N-3	Navy staff operations directorate
NAI	named area of interest
NCC	Navy component commander
NCIS	Naval Criminal Investigative Service
NCIX	National Counterintelligence Executive
NDP	naval doctrine publication
NECC	Navy Expeditionary Combat Command
NEIC	Navy Expeditionary Intelligence Command
NFC	numbered fleet commander
NGA	National Geospatial-Intelligence Agency
NGO	nongovernmental organization
NIC-C	National Intelligence Coordination Center
NIOC	Navy information operations command
NIPRNET	Nonsecure Internet Protocol Router Network
NIST	national intelligence support team
NLLS	Navy Lessons Learned System
NLT	no later than
NMAWC	Naval Mine and Anti-Submarine Warfare Command

NMETL		Navy mission-essential task list
NMET		Navy mission-essential task
NMMO		National Measurement and Signature Intelligence Management Office
NOSE		National Open Source Enterprise
NPP		Navy planning process
NSA		National Security Agency
NSAWC		Naval Strike and Air Warfare Center
NSW		naval special warfare
NTA		Navy tactical task
NTTL		Navy Tactical Task List
NTTP		Navy tactics, techniques, and procedures
NWDC		Navy Warfare Development Command
NWP		Navy warfare publication
OB		order of battle
OCO		offensive cyberspace operations
ODNI		Office of the Director of National Intelligence
OLW		operational level of war
ONI		Office of Naval Intelligence
OPAREA		operating area
OPCON		operational control
OPELINT		operational electronic intelligence
OPINTEL		operational intelligence
OPLAN		operation plan
OPNAV		Office of the Chief of Naval Operations
OPNAVINST		Chief of Naval Operations instruction
OPORD		operation order
OPSEC		operations security
OPTASK		operational tasking (message)

OSINT	open-source intelligence
PAX	personnel
PED	processing, exploitation, and dissemination
PIR	priority intelligence requirement
PMESII	political, military, economic, social, information, and infrastructure
PRISM	Planning Tool for Resource Integration, Synchronization, and Management
R2P2	rapid response planning process
RDBM	red track database management
RFI	request for information
RSTA	reconnaissance, surveillance, and target acquisition
S&TI	scientific and technical intelligence
SAG	surface action group
SAM	surface-to-air missile
SASO	stability and support operations
SECNAVINST	Secretary of the Navy instruction
SIGINT	signals intelligence
SIO	senior intelligence officer
SIPRNET	SECRET Internet Protocol Router Network
SIR	specific information requirement
SLOC	sea line of communications
SNOOPIE	ship's nautical or otherwise photographic interpretation and examination
SOF	special operations forces
SOP	standard operating procedure
SPOD	seaport of debarkation
SSG	surface strike group
SSN	attack submarine (nuclear propulsion)
TACAIR	tactical air

TAI	target area of interest
TAO	tactical action officer
TBMD	theater ballistic missile defense
TCPED	tasking, collection, processing, exploitation, and dissemination
TECHELINT	technical electronic intelligence
TECHINT	technical intelligence
TST	time-sensitive target
TTP	tactics, techniques, and procedures
TTW	territorial waters
U.S.	United States
UAS	unmanned aircraft system
UJTL	Universal Joint Task List
UNREP	underway replenishment
UNTL	Universal Naval Task List
URG	underway replenishment
USAFRICOM	United States Africa Command
USCENTCOM	United States Central Command
USCG	United States Coast Guard
USEUCOM	United States European Command
USMC	United States Marine Corps
USN	United States Navy
USSS	United States Signals Intelligence System
VTC	video teleconferencing
WARNORD	warning order
WARP	Web-based access and retrieval portal
WMD	weapons of mass destruction

NWP 2.0

LIST OF EFFECTIVE PAGES

Effective Pages	Page Numbers
MAR 2014	1 thru 24
MAR 2014	EX-1 thru EX-4
MAR 2014	1-1 thru 1-16
MAR 2014	2-1 thru 2-22
MAR 2014	3-1 thru 3-60
MAR 2014	4-1 thru 4-66
MAR 2014	5-1 thru 5-26
MAR 2014	A-1 thru A-4
MAR 2014	B-1 thru B-16
MAR 2014	C-1 thru C-4
MAR 2014	D-1 thru D-14
MAR 2014	E-1 thru E-12
MAR 2014	F-1 thru F-36
MAR 2014	G-1 thru G-6
MAR 2014	H-1 thru H-10
MAR 2014	I-1 thru I-6
MAR 2014	J-1 thru J-6
MAR 2014	Reference-1 thru Reference-4
MAR 2014	Glossary-1 thru Glossary-6
MAR 2014	LOAA-1 thru LOAA-10
MAR 2014	LEP-1, LEP-2

INTENTIONALLY BLANK

NWP 2-0
MAR 2014

Made in the USA
Coppell, TX
04 October 2020